Wavelet Applications in Engineering Electromagnetics

For a listing of recent titles in the *Artech House Electromagnetic Analysis Series*, turn to the back of this book.

Wavelet Applications in Engineering Electromagnetics

Tapan K. Sarkar
Magdalena Salazar-Palma
Michael C. Wicks

With contributions from
Raviraj Adve
Robert J. Bonneau
Russell D. Brown
Luis-Emilio García-Castillo
Yingbo Hua
Zhong Ji
Kyungjung Kim
Jinhwan Koh
Wonwoo Lee
Sergio Llorente-Romano
Rafael Rodriguez-Boix
Chaowei Su
Wenxun Zhang

Artech House
Boston • London
www.artechhouse.com

Library of Congress Cataloging-in-Publication Data
Sarkar, Tapan (Tapan K.)
  Wavelet applications in engineering electromagnetics / Tapan Sarkar, Magdalena
Salazar-Palma, Michael C. Wicks.
      p. cm. — (Artech House electromagnetic analysis series)
  Includes bibliographical references and index.
  ISBN 1-58053-267-5 (alk. paper)
    1. Signal processing—Mathematics. 2. Electric filters—Mathematical models.
  3. Wavelets (Mathematics). 4. Electromagnetism—Mathematical models. 5. Electro-
  magnetic theory. I. Salazar-Palma, Magdalena. II. Wicks, Michael C. III. Title.
  IV. Series.

TK5102.9 .S363 2002
621.382'2—dc21

                                                            2002023673

British Library Cataloguing in Publication Data
Sarkar, Tapan K.
  Wavelet applications in engineering electromagnetics. —
  (Artech House electromagnetic analysis series)
  1. Electromagnetism—Mathematics 2. Wavelets (Mathematics)
  I. Title II. Salazar-Palma, Magdalena III. Wicks, Michael C.
  621.3'015152433

  ISBN 1-58053-267-5

Cover design by Igor Valdman

© 2002 ARTECH HOUSE, INC.
685 Canton Street
Norwood, MA 02062

International Standard Book Number: 1-58053-267-5
Library of Congress Catalog Card Number: 2002023673

10 9 8 7 6 5 4 3 2 1

Contents

List of Figures

Chapter 5

Chapter 6

List of Tables

PREFACE

There has been an explosive growth of research on wavelets, particularly in the last decade, resulting in a plethora of books, journals, and research papers dealing with the topic. Methods based on wavelets have been particularly well suited for the area of image compression. Wavelet techniques have also been applied in many areas of mathematics and engineering. They have been depicted as flexible analysis tools as compared to conventional Fourier methodologies.

One of the objectives of this book is to apply this technique in computational electromagnetics and signal analysis. Primarily, it is shown to be an efficient tool in many cases for solving Maxwell's equations. The method has been applied to fast solutions of large dense complex matrix equations arising in the solution of the integral form of Maxwell's equations via the method of moments. The multiscale concept inherent in wavelet techniques has been applied to make dense complex impedance matrices arising in the solution of the method of moments problem sparse. In addition, the concept of dilation and shift inherent in multiresolution wavelet analysis can be applied to large system matrices when solving the differential form of Maxwell's equations via the finite element method, resulting in almost diagonal matrices. Finally, the multiscaling concept has been introduced for the adaptive refinement of the popular triangular patch basis extensively used in computational electromagnetics. This has resulted in the adaptive multiscale moment method.

The second application is in the design of waveforms that are strictly time limited and yet concentrated in a very narrow frequency band. This can be used as a possible window function in multiresolution analysis. This principle can be used to generate time-limited pulses that can be practically band limited in exciting selected frequency domain regions of the object. This principle can also be used in exciting substructure resonances.

A key feature of this book is that wavelet concepts are described from a filter theory point of view that is familiar to students with an electrical engineering background. Most of the presentations therefore are made from an engineering perspective without sacrificing mathematical rigor. Many examples have been presented to illustrate the principle of multiresolution. Hence, this book is suitable for researchers working primarily in the area of computational techniques with a junior level background in electrical engineering.

It is hoped that this book will provide procedures for the efficient solution of Maxwell's equations. Formulating wavelet-based solutions to electromagnetic fields and waves problems offers the user tremendous flexibility to address practical problems, especially when dealing with very large physical systems such as ships, or antennas on aircraft and spacecraft. Other important engineering problems that may benefit from these techniques include urban planning for the location and operation of wireless communications systems; ground-penetrating radar, where the analysis of buried objects must also include propagation attenuation; and reflection and scattering effects from nonhomogeneous media including soil, rocks, vegetated matter, or even moisture and pockets of water that surround the buried object. The benefits offered by wavelet techniques arise from their multifarious properties. For example, the wavelet basis is a function of two parameters, dilation and shift, as compared to a single parameter, dilation, when using Fourier-based methodologies. With two parameters, an additional degree of flexibility is afforded to the analyst who ultimately must deal with the challenge of solving numerical fields and waves problems. Of course, with the utilization of wavelet basis functions comes a natural reduction in computational complexity as well. This natural reduction in computational complexity is due to the redundancy that arises from scaling, which in turn arises again from the two parameters available given the form of the wavelet basis function.

Every attempt has been made to guarantee the accuracy of the materials in the book. We would, however, appreciate readers bringing to our attention any errors that may have appeared in the final version. These errors and any comments that you may have could be e-mailed to any of the authors. In addition, the contributions from the various authors are so interleaved that it is very difficult to delineate the precise contribution of each author. Hence they are all listed as primary authors.

ACKNOWLEDGMENTS

We gratefully acknowledge Professors Carlos Hartmann (Syracuse University, Syracuse, New York), Felix Perez-Martinez (Polytechnic University of Madrid, Madrid, Spain), and Gerard J. Genello (Air Force Research Laboratory, Rome, New York) for their continued support in this endeavor.

Thanks are also due to Ms. Brenda Flowers, Ms. Maureen Marano, and Ms. Roni Balestra (Syracuse University) for their expert typing of the manuscript.

We would also like to express sincere thanks to Mr. Seongman Jang, Mr. Raul Fernandez-Recio, Dr. Yongseek Chung and Dr. Sheyun Park for their help with the book.

Tapan K. Sarkar (tksarkar@syr.edu)
Magdalena Salazar-Palma (salazar@gmr.ssr.upm.es)
Michael C. Wicks (Michael.Wicks@rl.af.mil)
New York
April 2002

Chapter 1

ROAD MAP OF THE BOOK

1.1 INTRODUCTION

The objective of this chapter is to describe the contents of this book from the perspective of the various philosophies that exist in approximating a function arising in the numerical solution of electromagnetic scattering and radiation problems, wave propagation problems, and in signal analysis. The requirements for the component functions, which are called a basis, arising in the solution of these classes of problems are quite different. In the solution of operator equations, the approximating functions are termed the basis need to satisfy certain boundary and differentiability conditions, whereas for signal analysis these restrictions are nonexistent and even the basis can be linearly dependent.

In this book, the requirements of the basis in the solution of both matrix and operator equations arising in electromagnetics are studied using a wavelet-like expansion. The impact of such a basis on the condition number of the system matrix in the solution of boundary value problems is also presented.

A methodology is presented for the design of a T-pulse (i.e., a pulse that is strictly limited in time with most of its energy concentrated in a very narrow band; in addition, it may have a zero dc component and no intersymbol interference), which can be easily generated for many practical problems, including target identification. This is particularly important for dealing with strictly time-limited and practically band-limited waveforms. Here, the strength of the Fourier techniques is demonstrated. In addition, the shortcomings of a joint time-frequency representation are outlined and a method for effectively overcoming these shortcomings is discussed.

Finally, the use of wavelets in the choice of an optimum basis through matching pursuit or from the perspective of wavelet packets is discussed. In addition, a denoising procedure well suited to wavelet processing is also outlined.

1.2 WHY USE WAVELETS?

Conventional Fourier techniques are based on the principle of dilation. By dilation of a function $f(t)$ we mean $f(kt)$ for some constant k. In the Fourier techniques, we

1

use the normalized functions $\dfrac{1}{\sqrt{2\pi}}\exp(jnt)$ that are the dilated version of exp(jt).

Hence, in the Fourier techniques we use the integer dilations n to approximate any function between, say, $[-\pi, \pi]$. This amounts to approximating any function by a set of orthogonal functions. The problem associated with a Fourier basis is that when we approximate discontinuous functions, which are defined in the whole interval of interest by a global basis set, we run into Gibb's phenomenon. It is a problem of overshooting (and undershooting) the approximation to a function by an entire domain basis occurring near the point of discontinuity.

Wavelet analysis is more general in the sense that we employ not only an orthonormal set of functions (Schauder basis) as in the Fourier techniques, but we also employ a nonorthogonal, but linearly independent basis (Riesz basis) and a collection of functions that may not be linearly independent (frames) [1]. Since wavelets can use discontinuous functions, edge effects are reproduced much better in this methodology. Therefore, wavelets can approximate discontinuous functions with a fewer number of functions than can Fourier techniques. In fact, for image analysis the wavelets are quite suitable. We would also like to explore the suitability of such methodologies in numerical electromagnetics.

1.3 WHAT ARE WAVELETS?

A wavelet is described by the function $\psi_{a,b}(t)$, which is obtained by dilation and translation of a function $\psi(t)$ as defined by [1]

$$\psi_{a,b}(t) = \frac{1}{\sqrt{a}}\psi\left(\frac{t-b}{a}\right); \quad \text{with } a > 0, \text{ and } b \in \Re \qquad (1.1)$$

where $\Re$ is the set of real numbers. Here we have assumed that the functions ψ are real. The approximation of a function $f(t)$ by wavelets is carried out using the coefficients C given by

$$C(a,b) = <f; \psi_{a,b}> \qquad (1.2)$$

where $< \bullet ; \bullet >$ defines the inner product or scalar product.

The term *wavelets* is a literal translation of the French word *ondelettes* or *petites ondes*, that is, *small waves*. Moreover, *ondelettes* is a shorter form of the phrase *ondelettes à form constante*, that is, "wavelets of constant form." In fact, the term *ondelette* or *wavelet* has been in use for a long time to imply "small waves" as opposed to "waves of infinite duration." This implies that wavelets are waves, namely, functions that are localized in frequency around a central value and that are limited in time (i.e., they are of finite support and hence localized in time around a central value). In this sense, wavelets are different from the functions that are used in windowed Fourier analysis in that they do not have a constant waveform (i.e., they have the same envelope but their shape varies with frequency) nor are they of finite support. On the other hand, the family of functions $\psi_{a,b}$ defined by (1.1) can be of finite support in time and they can be generated by scaling and dilating the same function. Thus, wavelets have a constant shape because they are generated from only one function.

We now explain how they are used.

1.4 WHAT IS THE WAVELET TRANSFORM?

The wavelet transform defined by $Wf(a,b)$ of a function $f(t)$ is defined by

$$Wf(a,b) = \int\limits_{-\infty}^{+\infty} f(t)\, \frac{1}{\sqrt{a}} \psi(\frac{t-b}{a})\, dt \;=\; f \otimes \xi_a \qquad (1.3)$$

where $\otimes$ denotes the convolution between two functions, and

$$\xi_a(u) = \frac{1}{\sqrt{a}}\, \psi(\frac{-u}{a}) \qquad (1.4)$$

In (1.4) the shift parameter b is associated with the variable u through $u = b - t$. Hence, the wavelet transform is a function of two parameters a and b. The first parameter, a, denotes the scaling of the function ψ, whereas the second variable, b, denotes a shift of the function as defined in (1.1). Hence, the wavelet transform is carried out using a dilation and shift of the same function resulting in the coefficients in (1.2).

If we define the Fourier transform of the function ξ by the function Ξ, then

$$\Im[\xi_a(u)] \;=\; \Xi(\omega) \;=\; \int_{-\infty}^{+\infty} \xi_a(u)\exp(-j\omega u)\,du \;=\; \sqrt{a}\,\Psi(a\omega) \qquad (1.5)$$

where $\Psi(\omega)$ is the Fourier transform of the function $\psi(t)$, that is, $\Im\{\psi(t)\} = \Psi(\omega)$. Since the function ψ is assumed to have no dc value, we obtain

$$\Psi(\omega=0) \;=\; \int\limits_{-\infty}^{+\infty} \psi(t)\,dt = 0 \qquad (1.6)$$

In addition, if $\Psi(\omega)$ is continuously differentiable then the following admissibility condition is satisfied:

$$C = \int_0^\infty \frac{|\Psi(\omega)|^2}{\omega}\, d\omega < +\infty \tag{1.7}$$

where C is a finite constant. Therefore, $\Psi(\omega)$ has the characteristic of a bandpass transfer function. Equation (1.6) determines the low-frequency response, whereas (1.7) tells us that the function must decay rather rapidly to infinity.

Hence, the wavelet transform is carried out with a set of dilated bandpass filters determined by the parameter a. Because we are carrying out a filtering operation through a convolution, we can use the fast Fourier transform (FFT) techniques to perform these computations numerically. This is explained in detail in Chapter 2. Since a FFT of a sequence of length M can be computed in approximately $M \log (M)$ operations, the computation of the wavelet transform is quite fast.

We now illustrate how this technique can be used in computational electromagnetics and signal analysis.

1.5 USE OF WAVELETS IN THE NUMERICAL SOLUTION OF ELECTROMAGNETIC FIELD PROBLEMS

Analyses of electromagnetic radiation, scattering from material bodies and wave propagation in a dispersive medium are carried out using Maxwell's equations. Many methods have been proposed for the solution of these classes of problems, such as the expansion method [2], regularization method [3], Backus-Gilbert method [4], Galerkin method [2], or the moment method [5]. These are essentially low-frequency techniques. They can be grouped under the general heading of the well-known moment method (MM) [5], the finite element method (FEM) [6–9], the finite difference time domain (FDTD) method [10, 11], and so on. These techniques can conveniently be applied to analyze three-dimensional electromagnetic scattering from objects whose sizes are within several wavelengths. For electrically large structures, the number of unknowns becomes too large. The use of entire domain basis functions may partially alleviate the problem associated with using too many unknowns and the need for computers with very large memory, permitting solutions to be reached for relatively large systems [12] using modest computing resources.

Methodologies for computing electromagnetic scattering from electrically large material bodies can be divided into two categories. The first category is based on the physical characteristics of the bodies. The other is based on exploiting mathematical subtleties, rather than using the physical characteristics of the bodies derived from the principles of electromagnetics.

In the first category, we also have the hybrid techniques. An overview of these techniques may be found in [13–15]. A hybrid analysis combines a high-frequency method, like geometrical theory of diffraction (GTD) and the physical theory of diffraction (PTD) with the moment method, resulting in the GTD-MM [16–20] approach or the PTD-MM [13–15] approach, respectively. In addition, there are current-based hybrid methods [21], which combine the MM technique with Ansatz surface currents obtained from the physical optics (PO) approximation, PTD, GTD, or the Fock theory. According to the physical characteristics of induced currents on the surface of these bodies, the whole boundary surface of the scatterer has been subdivided into irregular and smooth surfaces. The current on the smooth surfaces can be separated into a PO term plus a term resulting from the surface-wave effects. The current on the irregular surface and the surface-wave term can be solved using the MM technique. The details of this technique and examples can be seen in [20–22]. Furthermore, Jakobus and Lansdorfer [23, 24] suggested an improved PO-MM hybrid formulation for the solution of scattering problems from three-dimensional perfectly conducting bodies of arbitrary shape. Another variation of the hybrid technique combines the high-frequency method known as the shooting-bouncing-ray method [25–27] and the FEM [28] or the FDTD method [29] for the analysis of electromagnetic scattering from a cavity with a complex termination. A high-frequency method called the generalized ray method (GRM) [30] has been used to track fields from the mouth of a cavity up to an arbitrarily defined planar surface close to the termination. The FDTD method in addition has been applied to the small region surrounding the termination. Although these hybrid techniques can deal with scattering and radiation from complex objects, the solution of scattering with high accuracy from electrically large complex objects is still a computational challenge.

The solution methodology based on mathematical principles in computational electromagnetics includes techniques like the multilevel or multigrid method and the wavelet-like methods. The multilevel or multigrid method has been used widely in solving differential equations and integral equations in the area of computational mathematics [31–35]. The multilevel method can be used to solve electromagnetic (EM) scattering problems. A multilevel algorithm utilizing the moment method has been developed by Kalbasi and Demarest [36]. A powerful multilevel algorithm for penetrable, volumetric scatterers has been developed by Chew and Lu [37]. Another multilevel algorithm named multilevel matrix decomposition algorithm (MLMDA) has been developed by Michelssen and Boag [38, 39] for analyzing scattering from electrically large surfaces. In addition, one can also mention the spatial decomposition technique [40] and the impedance matrix localization techniques [41–44].

Another technique for the solution of integral equations is the method of moments with a wavelet basis or wavelet-like basis [45, 46]. Alpert et al. [47] introduced wavelet-like basis functions in solving second-kind integral equations. The wavelet and wavelet-like basis functions have been chosen in the method of moments for the solution of integrodifferential equations arising in electromagnetic fields problems [48]. Steinberg and Leviatan [49] used the wavelet expansions for the unknown current (function) in the moment method, which is expressed as a twofold summation of shifted and dilated forms of properly chosen basis functions. Goswami et al. [50] proposed the use of compactly supported semi-orthogonal spline wavelets constructed for analyzing the two-dimensional electromagnetic scattering problems of metallic cylinders. Wang [51] proposed the hybrid wavelet expansion and boundary element method

(HWBM) to solve two-dimensional electromagnetic scattering problems over a curved computation domain. In HWBM, the unknown surface current was presented in terms of a basis of periodic, orthogonal wavelets in the interval [0,1] on the real axes. By the use of a wavelet basis or wavelet-like basis in the method of moments, the system matrix can be transformed into a sparse matrix due to the basis functions having local support and vanishing moment properties. Hence, it can reduce memory requirements and save CPU time for the solution of electromagnetic field problems. In addition, there are fast wavelet transform-based methods that compress the matrix equations [52] and fast multipole methods [53, 54].

Mathematically, a lossy compression methodology based on wavelets approximates the original function by using a limited number of multidimensional basis functions introduced through the tensor product of various one-dimensional functions. Wavelet processing offers an important technique to compress certain classes of data which exhibit isolated band-limited properties. They are not suitable for approximating impulsive functions exhibiting a broadband spectrum and therefore compression is not very efficient. Hence, many researchers have chosen the wavelet (or wavelet-like) basis function in the moment method with the hope that such a choice may make the impedance matrix sparse. Use of this powerful technique in the solution of operator equations is presented in Chapter 3.

In the solution of operator equations, wavelets can be introduced in three different ways:

- Use a subdomain basis and convert the operator equation to a matrix equation. Then use a wavelet-like decomposition with a threshold to produce a sparse matrix. The goal here is solve the sparse matrix equation in an efficient way. The wavelet-like approach is used directly on the impedance matrix in order to make it sparse rather than on the operator equations in the form of a set of basis functions to approximate the unknown. The result of making the matrix sparse changes its condition number. This technique is described in detail in Chapter 4 along with an efficient numerical implementation.

- Use wavelets as basis functions. In this case, they are entire domain-basis functions and are used directly in the solution of operator equations. One can apply this concept to either the integral form or the differential form of Maxwell's equations. The solution of integral equations using a wavelet-like basis is addressed in Chapter 3, while Chapter 5 describes the application of this methodology for the efficient solution of the differential form of Maxwell's equations.

- A multiscale basis may be used to make the solution sparse. This is accomplished a priori by making some of the unknown coefficients associated with a wavelet basis zero even though the solution is unknown. By applying this pruning of the unknown coefficients associated with the solution, one can eliminate rows and columns in the impedance matrix generated by a subdomain expansion and a thresholding operation. In a thresholding operation, elements whose magnitude is smaller than a prespecified value are set to zero. Once those elements in the matrix containing the unknowns are removed, the corresponding rows and columns of the impedance matrix are also eliminated along with that component of the unknown. From the point of view of combining the wavelet method with the multigrid method, use of the adaptive multiscale moment method (AMMM) [55–61] has been proposed to study the numerical solutions of Fredholm integral equations of the first kind. Using a special multiscale basis function on a bounded interval, which is similar to a wavelet-like basis function, AMMM has been introduced to overcome the limitation of a conventional moment method. The multiscale basis functions possess three important properties. First, they are equivalent to the triangular basis functions on the bounded interval. This means that there exists a matrix transformation within the array of the unknown coefficient matrix between the conventional MM basis and the AMMM basis. Second, when going from a lower to a higher scale, these functions are orthogonal among themselves on the new higher scale, and they are zero at the nodes of the previous scale. That means that adding new basis functions in AMMM will not change the values of the solution at the nodes corresponding to a previously computed solution at a lower scale. Third, the approximating function will not increase the number of terms when one scale is increased if the original function is linear. If the function is linear on some local interval, then this will not increase the number of terms in the approximating function on this interval. This means that we can use relatively few terms to represent the approximating function. This is described in Chapter 6.

This summarizes the application of wavelets for the solutions of operator equations arising in electromagnetics. In the first six chapters we deal with these classes of problems.

Next we look at the wavelet methodology for signal analysis. Here the goal is to characterize a time-varying phenomenon through wavelets. This is discussed in the following section.

1.6 WAVELET METHODOLOGIES COMPLEMENT FOURIER TECHNIQUES

The Fourier technique is a useful tool for analyzing and approximating functions using an orthonormal set of basis functions, which generally consist of the polynomials of trigonometric functions. In addition, the Fourier series converges in the mean when analyzing functions that are periodic. The convergence is pointwise if the function is continuous. However, the approximation methodology displays a Gibb's phenomenon when the Fourier technique is used to approximate a discontinuous function. Typically, we use this methodology to determine the extent of support of the spectrum of a given waveform. Hence, we are interested in characterizing the transform of the given wave shape and the region of support in the transform domain.

Here, we generally have a uniform subdivision of the axis dealing with the transform variable to characterize the spectrum. This is a serious limitation of the Fourier technique because it is not practical to observe the variations in the spectrum of a waveform, for example, over a 1-MHz window when the signal has 10 GHz of bandwidth. However, it may be useful to look at the spectrum of the signal with a 1-MHz window when it has a 10-MHz bandwidth. In addition, the resolution with which one can observe the spectrum is limited by the duration of the signal. For example, if a signal has two frequency components located at 10 and 20 Hz, then we need a signal of at least a $1/(20-10)$ seconds duration. Mathematically, this is termed the Heissenberg principle of uncertainty. This also tells us that if a signal is strictly time limited, it cannot simultaneously be band limited and vice versa. The resolution of the Fourier methodology (i.e., to resolve two spectral components from a finite time-limited description) is bounded by Heissenberg's principle of uncertainty. The wave shape that has the smallest time bandwidth spread has a Gaussian shape. Such a functional representation satisfies the uncertainty principle with equality, and all other waveforms have a larger time-bandwidth spread. Furthermore, to observe a time-varying phenomenon like the fields radiating from a structure as a function of time, we may want to know what the instantaneous bandwidth is or how effectively the frequency content changes as a function of time. One way to generate this information is to use a short time Fourier transform.

Wavelet methods, on the other hand, address similar issues and complements the Fourier techniques. First of all, wavelets are a function of two parameters, dilation and shift, whereas the Fourier methodology has only dilation. Wavelets essentially look at the spectrum with a constant Q window. The parameter Q is defined as the quality of a signal and is related to the relative bandwidth of the wave shape with respect to its center frequency. This parameter may be used to characterize the response of band-limited signals. A constant Q therefore implies that the ratio of the center frequency with respect to the bandwidth is constant. Thus, as we translate the window to a lower frequency, its support becomes smaller, whereas when we move it to a higher-frequency region, the window becomes wider, so that the ratio remains the same. Therefore, the bandwidth of the window is a function of the center frequency of observation for a wavelet, whereas for Fourier techniques the observation window is the same irrespective of the center frequency of the observation. In Fourier techniques one looks at the spectrum with a uniform bandwidth irrespective of where the window function is located (either in the high- or low-frequency regimes). For example, if one is observing a signal which has a spectrum of 10 MHz with a 1-MHz window, then at 1 GHz the Fourier technique observes the same signal through the same 1-MHz window. A wavelet method, on the other hand, observes the same spectrum of 10 MHz with a 1-MHz window, but when the signal has a center frequency of 1 GHz, it is observed with a 100-MHz window. In this way, the ratio of the bandwidth of the signal with respect to the center frequency is a constant, whereas with Fourier techniques, the spectrum of the observation window is the same irrespective of the center frequency. Thus, the width of the window through which we observe the signal changes as a function of location, whereas in the Fourier technique, it does not. However, the resolution in a wavelet-based technique is still limited by Heissenberg's principle of uncertainty just like for Fourier techniques. The representation of the window function through which we observe a signal in the time-frequency plane is illustrated through Figure 1.1.

Functional Representation

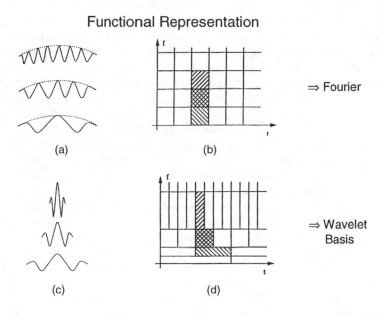

Figure 1.1   Philosophy of the wavelet and the Fourier methodology in characterizing a waveform in the time-frequency plane: (a) modulation and shift of a Gaussian window, (b) time-frequency representation, (c) shift and scaling of a wavelet, and (d) time-frequency representation.

In Figure 1.1(a) we see the window function in the time and in the frequency domain for the Fourier methodology. We observe a time-varying phenomenon with a resolution, which is independent of both time and frequency because the windows are uniform and square. So in a Fourier technique a function is approximated by modulation and shift of a function. This yields in the time-frequency plane regions that are of equal area.

For a wavelet methodology, the window function in the time-frequency plane is nonuniform and is a function of both time and frequency. Here, the approximation of a function is carried out by the principle of dilation and shift of the same function as shown in Figure 1.1(c). In the time-frequency plane this gives rise to approximating the time-frequency continuum in a set of nonuniform regions corresponding to the basis functions.

Mathematical properties and salient features of the wavelet techniques are described in Chapters 7 to 9.

1.7 OVERVIEW OF THE CHAPTERS

In Chapter 2, we describe the wavelet methodology utilizing the principles of filter theory. Using these principles, it is easy to visualize specifically for electrical engineers the principles of wavelets. For Fourier techniques, the approximating or basis functions are the continuous trigonometric functions, whereas for wavelets they can be either discontinuous like a pulse function (which is associated with the Haar wavelet described in Chapter 3) or they can be continuous. Therefore, in the wavelet methodology when discontinuous basis functions are used to approximate a discontinuous function, it does not display the Gibb's phenomenon, unlike its Fourier counterpart. However, if a continuous wavelet basis is used to represent a discontinuous function, then the approximation displays Gibb's phenomenon. Graphically the difference between representing a function in a Fourier or wavelet basis is illustrated in Figure 1.1.

Chapter 4 presents the wavelet methodology for compressing a system of matrix equations and thereby making the method of solution more efficient. In Chapter 5, we show how to employ the wavelet-like basis for efficient solution of the differential form of Maxwell's equations in rectangular regions. The adaptive multiscale moment method is described in Chapter 6 where the wavelet-like method is applied to the unknown solution instead of the impedance matrix, even though it is unknown. For the simultaneous time-frequency representation of a function, it is necessary that physical principles not be violated. Hence, it is necessary to develop a characterization that does not produce instantaneous power spectral densities that are negative. This issue is addressed in Chapter 7 where window functions are presented that do not produce physically meaningless solutions.

In Chapter 8, we show how to use numerical optimization to obtain a pulse that is strictly limited in time but whose energy is focused in a very narrow band. We call such a pulse the T-pulse. The shift orthogonality property can be used to design a pulse shape with zero intersymbol interference and with no dc component (i.e., no component at zero frequency or direct current). This may help in the design of window functions with better desired properties in the Fourier domain.

We present the concept of matching pursuits in Chapter 9 in terms of wavelet packets. Hence, a concept similar to singular value decomposition can be used in the optimum choice of a basis but with significantly reduced computational requirements. The concept of denoising is also introduced so as to improve the signal-to-noise ratio of signals through the use of wavelets. The advantage here is that the computational requirements are significantly less than that of comparable methodologies in linear algebra, which typically requires at least an order of magnitude more computations.

Finally, we provide a selected bibliography of various books and journal papers that have presented the various wavelet concepts for the topics that we have addressed in this book. The selected references have been chosen to illustrate where additional materials are available. No attempt has been made to provide earlier reference material.

REFERENCES

[1] P. G. Lemarie, "Introduction a la theorie des ondelettes," in *Les Ondelettes en 1989*, P. G. Lemarie, Ed., Lecture Notes in Mathematics, Springer-Verlag, New York, 1990, pp. 1–13.

[2] L. M. Delves, *Numerical Solution of Integral Equations*, Clarendon Press, Oxford, 1974.

[3] A. N. Tikhonov and V. Y. Arsenin, *On the Solution of Ill-Posed Problems*, John Wiley and Sons, NewYork, 1977.

[4] G. Backus and F. Gilbert, "Numerical applications of a formalism for geophysical inverse problems," *Geophys. J. Roy. Astron. Soc.*, Vol. 13, pp. 247–276, 1967.

[5] R. F. Harrington, *Field Computation by Moment Method*, Macmillan, New York, 1968.

[6] J. M. Jin, *The Finite Element Method in Electromagnetics*, Wiley, New York, 1993.

[7] M. Salazar-Palma et al., *Iterative and Self-Adaptive Finite-Elements in Electromagnetic Modeling*, Artech House, Norwood, MA, 1998.

[8] J. M. Jin and J. L. Volakis, "A finite element boundary integral formulation for scattering by three-dimensional cavity-backed apertures," *IEEE Trans. Antennas Propagat.*, Vol. AP-39, No.1, Jan. 1991, pp. 97–104.

[9] J. M. Jin and J. L. Volakis, "A hybrid finite element method for scattering and radiation by microstrip patch antennas and arrays residing in a cavity," *IEEE Trans. Antennas Propagat.*, Vol. AP-39, No. 11, Nov. 1991, pp. 1598–1604.

[10] K. S. Kunz and R. J. Luebbers, *Finite Difference Time Domain Method for Electromagnetics*, CRC Press, Boca Raton, FL, 1993.

[11] A. Taflove, *Computational Electrodynamics: The Finite Difference Time Domain Method*, Artech House, Norwood, MA, 1995.

[12] B. M. Kolundzija, J. Ognjanovic, and T. K. Sarkar, *WIPL-D Electromagnetic Modeling of Composite Metallic and Dielectric Structures*, Artech House, Norwood, MA, 2000.

[13] J. Bouche, F. A. Molinet, and R. Mittra, "Asymptotic and hybrid techniques for electromagnetic scattering," *Proc. IEEE*, Vol. 81, No. 12, Dec. 1993, pp. 1658–1684.

[14] G. A. Thiele, "Overview of selected hybrid method in radiating system analysis," *Proc. IEEE*, Vol. 80, No. 1, Jan. 1992, pp. 67–78.

[15] L. N. Medgyesi-Mitschang and D. S. Wang, "Hybrid methods in computational electromagnetics: A review," *Computer Physics Communications*, Vol. 68, May 1991, pp. 76–94.

[16] G. A. Thiele and T. H. Newhouse, "A hybrid technique for combining moment methods with a geometrical theory of diffraction," *IEEE Trans. Antennas Propagat.*, Vol. AP-23, Jan. 1975, pp. 62–69.

[17] W. D. Burnside, C. L. Lu, and R. J. Marhefka, "A technique to combine the geometric theory of diffraction and the moment method," *IEEE Trans. Antennas Propagat.*, Vol. AP-23, May 1975, pp. 551–558.

[18] E. P. Ekelman and G. A. Thiele, "A hybrid technique for combining the moment method treatment of wire antennas with the GTD for curved surfaces," *IEEE Trans. Antennas Propagat.*, Vol. AP-28, June 1980, pp. 831–839.

[19] J. N. Sahalos and G. A. Thiele, "On the application of the GTD-MM technique and its limitation," *IEEE Trans. Antennas Propagat.*, Vol. AP-29, June 1981, pp. 780–786.

[20] L. N. Medgyesi-Mitschang and D. S. Wang, "Hybrid solutions for scattering from large bodies of revolution with material discontinuities and coatings," *IEEE Trans. Antennas Propagat.*, Vol. AP-32, June 1984, pp. 717–723.

[21] D. S. Wang, "Current-based hybrid analysis for surface-wave effects on large scatterers," *IEEE Trans. Antennas Propagat.*, Vol. AP-39, June 1991, pp. 839–850.

[22] L. N. Medgyesi-Mitschang and D. S. Wang, "Hybrid solutions for scattering from perfectly conducting bodies of revolution," *IEEE Trans. Antennas Propagat.*, Vol. AP-31, May 1983, pp. 570–583.

[23] U. Jakobus and F. M. Lansdorfer, "Improved PO-MM hybrid formulation for scattering from three-dimensional perfectly conducting bodies of arbitrary shape," *IEEE Trans. Antennas Propagat.*, Vol. AP-43, Feb. 1995, pp. 162–169.

[24] U. Jakobus and F. M. Lansdorfer, "Improvement of the PO-MM hybrid method by accounting for effects of perfectly conducting wedges," *IEEE Trans. Antennas Propagat.*, Vol. AP-43, Oct. 1995, pp. 1123–1129.

[25] H. Ling, R. C. Chou, and S. W. Lee, " Ray versus modes: Pictorial display of energy flow in an open-ended waveguide," *IEEE Trans. Antennas Propagat.*, Vol. AP-35, No. 3, May 1987, pp. 605–607.

[26] H. Ling, R. C. Chou, and S. W. Lee, "Shooting and bouncing ray: Calculating the RCS of an arbitrary shaped cavity," *IEEE Trans. Antennas Propagat.*, Vol. AP-37, No. 2, Feb. 1989, pp. 194–205.

[27] J. Baldauf et al., "High frequency scattering from trihedral corner reflectors and other benchmark targets: SBR versus experiments," *IEEE Trans. Antennas Propagat.*, Vol. AP-39, No. 9, Sept. 1991, pp. 1345–1351.

[28] J. Jin, S. S. Ni, and S. W. Lee, "Hybridization of SBR and FEM for scattering by large bodies with cracks and cavities," *IEEE Trans. Antennas Propagat.*, Vol. AP-43, Oct. 1995, pp. 1130–1139.

[29] R. Lee and T. T. Chia, " Analysis of electromagnetic scattering from a cavity with a complex termination by means of a hybrid Ray-FDTD method," *IEEE Trans. Antennas Propagat.*, Vol. AP-41, No.11, Nov. 1993, pp. 1560–1564.

[30] R. J. Burkholder, "High-frequency asymptotic methods for analyzing the EM scattering by open-ended waveguide cavities," Ph.D dissertation, The Ohio State University, Columbus, OH, 1989.

[31] A. Brandt, "Multi-level adaptive solutions to boundary value problems," *Mathematics of Computation*, Vol. 31, 1977, pp. 330–390.

[32] W. Hackbusch, *Multigrid Methods and Applications*, Springer-Verlag, New York, 1985.

[33] S. F. McCormick, *Multigrid Methods: Theory, Applications, and Super-computing*, Marcel Dekker, New York, 1988.

[34] J. Mandel, "On multilevel iterative methods for integral equations of the second kind and related problems," *Numer. Math.*, Vol. 46, 1985, pp. 147–157.

[35] P. W. Hemker and H. Schippers, "Multiple grid methods for the solution of Fredholm integral equations of the second kind," *Mathematics of Computation*, Vol. 36, No. 153, 1981.

[36] K. Kalbasi and K. R. Demarest, "A multilevel formulation of the method of moments," *IEEE Trans. Antennas Propagat.*, Vol. AP-41, No. 5, May 1993, pp. 589–599.

[37] W. C. Chew and C. C. Lu, "The use of Huygen's equivalence principle for solving the volume integral equation of scattering," *IEEE Trans. Antennas Propagat.*, Vol. AP-41, No. 6, July 1993, pp. 897–904.

[38] E. Michelssen and A. Boag, "Multilevel evaluation of electromagnetic fields for the rapid solution of scattering problems," *Microwave Opt. Technol. Lett.*, Vol.7, Dec. 1994, pp. 790–795.

[39] E. Michelssen and A. Boag, "A multilevel matrix decomposition algorithm for analyzing scattering from large structures," *IEEE Trans. Antennas Propagat.*, Vol. AP-44, No. 8, Aug. 1996, pp. 1086–1093.

[40] K. R. Umashankar, S. Nimmagadda, and A. Taflove, "Numerical analysis of electromagnetic scattering by electrically large objects using spatial decomposition technique," *IEEE Trans. Antennas Propagat.*, Vol. AP-40, No. 8, 1992, pp. 867–877.

[41] F. X. Canning, "The impedance matrix localization method (IML)," *IEEE Trans. Antennas Propagat.*, Vol. AP-41, No. 5, 1993, pp. 659–667.

[42] F. X. Canning, "The impedance matrix localization method (IML) permits solution of large scatterers," *IEEE Trans. Magnetics*, Vol. 27, Sept. 1991, pp. 4275–4277.

[43] F. X. Canning, "The impedance matrix localization method (IML) for MM calculation," *IEEE Antennas Propagat. Magazine*, Vol. 32, Oct. 1990, pp. 18–30.

[44] F. X. Canning, "Transformations that produce a sparse moment method matrix," *J. Electromag. Wave Applicat.*, Vol. 4, No. 9, 1990, pp. 893–913.

[45] V. Rohklin, "Rapid solution of integral equations of scattering in two dimensions," *J. Comput. Phys.*, Vol. 86, 1990, pp. 414–439.

[46] V. Rokhlin, "Rapid solution of integral equations of classical potential theory," *J. Comput. Phys.*, Vol. 60, 1985, pp. 187–207.

[47] B. K. Alpert et al., "Wavelet-like bases for the fast solution of second-kind integral equations," *SIAM J. Sci. Comp.*, Vol. 14, Jan. 1993, pp. 159–184.

[48] R. L. Wagner, P. Otto, and W. C. Chew, "Fast waveguide mode computation using wavelet-like basis functions," *IEEE Microwave Guided Wave Lett.*, Vol. 3, July 1993, pp. 208–210.

[49] B. Z. Steinberg and Y. Leviatan, "On the use of wavelet expansions in the method of moments," *IEEE Trans. Antennas Propagat.*, Vol. AP-41, No. 5, 1993, pp. 610–619.

[50] C. Goswami, A. K. Chan, and C. K. Chui, "On solving first-kind integral equations using wavelets on a bounded interval," *IEEE Trans. Antennas Propagat.*, Vol. AP-43, No. 6, June 1995, pp. 614–622.

[51] G. Wang, "A hybrid wavelet expansion and boundary element analysis of electromagnetic scattering from conducting objects," *IEEE Trans. Antennas Propagat.*, Vol. AP-43, No. 2, Feb. 1995, pp. 170–178.

[52] H. Kim and H. Ling, "On the application of fast wavelet transform to the integral equation of electromagnetic scattering problems," *Microwave Opt. Technol. Lett.*, Vol. 6, No. 3, March 1993, pp. 168–173.

[53] R. Coifman, V. Rokhlin, and S. Wandzura, "The fast multipole method for the wave equation: A pedestrian prescription," *IEEE Antennas Propagat. Mag.*, Vol. 35, 1993, pp. 7–12.

[54] W. C. Chew et al., "Fast solution methods in electromagnetics," *IEEE Trans. Antennas Propagat.*, Vol. AP-45, No. 3, March 1997, pp. 533–543.

[55] C. Su and T. K. Sarkar, "A multiscale moment method for solving Fredholm integral equation of the first kind, " *J. Electromag. Waves Appl.*, Vol. 12, 1998, pp. 97–101.

[56] C. Su and T. K. Sarkar, "Scattering from perfectly conducting strips by utilizing an adaptive multiscale moment method," *Progress in Electromagnetics Research,* PIER 19, 1998, pp. 173–197.

[57] C. Su and T. K. Sarkar, "Electromagnetic scattering from coated strips utilizing the adaptive multiscale moment method," *Progress in Electromagnetics Research*, PIER 18, 1998, pp. 173–208.

[58] C. Su and T. K. Sarkar, "Electromagnetic scattering from two-dimensional electrically large perfectly conducting objects with small cavities and humps by use of adaptive multiscale moment methods (AMMM)," *J. Electromag. Waves Appl.*, Vol. 12, 1998, pp. 885–906.

[59] C. Su and T. K. Sarkar, "Adaptive multiscale moment method for solving two-dimensional Fredholm integral equation of the first kind," *J. Electromag. Waves Appl.*, Vol. 13, No. 2, 1999, pp. 175–176.

[60] C. Su and T. K. Sarkar, "Adaptive Multiscale Moment Method (AMMM) for analysis of scattering from perfectly conducting plates," *IEEE Trans. Antennas Propaga.*, Vol. 48, No. 6, June 2000, pp. 932–939.

[61] C. Su and T. K. Sarkar, "Adaptive multiscale moment method (AMMM) for analysis of scattering from three-dimensional perfectly conducting structures," *IEEE Trans. Antennas Propagat.*, Vol. 49, No. 3, March 2002 (to be published).

Chapter 2

WAVELETS FROM AN ELECTRICAL ENGINEERING PERSPECTIVE

The objective of this chapter is to present the subject of wavelets from an electrical engineering perspective. This is carried out using filter theory concepts, which are quite familiar to electrical engineers. Such a presentation provides both physical and mathematical insights into the problem [1–3]. It is shown that taking the discrete wavelet transform of a function is equivalent to filtering it by a bank of constant Q filters whose nonoverlapping bandwidths (for a set of ideal filters) differ by an octave. The discrete wavelets are presented and a recipe is provided describing how to generate such entities. This chapter presents wavelet decomposition starting with the fundamentals and shows how the scaling functions and wavelets are generated from the filter theory perspective. Finally, we illustrate how a function is approximated in terms of the scaling functions and wavelets and how this representation can be computationally determined in an efficient fashion. Examples are presented to illustrate the class of problems for which the discrete wavelet techniques are ideally suited. It is interesting to note that it is not necessary to generate the wavelets or the scaling functions in order to implement the discrete wavelet transform.

2.1 INTRODUCTION

Wavelets are a set of functions that can be used effectively in a number of situations to represent naturally occurring highly transient phenomena, which result from a dilation and shift of the original waveform. For example, when a pulse propagates through a layered medium, where different layers have different electrical properties, the pulse gets dilated and is delayed due both to dispersion and the finite velocity of propagation. The application of wavelets was first carried out in the area of geophysics [4] in 1980 by the French geophysicist J. Morlet of Elf-Aquitane and his coworkers. A development of the wavelet theory from the mathematical perspective is available in a special issue of the *IEEE Proceedings* [5]. In electrical engineering [6–9], however, it was popular for some time under the various names of multirate sampling, quadrature mirror filters, and so on. In this chapter, the theory of the discrete wavelet transform is developed from a filter theory perspective to demonstrate that the discrete wavelet transform is equivalent to filtering a signal by a band of constant Q filters whose nonoverlapping bandwidths (when dealing with a set of ideal constant Q filters) differ by an octave, that is, by a factor of 2. It is hoped that this mode of presentation will make it easier to visualize, conceptualize, and apply the theory to the problem at hand, if the wavelet theory is relevant! Then, a connection is made with scaling functions and wavelets and how they relate to the filters. A basic understanding of these principles can help one design a wavelet specially tailored to one's needs.

Consider a signal $x(t)$ whose Fourier transform is $X(\omega)$. Here t represents time in seconds, and ω stands for the angular frequency in radians per second. If f represents frequency in hertz (Hz), then $\omega = 2\pi f$. In this chapter, we only deal with discrete wavelet techniques. The discrete wavelet techniques are quite suitable for discrete signal processing, for example, in speech and image processing. Their applications are particularly desirable in data compression.

2.2 DEVELOPMENT OF THE DISCRETE WAVELET METHODOLOGY FROM FILTER THEORY CONCEPTS

2.2.1 Preliminaries

Let us assume that the signal $x(t)$ has been sampled in order to obtain the discrete signal $x(n)$ represented by the sequence

$$x(n) \text{ for } n = 0, 1, 2, \ldots \tag{2.1}$$

where n stands for the sample number taken at the time instance $n\Delta t$, where Δt stands for the time interval between samples. Without loss of generality, we can set $\Delta t = 1$. Then the Fourier transform of $x(n)$ is best handled by the Z transform [1] (lowercase letters represent the function in the original domain and uppercase letters are used for the Z transform), namely,

$$X(z) = X(e^{j2\pi f}) = X(e^{j\omega}) = \sum_n x(n) z^{-n} \tag{2.2}$$

where $\omega = 2\pi f$, and $z = e^{j\omega} = e^{j2\pi f}$. The continuous signal $x(t)$ is assumed to be band limited and sampled at a frequency of 1 Hz to produce the discrete signal $x(n)$. Thus the sampling interval is $\Delta t = 1$ second. From the Nyquist sampling criteria, it is then necessary for the signal $x(n)$ to be band limited to ½ Hz so that it can be sampled at two times its bandwidth at the Nyquist frequency [1], without aliasing. The bandwidth of the signal (in angular frequency) B_ω is

$$B_\omega = 2\pi B_f = 2\pi \cdot \tfrac{1}{2} = \pi \tag{2.3}$$

and the ideal sampling angular frequency ω_{samp} is

$$\omega_{samp} = 2\pi \qquad (2.4)$$

Now let us filter the signal $x(n)$ through a lowpass digital filter with a transfer function $H(z)$ of bandwidth $\pi/2$ (i.e., $0 \le \omega \le \pi/2$) and a highpass digital filter $G(z)$ with a bandwidth from $\omega = \pi/2$ to π. Let $u'(n)$ be the lowpass filtered signal [i.e., $u'(n)$ has been obtained by passing $x(n)$ through the lowpass filter of impulse response $h(n)$] and $v'(n)$ be the highpass filtered signal [i.e., $v'(n)$ has been obtained by passing $x(n)$ through the highpass filter of impulse response $g(n)$]. These two filters are called analysis filters. This is shown in Figure 2.1. The bandwidth of the original signal is the sum of the bandwidths of the lowpass and highpass filters. The bandwidth of the signals $u'(n)$ and $v'(n)$ has now been reduced by a factor of 2, since they are the result of filtering the original signal by a filter of half the bandwidth. Then, according to the Nyquist sampling criteria, the sampling rate which forms the digitized signal can be reduced by a factor of 2 over the rate at which $x(t)$ has been sampled to produce $x(n)$. A reduction of the sampling rate can be achieved for $u'(n)$ and $v'(n)$ by decimating them by a factor of 2. This reduction in the sampling rate does not produce any aliasing of the spectrum of these two signals. The principle of decimation is equivalent to reducing the sampling rate. Decimation or downsampling by a factor of 2 implies that alternate samples of the signal are dropped. So now the number of data samples for each of the low- and highpass signals is reduced by 2 as illustrated in Section 2A.1 of Appendix 2A.

The purpose of decimation is to reduce the effective sampling rate and thereby the bandwidth of the signal. However, the total number of samples remains the same. The difference is seen in how they are distributed. This downsampling is possible because both $u'(n)$ and $v'(n)$ have an effective bandwidth $f = \frac{1}{4}$ or $\omega = \pi/2$, because they have been filtered. Thus, both $u'(n)$ and $v'(n)$ are downsampled by a factor of 2 resulting in $u(n)$ and $v(n)$. This subsampling can continue further as we shall see later on. Here we will restrict ourselves to the two stages of filtering $x(n)$ by $h(n)$ and $g(n)$ for illustration purposes. In the transmitter block of Figure 2.1, the goal is to generate the signals $u(n)$ and $v(n)$ from $x(n)$. These signals have a smaller bandwidth, exactly half, of the original signal. In the receiver block of Figure 2.1, the two signals $u(n)$ and $v(n)$ are to be processed in order to recover the original signal $x(n)$. The downsampled versions of $u(n)$ and $v(n)$ can be transmitted at a much lower bit rate than the original signal without any loss of information because they have smaller bandwidths than the original signal [1].

The problem in the receiver block is how to process signals $u(n)$ and $v(n)$ in order to have a perfect reconstruction of the original signal $x(n)$. The perfect reconstruction is carried out in the receiver through upsampling signals $u(n)$ and $v(n)$ by a factor of 2 and then filtering the signal through the synthesis filters of impulse responses $h'(n)$ and $g'(n)$. The principle of upsampling involves introducing a zero between the sampled values of the signal and is described in Section 2A.2 of Appendix 2A. The outputs from the synthesis filters are combined to obtain the reconstructed signal $\tilde{x}(n)$.

The next section describes how to design the filters $H(z)$, $G(z)$, $H'(z)$, and $G'(z)$ [where $H'(z)$ and $G'(z)$ represent the transfer functions of the synthesis or reconstruction filters of impulse responses $h'(n)$ and $g'(n)$, respectively] in order to obtain a perfect reconstruction of $x(n)$ from its approximation $\tilde{x}(n)$. Splitting the

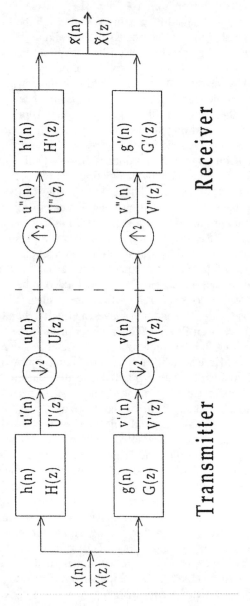

Figure 2.1 The principles of subband filtering.

original signal $x(n)$ into two bands comprising the low- and high-frequency components is termed subband splitting. Such a separation of the signal into low- and high-frequency parts has many advantages:

1. This methodology results in a decimated signal (i.e., some of the sample values of the signal have been deleted or set to zero), namely, the sampling rate can be reduced without any loss of information. [Note that $u(n)$ and $v(n)$ have a lower bandwidth than the original signal $x(n)$ and hence the sample rate has been reduced by a factor of 2.] This is equivalent to saying that $u(n)$ and $v(n)$ have been decimated by a factor of 2.

2. In the transmitter block of a real system, we usually take the signals $u(n)$ and $v(n)$, quantize their amplitudes, and then transmit them. In the receiver these quantized signals are received and they are used to reconstruct the original signal. It has been observed in such systems that even if the two signals are quantized in a very rough fashion (say, quantized into 2 bits rather than the conventional 8 bits or 16 bits), the accuracy of the reconstruction is quite high even in the presence of large quantization errors in $u(n)$ and $v(n)$ [1–9]. This has been demonstrated in analysis of images. Since a matrix is a digitized version of a two-dimensional image, we will demonstrate that this same phenomenon also happens when dealing with large matrices arising in the solution of electromagnetic field problems. Hence, this principle can be applied to make the large complex dense matrices sparse. Also, can we apply this methodology to the efficient solution of operator equations? We will address this issue later.

2.2.2   Development of the Quadrature Mirror Filters

In this section, we develop the mathematical properties necessary for the four filters $h(n)$, $g(n)$, $h'(n)$, and $g'(n)$ to obtain a perfect reconstruction for the signal $x(n)$. At the first step, a transmitter generates the decimated signals $u(n)$ and $v(n)$. These signals are transmitted. In the receiver, the signals are first upsampled. Then, they are filtered by the two receiving filters $g'(n)$ and $h'(n)$. Their outputs are combined to form $\tilde{x}(n)$. Now let us see how $x(n)$ is related to its estimate $\tilde{x}(n)$ and then the methodology to extract $x(n)$ will be obvious. Observe that

$$U'(z) = H(z)\ X(z) \tag{2.5}$$

$$V'(z) = G(z)\ X(z) \tag{2.6}$$

and from Section 2A.1 [since $u(n)$ and $v(n)$ have been decimated by a factor of 2], the following expressions hold:

$$U(z) = 0.5 \times \left[U'(\sqrt{z})\ + U'(-\sqrt{z}) \right] \tag{2.7}$$

$$V(z) = 0.5 \times \left[V'(\sqrt{z})\ + V'(-\sqrt{z}) \right] \tag{2.8}$$

and by using the results of Section 2A.2 [where $u''(n)$ and $v''(n)$ have been upsampled by a factor of 2], one may write

$$U''(z) = U(z^2) = 0.5 \times [U'(z) + U'(-z)] \tag{2.9}$$

$$V''(z) = V(z^2) = 0.5 \times [V'(z) + V'(-z)] \tag{2.10}$$

Therefore,

$$\tilde{X}(z) = \frac{1}{2}[\{G(z)X(z) + G(-z)X(-z)\}G'(z) \tag{2.11}$$

$$+ \{H(z)X(z) + H(-z)X(-z)\}H'(z)]$$

$$= \frac{1}{2}[\{G(z)G'(z) + H(z)H'(z)\}X(z)$$

$$+ \{G(-z)G'(z) + H(-z)H'(z)\}X(-z)] \tag{2.12}$$

where $\tilde{X}(z)$, $G'(z)$, $H'(z)$, $G(z)$, $H(z)$, and $X(z)$ are the Z transforms of $\tilde{x}(n)$, $g'(n)$, $h'(n)$, $g(n)$, $h(n)$, and $x(n)$, respectively. The Z transform has been defined by (2.1).

Therefore, the estimated signal $\tilde{X}(z)$ contains the original signal, which is given by the first term, and an aliased part, which is given by the second term of (2.12). Now to remove the aliasing effect, the second term must be zero, that is,

$$H(-z)H'(z) + G(-z)G'(z) = 0 \tag{2.13}$$

Let $H(z)$ be a FIR (finite impulse response) filter of order N and of length $N + 1$ [1]. Therefore, $h(n)$ would have $N + 1$ terms. We consider N to be always odd. For N even, the derivation that follows would be done in a different way. Then

$$H(z) = h(0) + h(1)z^{-1} + \ldots + h(N)z^{-N} \tag{2.14}$$

Without loss of generality and for convenience we choose

$$H'(z) = z^{-N}H(z^{-1}) \quad \text{so that} \quad h'(n) = h(N - n) \tag{2.15}$$

The factor z^{-N} is used to guarantee causality of the filters $H'(z)$, that is, $h'(n) = 0$ for $n < 0$. The highpass filter $g'(n)$ is chosen in such a way that

$$G'(z) = z^{-N}G(z^{-1}) \quad \text{with} \quad g'(n) = g(N - n) \tag{2.16}$$

In addition, we define the highpass filters $g'(n)$ and $g(n)$ in terms of the lowpass filter coefficients $h'(n)$ and $h(n)$, respectively, by relating

$$g'(n) = -(-1)^n h'(N-n) = -(-1)^n h(n)$$
$$g(n) = (-1)^n h(N-n) = (-1)^n h'(n)$$
$$(2.17)$$

so that

$$G'(z) = z^{-N} H'(-z^{-1}) = -H(-z)$$
$$G(z) = -z^{-N} H(-z^{-1}) = H'(-z)$$
$$(2.18)$$

Substitution of (2.18) into (2.13) demonstrates that all four equations of (2.15) to (2.18), are consistent and the aliased component due to $X(-z)$ is zero. Furthermore, substituting (2.13), (2.15), and (2.16) into (2.12) and taking into account (2.18), (2.12) simplifies to

$$\tilde{X}(z) = 0.5 \times \left[z^{-N} G(z)G(z^{-1}) + z^{-N} H(z)H(z^{-1}) \right] X(z)$$
$$= 0.5 \times \left[H(-z^{-1}) H(-z) + H(z) H(z^{-1}) \right] z^{-N} X(z)$$
$$(2.19)$$

If the filter $H(z)$ is chosen in such a way that

$$H(z)H(z^{-1}) + H(-z)H(-z^{-1}) = 2 \qquad (2.20)$$

or

$$\left| H(e^{j\omega}) \right|^2 + \left| H[e^{j(\omega+\pi)}] \right|^2 = 2 \qquad (2.21)$$

then one would have a perfect reconstruction property as

$$\tilde{X}(z) = z^{-N} X(z); \quad \text{or equivalently} \quad \tilde{x}(n) = x(n-N) \qquad (2.22)$$

so that the reconstructed signal is exactly the original signal $x(n)$, but delayed by N samples.

The filters $H(z)$ and $G(z)$ are called quadrature mirror filters (QMF). Moreover, we would like $H(z)$ and $G(z)$ to be FIR filters as opposed to infinite impulse response (IIR) filters [1] for ease of numerical computations. To illustrate the nature of the various FIR filters given by the various equations, consider as an example $N = 3$ [9]. Then the four filters $H(z)$, $G(z)$, $H'(z)$, and $G'(z)$ of order 3 will have four nonzero coefficients in their expansion. Or equivalently $h(n)$, $g(n)$, $h'(n)$, and $g'(n)$ will have four entries. If we choose

$$H(z) = \sum_{n=0}^{N=3} h(n) z^{-n} = h(0) + h(1)z^{-1} + h(2)z^{-2} + h(3)z^{-3} \qquad (2.23)$$

Then, from (2.18),

$$G(z) = -z^{-3} H(-z^{-1}) = -z^{-3} \left[h(0) - h(1)z^{+1} + h(2)z^{+2} - h(3)z^{+3} \right]$$
$$= +h(3) - h(2)z^{-1} + h(1)z^{-2} - h(0)z^{-3} \qquad (2.24)$$

Figure 2.2 shows the amplitude response of typical highpass $G(z)$ and lowpass $H(z)$ filters. These filters are called quadrature mirror filters because they have symmetry around the point $\pi/2$ as shown in Figure 2.2. Figure 2.3 shows the response of two ideal nonoverlapping high- and lowpass filters. For a real filter, the two responses will always overlap.

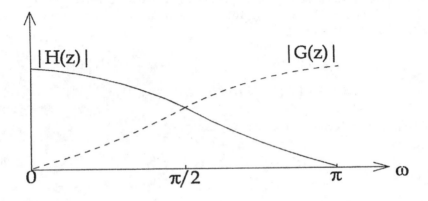

Figure 2.2 Amplitude response of the quadrature mirror filters.

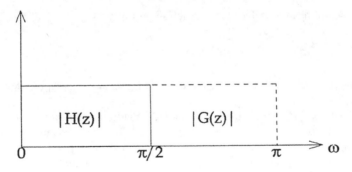

Figure 2.3 Amplitude response of nonoverlapping ideal filters.

From (2.15) one obtains

$$H'(z) = z^{-N} H(z^{-1}) = h(3) + h(2)z^{-1} + h(1)z^{-2} + h(0)z^{-3} \qquad (2.25)$$

and from (2.16) and (2.18) one gets

$$G'(z) = z^{-N}G(z^{-1}) = -H(-z) = -h(0) + h(1)z^{-1} - h(2)z^{-2} + h(3)z^{-3}$$
$$(2.26)$$

So if we solve only for $H(z)$, the other three filters will then be given by (2.24), (2.25), and (2.26). Now we show how to solve for $H(z)$. In this example, since we have $N = 3$, the length of the filter L is given by

$$L = \text{length of filter} = \text{order of filter} + 1 = N + 1 = 4 \qquad (2.27)$$

and utilizing (2.20) we have

$$\left[h(0) + h(1)z + h(2)z^2 + h(3)z^3\right]\left[h(0) + h(1)z^{-1} + h(2)z^{-2} + h(3)z^{-3}\right] +$$
$$\left[h(0) - h(1)z + h(2)z^2 - h(3)z^3\right]\left[h(0) - h(1)z^{-1} + h(2)z^{-2} - h(3)z^{-3}\right] = 2$$
$$(2.28)$$

Equating all coefficients related to the individual powers of z in (2.28) in order to fulfill (2.20) leads to

$$h(0)^2 + h(1)^2 + h(2)^2 + h(3)^2 = 1 \qquad (\text{for } z^0) \qquad (2.29)$$

$$h(0)\,h(2) + h(1)\,h(3) = 0 \qquad (\text{for } z^2 \text{ and } z^{-2}) \qquad (2.30)$$

Thus, we need two more equations to solve for (2.23), that is, the coefficients of $H(z)$. Since $G(z)$ is a highpass filter, then at $\omega = 0$

$$G(e^{j0}) = G(1) = 0 = H(e^{j\pi}) = H(-1) \qquad (2.31)$$

and therefore using (2.24) for $z = 1$, we obtain

$$h(3) - h(2) + h(1) - h(0) = 0 \qquad (2.32)$$

In addition we have from (2.21) and (2.31) at $\omega = 0$

$$H(1) = H(e^{j0}) = \sqrt{2} \qquad (2.33)$$

From (2.23) and (2.33) we get

$$H(e^{j0}) = H(1) = h(0) + h(1) + h(2) + h(3) = \sqrt{2} \qquad (2.34)$$

Thus, from (2.32) and (2.34) we get

$$h(0) + h(2) = \frac{1}{\sqrt{2}} \qquad (2.35)$$

$$h(1) + h(3) = \frac{1}{\sqrt{2}} \qquad (2.36)$$

in addition to (2.29) and (2.30). We need one more equation because these four equations, (2.29), (2.30), (2.35), and (2.36), are linearly dependent since (2.35) and (2.36), in conjunction with (2.30), lead to (2.29). The question is how to find the fourth equation. Without the fourth equation it is not possible to obtain the complete solution. Here the various methodologies differ and several researchers have come up with varying procedures.

For example, Daubechies [3] originally developed this methodology by mandating the constraint that the filter $H(z)$ be smooth and have a finite impulse response. Daubechies accomplished this by enforcing all derivatives up to order p at $\omega = 0$ to be zero, or equivalently $\left. \dfrac{d^P G'}{dz^P} \right|_{z=1} = 0$ where the superscript p denotes the pth derivative of G'. So, first of all, the Daubechies filters are FIR filters, and if p of the derivatives at $\omega = 0$ are zero, then they are at least of length $2p$. This is in contrast to the filters developed earlier by electrical engineers that were infinite impulse response. The actual value for p is determined from the number of equations needed to solve for the values of $h(n)$, $n = 0, 1, \ldots, N$. This leads to taking the various moments of $g'(n)$ and setting them equal to zero. Daubechies chose this procedure because enforcing the above conditions guarantees smooth wavelets, which we will define later.

The magnitude and the phase responses of $H(z)$, $H'(z)$, $G(z)$, and $G'(z)$ are shown in Figure 2.4(a) and (b), respectively. Note that these filters have no ripples, with multiple transfer function zeros at π. Filters for other higher orders have the following characteristics: The slope at the center of the transition band is proportional to $\sqrt{N}$ (for the results of Figure 2.4 that would be $\sqrt{N} = \sqrt{3}$). The transition band of the filter is defined from the change of the amplitude response from $|H(z)| = 0.98\sqrt{2}$ of the peak value to a level of $|H(z)| = 0.02\sqrt{2}$ of the maximum amplitude. The length of the transition band is given by $4/\sqrt{N} = 2.3$ Hz approximately. In mathematical terms, setting equal to zero the derivative of $G'(1)$, that is, $\left[\left. \dfrac{dG'}{dz} \right|_{z=1} = 0 \right]$ leads to

$$-0 \times h(0) - 1 \times h(1) + 2 \times h(2) - 3 \times h(3) = 0 \qquad (2.37)$$

Solution of (2.30), (2.35), (2.36), and (2.37) leads to

$$h'(0) = h(3) = \frac{1 + \sqrt{3}}{4\sqrt{2}} \ ; \ \ h'(1) = h(2) = \frac{3 + \sqrt{3}}{4\sqrt{2}} \ ; \ \ h'(2) = h(1) = \frac{3 - \sqrt{3}}{4\sqrt{2}}$$

$$\text{and } h'(3) = h(0) = \frac{1 - \sqrt{3}}{4\sqrt{2}}$$

$$(2.38)$$

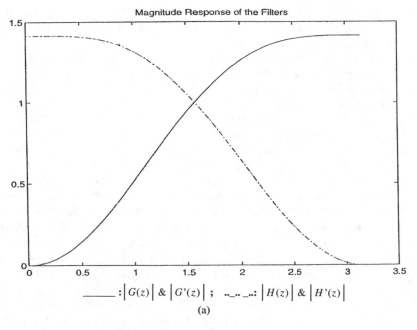

$$\underline{\quad\quad} : \left| G(z) \right| \, \& \, \left| G'(z) \right| \, ; \quad \cdot\cdot\_\cdot\cdot\_\cdot\cdot: \left| H(z) \right| \, \& \, \left| H'(z) \right|$$

(a)

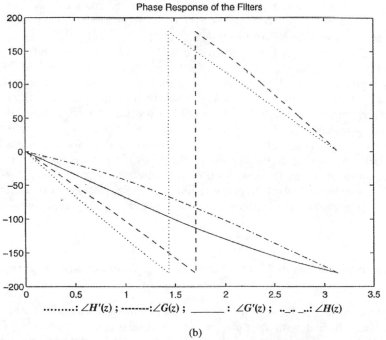

$$\cdots\cdots\cdots: \angle H'(z) \, ; \quad \texttt{-------} : \angle G(z) \, ; \quad \underline{\quad\quad} : \angle G'(z) \, ; \quad \cdot\cdot\_\cdot\cdot\_\cdot\cdot: \angle H(z)$$

(b)

Figure 2.4 (a) The magnitude responses of a third-order Daubechies filter versus ω and (b) the phase responses of a third-order Daubechies filter as a function of ω.

Another criterion, instead of using the derivatives, may be to keep the signal-to- noise ratio of the truncated wavelet coefficients as large as possible. This methodology has been presented in [10, 11].

Instead of having a two-stage decomposition of the signal $x(n)$ to $u(n)$ and $v(n)$, one can perform a multistage decomposition by applying the filters successively to each stage of the decomposition, as shown in Figure 2.5. In this figure a three-stage decomposition is shown. In this case it is the lowpass signal, which is further filtered by both a lowpass and a highpass filter. The highpass filtered components are left unprocessed. The assumption here is that there is more information in the low-frequency part of the signal and therefore we need to subdivide it further into various smaller bands. However, this is not the only strategy to subdivide the signal into various bands. One could apply the same high- and lowpass filtering procedure to the highpass channel leaving the lowpass filtered signals intact. Or one could carry out a combination of some of these filters, as we shall see in Chapter 9. For our present discussion, we will consider that at each stage only the low-frequency part of the signal is further filtered, leaving the high-frequency part intact. The electrical filter theory equivalent of this example for a three-stage decomposition is shown in Figure 2.6. The meaning of the coefficients $c_{k,2}$, $d_{k,0}$, $d_{k,1}$ and $d_{k,2}$ at the output of Figure 2.5 will be clear when we present the scaling functions and the wavelets in the next section. Now, these filtered and downsampled coefficients are thresholded. This is equivalent to keeping only those coefficients that are bigger in magnitude than some prespecified constant ε. The magnitudes of the coefficients smaller than ε are set equal to zero resulting in approximate outputs.

To reconstruct the signal from these coefficients, we follow the procedure shown in Figure 2.1. Namely, we up sample the data that one receives and then filter it through a set of synthesis filters. The reconstruction algorithm is depicted in Figure 2.7, where the approximated coefficients are upsampled and then filtered in the way depicted in the receiver of Figure 2.1. Now the interesting part is that the "filtered" thresholded coefficients received at the receiver can now be used to recover the original signal with reasonably good accuracy [9]. We will see this feature later. The type of decomposition outlined in Figure 2.5 is identical to a wavelet decomposition, as the next section will illustrate. On the other hand, Figure 2.7 represents the wavelet reconstruction methodology. In the next section, we illustrate the connection between the filters and the scaling functions and wavelets.

2.2.3   Connection Between Filter Theory and the Mathematical Theory of Wavelets

To establish a connection between the filter theory and the mathematical theory of wavelets, we will deal with functions of a real continuous variable. Discrete samples will then be viewed as the result of "sampling and hold" of a continuous variable. Such an approach is absolutely necessary because the wavelets cannot

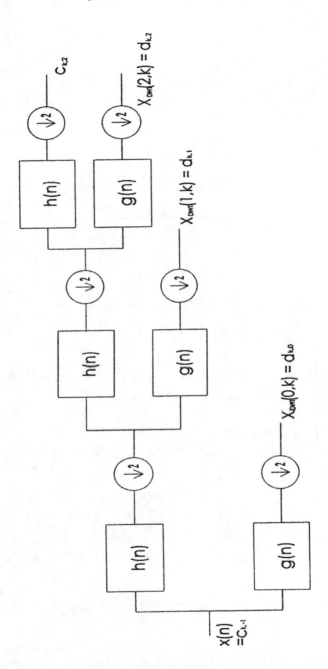

Figure 2.5 A multistage decomposition of the signal $x(n)$.

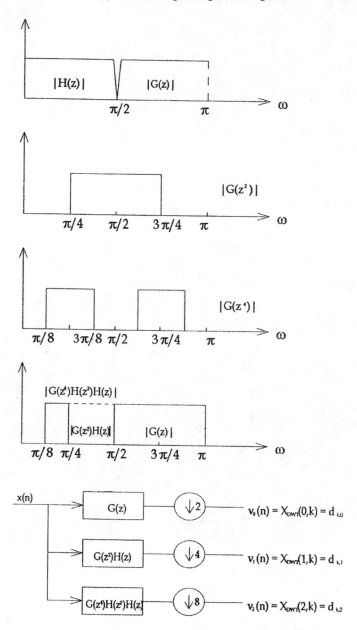

Figure 2.6 A filter theory representation of the discrete wavelet transform.

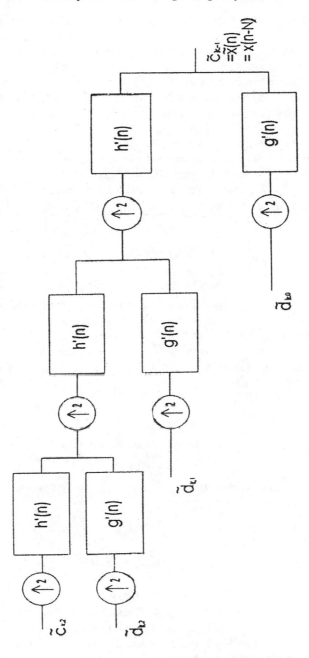

Figure 2.7 The reconstruction of the original signal, $x(n)$, from the approximate coefficients of the discrete wavelet transform.

be expressed in terms of discrete samples. Moreover, the dilation equation, which is at the heart of the wavelet theory, has no solution for discrete samples. So the discrete wavelet transform implies that the functions we are dealing with are functions of a continuous variable. These functions can have integer shifts, that is, they can be delayed in time by an integer number of time intervals. In addition, their spread along the time axis can be scaled up and down by integer multiples. We now illustrate that filtering a function by $h(n)$ is equivalent to fitting a scaling function at a certain scale, and filtering by $g(n)$ is equivalent to curve fitting $x(n)$ by wavelets at the same scale of the scaling function. The mathematical connection is now established between wavelet theory and filter theory.

We consider the original signal $x(t)$ with bandwidth 0 to π. Also consider some function ϕ. We assume that the integer shifts m of $\phi(t)$, namely $\phi(t - m)$ form a Riesz basis [1] and would lead to a perfect approximation for $x(t)$. Let us denote

$$\phi_m(t) = \phi(t - m) \tag{2.39}$$

which is the function obtained by shifting $\phi(t)$ along the time axis by m time intervals, where m is an integer. The point here is that since the functions $\phi(t - m)$, for $m = 0, 1, 2, \ldots, N$ form a complete set for a particular space, any finite combination of these functions should be able to represent a dilated version of the function $\phi(t)$, which is chosen here to be $\phi(t/2)$. Hence, from a mathematical standpoint, one can represent a dilated version of $\phi(t)$ [e.g., $\phi(t/2)$] by a combination of the functions $\phi(t - m)$ [or $\phi_m(t)$] with some coefficients $h'(m)$ resulting in the dilation equation

$$\phi\left(\frac{t}{2}\right) = \sqrt{2} \sum_{m=0}^{N} h'(m)\, \phi(t - m) \tag{2.40}$$

or equivalently, with $h'(m) = h(N - m)$ as defined by (2.15), one may write the dilation equation as

$$\phi(t) = \sqrt{2} \sum_{m=0}^{N} h'(m)\, \phi(2t - m) = \sqrt{2} \sum_{m=0}^{N} h(N - m)\phi(2t - m) \tag{2.41}$$

The constant factor $\sqrt{2}$ in (2.40) and (2.41) has been included for convenience. This will simplify certain constant factors later. It is interesting to note that the coefficients $h'(m)$ turn out to be the same lowpass filter coefficients for the synthesis filters that are shown in the receiver part of Figure 2.1. These are the same filters that have been shown in Figure 2.7. A solution of the dilation equation in (2.40) or (2.41) for $\phi(t)$, using a given set $h'(m)$, for $m = 0, \ldots, N$ is called a scaling function, or the father of wavelets. This is because the wavelets are generated from the scaling functions. In addition, we assume the functions $\phi(t-m)$ to be normalized, that is,

$$\int_t \phi(t-m)\, dt = 1 \qquad (2.42)$$

Here dt can be represented as the length of the incremental time interval. Therefore, from (2.41) utilizing (2.42) we get

$$\int_t \phi(t-k)\, dt = 1 = \sqrt{2} \sum_{m=0}^{N} h'(m) \int_t \frac{\phi(2t-2k-m)}{2}\, d(2t) = \frac{\sqrt{2}}{2} \sum_{m=0}^{N} h'(m) \quad (2.43)$$

or equivalently

$$\sum_{m=0}^{N} h'(m) = \sqrt{2} \qquad (2.44)$$

Hence the coefficients $h'(m)$ satisfying the dilation equation must satisfy (2.44). This is the same equation as (2.34) where $L = N + 1 = 4$ terms.

The corresponding wavelet is now generated from the highpass filter coefficients and the scaling function. The wavelet $\psi(t)$ is defined in terms of the dilated versions of the scaling functions through

$$\psi(t) = \sqrt{2} \sum_{m=0}^{N} g'(m)\, \phi(2t-m)$$

$$= \sqrt{2} \sum_{m=0}^{N} -(-1)^m\, h'(N-m)\, \phi(2t-m) \qquad (2.45)$$

where the $N + 1$ coefficients of the highpass synthesis filter $g'(m)$ are related to the lowpass synthesis filter $h'(m)$ through (2.17). In addition, $N + 1$ is the length of the filter (which is always even). Here the filter coefficients $h'(m)$ and $g'(m)$ are assumed to be real. The function $\psi(t)$ for a given $g'(m)$ is called the mother of wavelets. [This is the traditional Judeo-Christian concept of the mother—where the mother (Eve) is generated from the father (Adam)!]

So from an electrical engineering perspective, if we have the filter coefficients $h'(m)$, then the scaling function $\phi(t)$ can be obtained by solving dilation equation (2.41). Once the scaling function is known, the wavelet is given by (2.45). The relationship between the scaling function, wavelet, and the signal $x(t)$ will be described in the next section.

The scaling function can be more easily solved for in the transform domain. If we define $\Phi(\omega)$ to be the Fourier transform of $\phi(t)$, then

$$\Phi(\omega) = \int_{-\infty}^{\infty} \phi(t)\, e^{-j\omega t}\, dt$$

Now taking the Fourier transform of both sides of (2.41), one can rewrite it in the Fourier domain (ω) as

$$\Phi(\omega) = \frac{\sqrt{2}}{2}\, H'(e^{-j\frac{\omega}{2}})\, \Phi(\omega/2) \qquad (2.46)$$

where $H'(z)$ is the Z transform of $h'(n)$ [see (2.2)]. One can repeatedly apply (2.46) to obtain

$$\Phi(\omega) = \left(\prod_{k=1}^{K} 2^{-1/2}\, H'(e^{-j\frac{\omega}{2^k}}) \right) \Phi\left(\frac{\omega}{2^K} \right) \qquad (2.47)$$

Owing to (2.42), $\Phi(0) = 1$. Therefore, as $K \to \infty$, (2.47) becomes

$$\Phi(\omega) = \left\{ \prod_{k=1}^{\infty} 2^{-1/2}\, H'(e^{-j\frac{\omega}{2^k}}) \right\} = \prod_{k=1}^{\infty} \left[2^{-1/2}\, H'(e^{-j\frac{\omega}{2^k}}) \right] \qquad (2.48)$$

The function of (2.48), $\Phi(\omega)$, is pointwise convergent provided the infinite series converge. In the original domain, this is equivalent to $\phi(t) = \prod_{k=1}^{\infty} [\otimes\, 2^{-1/2}\, h'(2^k t)]$ where $\otimes$ denotes a convolution. Note that the sequence of convolutions is carried out by various compressed, scaled versions of the same filter. When the sequence of convolutions converges, it yields the function $\phi(t)$. The evolution of the scaling function for a third-order ($N = 3$) Daubechies filter [1] at different scales k is shown in Figure 2.8.

As an example, consider the solution of the dilation equation for $N = 0$. Then (2.44) yields $h'(0) = \sqrt{2}$, and dilation equation (2.41) becomes $\phi(t) = 2\phi(2t)$ and the scaling function becomes a delta function, that is, for $N = 0$, $\phi(t) = \delta(t)$.

If we choose $N = 1$, then the length of the filter $L = N + 1 = 2$ and (2.44) lead to

$$h'(0) = \frac{1}{\sqrt{2}} = h'(1) \qquad (2.49)$$

Hence, in this case the dilation equation becomes $\phi(t) = \phi(2t) + \phi(2t - 1)$. One possible solution of this dilation equation is the pulse function

$$\phi(t) = \begin{cases} 1; & \text{if } 0 \le t < 1 \\ 0; & \text{otherwise} \end{cases} \qquad (2.50)$$

The wavelet is generated from (2.45), that is, $\psi(t) = -\phi(2t) + \phi(2t - 1)$. Thus, it is given by

$$\psi(t) = \begin{cases} -1 & \text{for} \quad 0 \le t < 1/2 \\ +1 & \text{for} \quad 1/2 \le t < 1 \\ 0 & \text{otherwise} \end{cases} \tag{2.51}$$

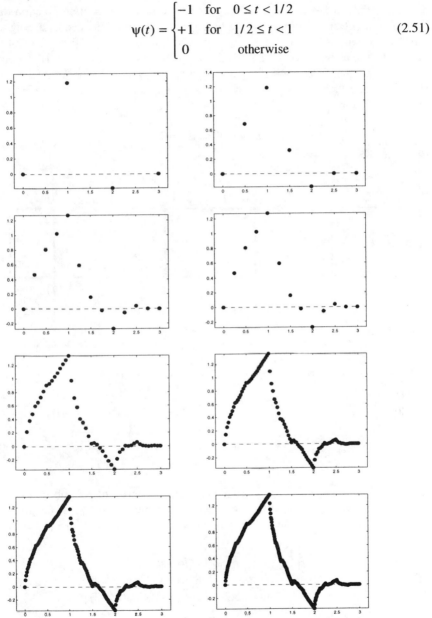

Figure 2.8 Evolution of the scaling function for $N = 3$ Daubechies filters.

In this case, $\phi(t)$ is a pulse function of magnitude 1 defined between $t = 0$ and 1 as shown in Figure 2.9. It is also a combination of the functions $\phi(2t)$ and $\phi(2t - 1)$ [i.e., the latter is a shifted version of $\phi(2t)$] where these functions are defined between 0 to ½ and ½ to 1, respectively. The term $\phi(t)$ is the scaling function for a Haar wavelet. Here we observe that a function can be approximated by a weighted sum [the weighting factors being $h'(m)$] of the dilated and shifted versions of the same function as observed from (2.41). The Haar wavelet function $\psi(t)$ is a pulse doublet generated from the scaling functions and is shown in Figure 2.9. For the Haar wavelet, the scaling function $\phi(t)$ is orthogonal with respect to its own translates, that is,

$$\int_{-\infty}^{\infty} \phi(t)\, \phi(t-m)\, dt = 0 \qquad\qquad (2.52)$$

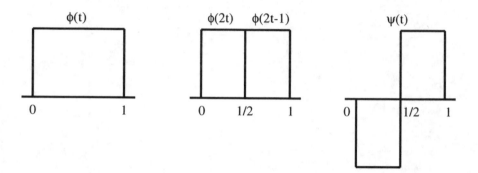

Figure 2.9 Solution of the dilation equation for $N = 1$, that is, $L = 2$, resulting in the Haar wavelet.

This is also true for the wavelets

$$\int_{-\infty}^{\infty} \psi(t)\, \psi(t-m)\, dt = 0 \qquad\qquad (2.53)$$

For $N = 3$, we obtain the case presented earlier, the third-order Daubechies filter. Note that for this case, the filter coefficients $h'(i)$, for $i = 0, \ldots, 3$ are given by (2.38). Hence the scaling function can be obtained from the dilation equation as

$$\phi(t) = \frac{1}{4}\left[(1+\sqrt{3})\,\phi(2t) + (3+\sqrt{3})\,\phi(2t-1)\right]$$
$$+ \left[(3-\sqrt{3})\,\phi(2t-2) + (1-\sqrt{3})\,\phi(2t-3)\right] \qquad (2.54a)$$

The wavelets are generated in an analogous fashion by using (2.45) and (2.38) resulting in

$$\psi(t) = \frac{-1}{4}\left[(1-\sqrt{3})\ \phi(2t) - (3-\sqrt{3})\ \phi(2t-1)\right]$$
$$+\left[(3+\sqrt{3})\ \phi(2t-2) - (1+\sqrt{3})\ \phi(2t-3)\right] \tag{2.54b}$$

The scaling function and wavelet and their Fourier transforms for an $N = 3$ Daubechies filter are shown in Figure 2.10. In Figure 2.11 the scaling function and the wavelet for $N = 1, 3, 5, 7,$ and 9 are illustrated.

We are not going to delve further into the solution of the dilation equation because for the electromagnetics problems in which we are interested—namely, solution of large matrix equations—the scaling functions and the wavelets are really not necessary because the discrete wavelet representation can be carried out from the knowledge of only $h(m)$! However, we present some other wavelets for illustration purposes.

For example, consider the Shannon wavelet, which is the dual of the Haar wavelet. The scaling function is given by

$$\phi(t) = \frac{\sin \pi t}{\pi t} \tag{2.55}$$

and its transform is given by

$$\Phi(\omega) = \begin{cases} 1 & \text{for } 0 \leq |\omega| < \pi \\ 0 & \text{otherwise} \end{cases} \tag{2.56}$$

The wavelet can be generated from (2.45) and is given by

$$\psi(t) = \frac{\sin \pi t}{\pi t}\ \cos\frac{3\pi}{2}t \tag{2.57}$$

and its transform is

$$\Psi(\omega) = \begin{cases} 1 & \text{for } \pi \leq |\omega| < 2\pi \\ 0 & \text{otherwise} \end{cases} \tag{2.58}$$

Sometimes, one may need a filter that has only a positive response. In that case, one can use the Lagrange half-band filters. They provide a nonnegative frequency response for the filters $h(n)$. A closed-form expression was given by Ansari et al. [12] starting from

$$H'(z) = \frac{1}{2} + \sum_{n=1}^{i} h'\ (2n-1)\ \left[z^{-2n+1} + z^{2n-1}\right] \tag{2.59}$$

with $4i - 1$ coefficients. The coefficients $h'(k)$, with $k = 2n - 1$, are determined using the Lagrange interpolation formula:

$$h'(2n-1) = \frac{(-1)^{n+i-1} \displaystyle\prod_{k=1}^{2i} (i-k+1/2)}{(i-n)! \; (i-1+n)! \; (2n-1)} \qquad (2.60)$$

These filters are very regular as they have a $2i$-fold zero at $z = -1$.

Therefore, in the development of the wavelets for the discrete case, one has the choice of either of the two following procedures:

1. Do we start with $\phi(t)$ and construct $h'(n)$ and then generate $\psi(t)$, from (2.45)?
2. Do we start with $h'(n)$ and then create $\phi(t)$ and $\psi(t)$?

For the discrete case with which we are dealing, the answer is straightforward, that is, we design $h'(n)$ and then obtain $\phi(t)$ and $\psi(t)$. This is also much simpler in practice. However, for the discrete wavelet transform, as we shall see, $\phi(t)$ and $\psi(t)$ are not at all required in the numerical computation!

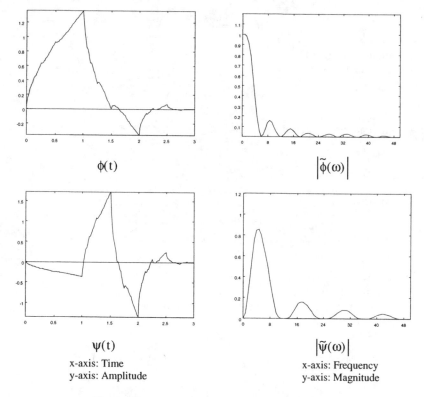

$\phi(t)$ $\left|\tilde{\phi}(\omega)\right|$

$\psi(t)$ $\left|\tilde{\psi}(\omega)\right|$

x-axis: Time   x-axis: Frequency
y-axis: Amplitude   y-axis: Magnitude

Figure 2.10 Scaling function and wavelet and their transforms corresponding to the third-order Daubechies filter.

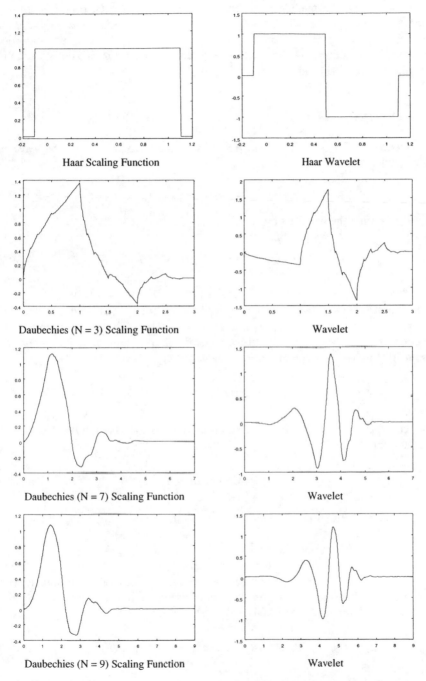

Figure 2.11 Scaling functions and wavelets for different orders of Daubechies filters.

In summary, the mathematical basis of wavelets has been presented from a filter theory perspective. We have shown how to construct scaling functions ϕ and wavelets ψ starting from filters $h'(m)$ and utilizing the perfect reconstruction argument presented in subband filtering techniques. Once $h(m)$ is known, ϕ can be generated from (2.41) and ψ from (2.45).

2.3 APPROXIMATION OF A FUNCTION BY WAVELETS

Consider a function $x(t)$. The objective now is to approximate it by the wavelets $\psi_{n,k}(t)$ so that

$$x(t) = \sum_{n=-\infty}^{\infty} \sum_{k=-\infty}^{\infty} d_{k,n} \, \psi_{n,k}(t) \tag{2.61}$$

where we define the wavelets by

$$\psi_{n,k}(t) = 2^{-n/2} \, \psi(2^{-n}t - k) \tag{2.62}$$

Note that $\psi(2^{-n}t)$ represents dilated versions of $\psi(t)$. For $n = -1$, we say the scale is the finest. This is because for $n > -1$, the function gets dilated and becomes wider. The term $\psi(t - k)$ represents the shifts. Therefore we approximate the function $x(t)$ by a dilated and shifted version of the same function $\psi(t)$. We assume that we are dealing with orthogonal wavelets; hence,

$$\int \psi_{j,k}(t) \, \psi_{n,i}(t) \, dt = 0, \text{ if } j \neq n \text{ or } k \neq i \tag{2.63}$$

The scale factor $2^{-n/2}$ appearing in (2.62) is there so as to make the functions $\psi_{n,k}(t)$ orthonormal, that is,

$$\int \psi_{j,k}(t) \, \psi_{n,i}(t) \, dt = \delta_{jn} \, \delta_{ki} \tag{2.64}$$

where δ_{pq} represents a Dirac delta function, so that its value is unity when $p = q$ and zero otherwise.

It is interesting to note that $\psi_{n,k}(t)$ has zero dc value $\{\psi_{n,k}(\omega = 0) = 0\}$, whereas $x(t)$ may not! Therefore, if we have to talk about convergence of the series (2.61), we can only talk about mean square convergence because the constant term is missing. This dichotomy, however, does not arise when the sum in (2.61) is finite, as we will see later. Utilizing (2.64) and (2.61), we observe that

$$d_{k,n} = \int x(t) \, \psi_{n,k}(t) \, dt = \langle x; \psi_{n,k} \rangle = X_{DWT}(n,k) \tag{2.65}$$

where $\langle \bullet, \bullet \rangle$ defines the inner product. The terms $X_{DWT}(n,k)$ are the kth discrete wavelet coefficients at scale n of the function $x(t)$ and are symbolically denoted by $d_{k,n}$. The isomorphism between the discrete wavelet coefficients and the output from the highpass filters is shown in Figure 2.5. In fact, they are the same.

When the number of scales in the wavelet decomposition goes to infinity, the scaling functions are not present in the representation of x as may be deduced from (2.61). However, for a finite number of terms in the summation, the scaling functions are always explicitly present. This is shown in Figure 2.5 as the coefficients from the output of the lowpass filter. Therefore, we now assume for practical considerations that the value of n is finite.

Now if we have to carry out the inner products in (2.65), it will be extremely time-consuming because the inner products have to be computed for all values of n and k. Here the strength of the wavelet techniques is seen because they provide a fast and accurate way to recursively evaluate the inner products. This is accomplished through the introduction of the scaling functions

$$\phi_{i,j}(t) = 2^{-1/2} \phi(2^{-i}t - j) \tag{2.66}$$

We further assume/utilize the orthogonality relationships between the scaling functions and the wavelets and between the scaling functions themselves:

$$\int \phi_{j,k}(t)\, \psi_{i,n}(t)\, dt = 0 \quad \text{if} \quad i \le j \tag{2.67}$$

and

$$\int \phi_{j,k}(t)\, \phi_{j,n}(t)\, dt = \delta_{k,n} \tag{2.68}$$

We now define coefficients $c_{k,n}$:

$$c_{k,n} = \int x(t)\, \phi_{n,k}(t)\, dt \tag{2.69}$$

We now illustrate how the wavelet coefficients $d_{k,n}$ are evaluated recursively through $c_{k,n}$.

We have from (2.64), (2.67), and (2.68) that the following orthonormal set

$$\{\phi(t-k);\ \psi(t-k)\}_{k=-\infty}^{\infty} \tag{2.70}$$

can represent any function in the space they span, because they are orthogonal. Hence the set of (2.70) may be termed as a basis for that finite dimensional space of dimension n. From dilation equation (2.41) and from (2.45) we note that the set

$$\left\{\sqrt{2}\ \phi(2t-k)\right\}_{k=-\infty}^{\infty} \tag{2.71}$$

also forms an orthonormal set for the same space. Therefore at scale $n = -1$ we can expand any function $p(t)$ by

$$p(t) = \sum_k a^{(-1)}(k) \sqrt{2}\ \phi(2t-k) \tag{2.72a}$$

$$= \sum_k \left[a^{(0)}(k)\ \phi(t-k) + b^{(0)}(k)\ \psi(t-k) \right] \tag{2.72b}$$

At scale $n = 0$, we have $a^{(-1)}(k)$ as the coefficients weighting the function $\phi(2t-k)$. The superscripts on the coefficients $a(k)$ and $b(k)$ represent the value of the scale n at which they are defined. We have from (2.41)

$$\phi(t-n) = \sqrt{2} \sum_k h'(k)\ \phi(2t-2n-k) \tag{2.73}$$

and from (2.45)

$$\psi(t-n) = \sqrt{2} \sum_k g'(k)\ \phi(2t-2n-k) \tag{2.74}$$

The limits for the summation over k are determined from the number of stages of decompositions. It is interesting to note that a useful result can be derived from (2.73) and (2.74) in the form of

$$\left\langle \phi_{0,k} ; \phi_{-1,j} \right\rangle = h'(j-2k)$$

$$\left\langle \psi_{0,k} ; \phi_{-1,j} \right\rangle = g'(j-2k)$$

which can be generalized to yield

$$\left\langle \phi_{n,k} ; \phi_{n-1,j} \right\rangle = h'(j-2k)$$

$$\left\langle \psi_{n,k} ; \phi_{n-1,j} \right\rangle = g'(j-2k)$$

Next, observe that

$$a^{(0)}(n) = \int_{-\infty}^{\infty} p(t)\ \phi(t-n)\,dt = \int_{-\infty}^{\infty} p(t) \sum_k h'(k) \sqrt{2}\ \phi(2t-2n-k)\,dt \tag{2.75}$$

$$= \sum_k h'(k)\ a^{(-1)}(2n+k) = \sum_k a^{(-1)}(k)\ h'(k-2n)$$

and

$$b^{(0)}(n) = \sum_k a^{(-1)}(k)\ g'(k-2n) \tag{2.76}$$

Given the existence of relationships like (2.75) and (2.76) and drawing the isomorphism between

$$c_{k,j} \leftrightarrow a^{(j)}(k)$$

$$d_{k,j} \leftrightarrow b^{(j)}(k)$$

we can generalize the expressions (2.75) and (2.76) as follows:

$$c_{j,k+1} = \sum_n c_{n,j} h'(n-2k) = \sum_n c_{n,j} h(N+2k-n) \quad for \ j \geq -1 \quad (2.77a)$$

$$d_{j,k+1} = \sum_n c_{n,j} g'(n-2k) = \sum_n c_{n,j} g(N+2k-n) \quad for \ j \geq -1 \quad (2.77b)$$

These results have been derived utilizing (2.15) and (2.16). The above recursive relations show that we need to compute the inner product of (2.64) at the highest scale, $n = -1$, only once [instead of using (2.65)] and then the wavelet coefficients $d_{k,n}$, that is, $X_{DWT}(n,k)$ ($n = 0, 1, 2, \ldots$) are computed recursively from (2.77a) and (2.77b). From the filter theory point of view, (2.77a) and (2.77b) show that the c_{nj} need to be convolved with $h(n)$ and $g(n)$ and then downsampled by a factor of 2. Through Figure 2.5 and from the above development it is clear that for the computation of the DWT it is not necessary to even know what the scaling functions and wavelets are because one can use (2.77a) and (2.77b) directly without going through the mathematical derivations as Figures 2.5 to 2.7 illustrate. We start with the computation of $c_{n,-1}$ through (2.68) and (2.66) and then take into account the scaling factor in (2.71), which is the coefficient generated by correlating $\phi(2t)$ with the function $x(t)$. Then we recursively compute the discrete wavelet transform mathematically through (2.77a) and (2.77b) and graphically using Figure 2.5, which is easier to visualize from a filter theory perspective. The methodology is the same. The process described so far is similar to the transmitter part marked in Figure 2.1. If $\phi(2t)$ are the impulse scaling functions as we show later, then $c_{k,-1}$ will be equivalent to the sampled version of $x(t)$, namely, $x(n)$.

There is another subtle point, which we should introduce now. So far in the approximation in (2.61), the limits are infinity. This is good from a mathematical perspective. However, from a practical reality the limits have to be finite. Hence, from Figure 2.5 it is seen that in addition to the coefficients $d_{k,n}$ we also need the coeffcients $c_{k,n}$. To be more precise, we need the coefficients $c_{k,M}$, where M represents the highest scale in which one is interested. The approximation of $x(t)$ from a practical standpoint is done by

$$x(t) = \sum_{n=0}^{M} \sum_k d_{k,n} \Psi_{n,k}(t) + \sum_k c_{k,M} \phi_{M,k}(t) \quad (2.78)$$

Hence, we have both wavelets and scaling functions in the representation of $x(t)$ whenever the scale n is finite. However, when $M \rightarrow \infty$, then we have only the wavelets in the expansion and not the scaling functions. Now observe that (2.78) is the practical approximation of $x(t)$ in a finite dimensional space, instead of the representation used by (2.61), where only the wavelet functions are used.

In summary, the evaluation of the discrete wavelet coefficients in (2.61) is equivalent to filtering the signal $x(t)$ [or equivalently the sampled values of $x(t)$ for a certain class of scaling functions] by a cascade of mutually orthogonal bandpass filters as shown in Figure 2.5. The filter output at each stage represents the discrete wavelet coefficients. The bandwidths of the bandpass filters are reduced by a factor of 2 as one goes toward the dc value, so that the ratio between the center frequency and its bandwidth defined by the symbol Q to represent the quality factor of the filters remains the same. For the infinite sum in (2.61), the scaling functions do not enter in the final sum except in the intermediate computations as can be observed from (2.77a) and (2.77b). However, if the sum is finite in (2.61), then one obtains (2.78) and the scaling functions are needed in the summation.

In practice, once the coefficients $d_{k,n}$ and $c_{k,M}$ have been obtained, those coefficients whose magnitudes are below a certain threshold value of ε (e.g., $\varepsilon = 10^{-3}$) are set equal to zero. Thus, the final result of the transform is the approximate coefficients $\tilde{d}_{k,n}$ and $\tilde{c}_{k,M}$, which are the values obtained after thresholding $d_{k,n}$ and $c_{k,M}$ by ε. Now the problem is how does one recover $x(t)$ from these approximate coefficients in a fast efficient way.

By utilizing (2.73) and (2.74) in (2.72b), we get

$$p(t) = \sum_k \left[a^{(0)}(k) \sqrt{2} \sum_n h'(n) \phi(2t - 2k - n) + b^{(0)}(k) \sqrt{2} \sum_n g'(n) \phi(2t - 2k - n) \right]$$

$$= \sum_k \sqrt{2} \left[a^{(0)}(k) \sum_n h'(n - 2k) \phi(2t - n) + b^{(0)}(k) \sum_n g'(n - 2k) \phi(2t - n) \right]$$

(2.79)

By equating (2.79) to (2.72a) we find

$$a^{(-1)}(k) = \sum_n \left[h'(k - 2n) a^{(0)}(n) + g'(k - 2n) b^{(0)}(n) \right]$$

$$= \sum_n \left[h(N + 2n - k) a^{(0)}(n) + g(N + 2n - k) b^{(0)}(n) \right]$$

(2.80)

Hence we start with the approximate coefficients $\tilde{c}_{k,M}$ and $\tilde{d}_{k,n}$, and we then recursively generate, through the use of a generalization of (2.80), the coefficients

$$\tilde{c}_{k,m+1} = \sum_n \left[h(N + 2n - k) \tilde{c}_{n,m} + g(N + 2n - k) \tilde{d}_{n,m} \right] \qquad (2.81)$$

Note that $\tilde{c}_{k,-1}$ are the estimates of the discrete values of $x(n)$ for the impulse scaling function. Also observe that (2.81) is equivalent to Figure 2.7 from a filter theory perspective and is identical to the receiver part in Figure 2.1 when the number of the scale is equal to unity. The most remarkable point here is that even though the original coefficients have been thresholded by ε to produce the approximate coefficients, $\tilde{c}_{k,M}$ and $\tilde{d}_{k,n}$, the original sampled function $x(n)$ can be reconstructed through $\tilde{x}(n)$ with an accuracy better than ε. This we will illustrate later through numerical examples.

In summary, in order to implement and carry out the discrete wavelet transform it is not even necessary to introduce the concept of scaling functions and wavelets. The filter theory approach essentially provides the same methodology in a simpler practical fashion. Finally, we have to be careful that we do not commit any wavelet crimes [13]. Wavelet crimes are related to the question of whether one is approximating a continuous function or a discrete function. The problem here is that the wavelets are always continuous functions. So how to correlate the approximation of a discrete function by a set of continuous basis functions is a problem as the dilation equation that produces wavelets has no solution for the discrete case.

2.4 EXAMPLES

As an illustration of how to utilize the discrete wavelet transform, we consider the compression of an image. For the case of the images, one is dealing with the 2-D discrete wavelet transform, which is a generalization of the 1-D case.

In the following examples, we consider the image as a 2-D array. We take the discrete wavelet transform in 2-D by utilizing a recursive relationship similar to that of (2.77a) and (2.77b). We utilize an eighth-order filter, which has been designed to match the signal [14] for an efficient decomposition. Once the discrete wavelet coefficients are obtained, they are thresholded and then the original image is reconstructed utilizing the recursive relation of (2.81). The objective is to illustrate that even though only a few wavelet coefficients are utilized to represent the original image, the reconstruction is still better than the conventional JPEG algorithm [14].

As an example consider the original image of "Lena" shown in Figure 2.12(a). Figures 2.12(b) and (c) show the result of reconstructing the image after a 40:1 compression of "Lena" using both wavelet techniques utilizing signal-dependent QMF filters and JPEG compression, respectively. JPEG is the current standard for image compression. The coefficients of the filters $h(m)$ are determined from the picture [14]. Here compression refers to the total number of bits required to store the wavelet coefficients of the image after eliminating the small wavelet coefficients as opposed to the total number of bits of the original image.

Even though noticeable degradation is seen in both of the images, the two methods perform reasonably in the reconstruction of the image. This is because the image is processed by blocks. The image processed utilizing the signal-dependent QMF decomposition loses detail in the local (i.e., high-frequency) edge information [14]. In particular, notice that the sharpness in the eyes and the detail in the feather cap are blurred. Unlike the image obtained through the current standard JPEG, however, there are no objectionable artifacts such as blockiness.

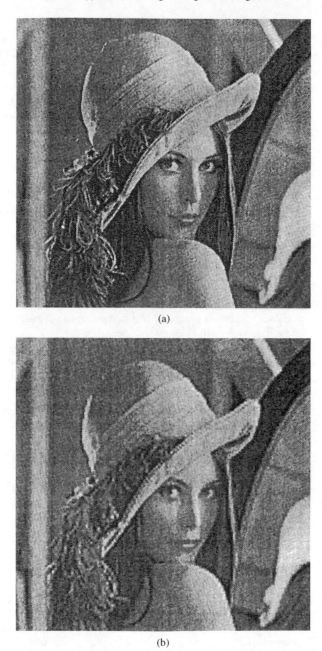

(a)

(b)

Figure 2.12   (a) Original picture of "Lena," (b) compressed version of "Lena" utilizing QMF and a compression ratio of 40:1, and (c) compressed version of "Lena" utilizing JPEG and a compression ratio of 40:1.

(c)

Figure 2.12 Continued.

For another example, consider the application processing halftone images utilizing wavelets. Unlike continuous-tone images such as photographs, a halftone image is digital in nature; all pixels are either on or off. These digital pixels are densely placed on the paper typically with a density of 300–600 dots per inch. Halftone images are the output produced by the most common printers: laser jet and dot matrix. In many instances, these images are scanned into a computer using an 8 to 16 bits per pixel optical input scanner with the resulting image placed into memory. A halftone image is shown in Figure 2.13(a). This figure has been drawn with a magnification of 3 for demonstration purposes. Notice the dot pattern of the background, which is not present in most continuous-tone images. This image was printed by a 400-dpi printer. Figures 2.13(b) and (c) show the printed results after a 30:1 compression of Figure 2.13(a) using both the signal-dependent QMF filters [14] and the JPEG compression, respectively. Notice that the QMF compression attenuates the dot pattern associated with the original halftone picture. This is because the high detail content is actually improving the appearance of the image. Again the term *compression* implies that 30 times less storage is required to store the dominant wavelet coefficients as opposed to that of the original image. The JPEG compressed image seems to produce a noise pattern in the output. This is indicative of the problems faced when using JPEG compression on halftone images.

(a)

(b)

Figure 2.13 (a) Original, simulated halftone picture printed on a 400-dpi digital printer, (b) compressed version of the simulated halftone picture utilizing QMF and a compression ratio of 30:1, and (c) compressed version of the simulated halftone picture utilizing JPEG and compression ratio of 30:1.

(c)

Figure 2.13 Continued.

By utilizing a wavelet technique, it is possible to quantize images with bit rates as low as 0.4 bits per pixel while maintaining sufficiently high-quality reconstructions [9].

2.5 CONCLUSION

The discrete wavelet transform is presented from the first principles utilizing the basic concepts of filter theory. We show how to construct the high- and lowpass filters that produce the wavelets and the scaling functions. We also show how these filters relate to the scaling functions and wavelets. An efficient method for implementing the discrete wavelet transform has also been described. However, for the discrete wavelet transform, the introduction of wavelets and scaling functions is not at all necessary because all the computations can be done using the coefficients of the various high- and lowpass filters.

REFERENCES

[1] P. P. Vaidyanathan, *Multirate Systems and Filter Banks*, Prentice-Hall, Englewood Cliffs, NJ, 1993.

[2] C. K. Chui, *An Introduction to Wavelets*, Academic Press, San Diego, 1992.

[3] I. Daubechies, *Ten Lectures on Wavelets*, CBMS-NSF Regional Conference Series in Applied Mathematics, Philadelphia, SIAM, 1992.

[4] J. Morlet et al., "Wave propagation and sampling theory," *Geophysics*, Vol. 47, 1982, pp. 203–236.

[5] *Special Issue of IEEE Proceedings*, Vol. 84, No. 4, April 1986.

[6] D. Esteban and C. Galland, "Application of quadrature mirror filters to split-band voice coding schemes," *Proc. IEEE Int. Conf. Acoustic Speech & Signal Process.*, Hartford, 1977, pp. 191–195.

[7] M. J. T. Smith and T. P. Barnwell III, "Exact reconstruction techniques for true structured subband coders," *IEEE Trans. Acoustic Speech and Signal Process.*, Vol. 39, 1986, pp. 434–441.

[8] M. Vettereli, "Multidimensional subband coding: Some theory and algorithms," *Signal Processing*, Vol. 6, 1984, pp. 97–112.

[9] T. H. Koornwinder, Ed., *Wavelets: An Elementary Treatment of Theory and Applications*, World Scientific, River Edge, NJ, 1993.

[10] S. Schweid and T. K. Sarkar, "A sufficiency criteria for orthogonal QMF filters to ensure smooth wavelet decomposition," *Applied and Computational Harmonic Analysis*, Vol. 2, 1995, pp. 61–67.

[11] S. Schweid and T. K. Sarkar, "Iterative calculation and factorization of the autocorrelation function of orthogonal wavelets with maximal vanishing moments," *IEEE Trans. Circuits and Systems*, Vol. 42, No. 11, Nov. 1995, pp. 694–701.

[12] R. Ansari, C. Guillemot, and J. F. Kaiser, "Wavelet construction using Lagrange half-band filter," *IEEE Trans. Circuits and Systems*, Vol. 38, 1991, pp. 1116–1118.

[13] G. Strang and T. Nguyen, *Wavelet and Filter Banks*, Wellesley-Cambridge Press, Wellesley, MA, 1996.

[14] S. Schweid, "Projection method minimization techniques for smooth multistage QMF filter decomposition," Ph.D. Thesis, Syracuse University, Syracuse, NY, May 1994.

APPENDIX 2A   Principles of Decimation and Expansion

2A.1   Principle of Decimation by a Factor of 2

Pictorially the principle of downsampling or decimation of a signal is represented by the symbol in Figure 2A.1 where the decimated signal $y_D(n)$ has been generated from the original signal $x(n)$ as shown in Figure 2A.2. Note that alternate sample values have been dropped, therefore from Figure 2A.2, in the sampled domain one has $y_D(n) = x(2n)$, or in the Z-transform domain

$$Y_D(z) = \sum_n y_D(n)\, z^{-n} = \sum_{\text{for } m \text{ even}} x(m)\, z^{-\frac{m}{2}}$$

$$= \sum_m \frac{x(m)}{2} [1 + (-1)^m]\, z^{-\frac{m}{2}} = \frac{1}{2}[X(\sqrt{z}) + X(-\sqrt{z})]$$

The original signal x has the spectrum given by Figure 2A.3. Once the signal is downsampled, the spectrum is given by Figure 2A.4. Hence, the spectrum of $Y_D(e^{j\omega})$ is aliased because it is the sum of the two terms given by

$$Y_D(z) = \frac{X(\sqrt{z}) + X(-\sqrt{z})}{2}$$

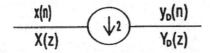

Figure 2A.1 The symbol used to represent decimation by a factor of 2.

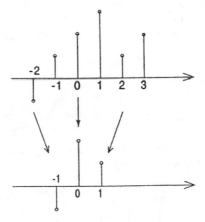

Figure 2A.2 The waveform obtained in the sampled domain from a decimation by a factor of 2. *Top*: The original sampled signal. *Bottom*: The decimated discrete signal.

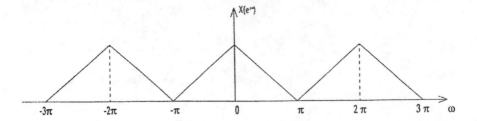

Figure 2A.3 The spectrum of the original signal.

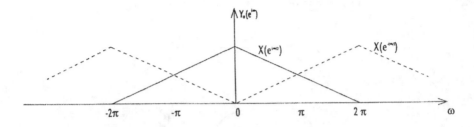

Figure 2A.4 The spectrum of the downsampled signal.

2A.2   Principle of Expansion by a Factor of 2

Pictorially, the upsampling or expansion is represented by the symbol in Figure 2A.5. In the sampled domain, this is equivalent to inserting a zero between the sampled signals as shown in Figure 2A.6. Mathematically this is equivalent to $y_L(n) = x\left(\dfrac{n}{2}\right)$, or, in the Z-transform domain

$$Y_L(z) = \sum_n y_L(n)\, z^{-n} = \sum_n x\left(\frac{n}{2}\right) z^{-n} = \sum_m x(m)\, z^{-2m} = X(z^2)$$

If the spectrum of the original signal x is that of Figure 2A.7, then the spectrum of $Y_L(z)$ would be that of Figure 2A.8.

$$\frac{x(n)}{X(z)} \quad \left(\uparrow 2 \right) \quad \frac{y_L(n)}{Y_L(z)}$$

Figure 2A.5 The symbol used to represent upsampling by a factor of 2.

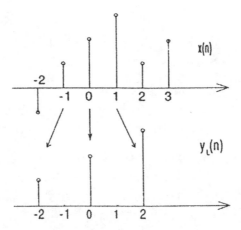

Figure 2A.6 Upsampling by a factor of 2 is equivalent to inserting a zero between the sampled signals in the samples domain. *Top*: The original sampled signal. *Bottom*: The expanded discrete signal.

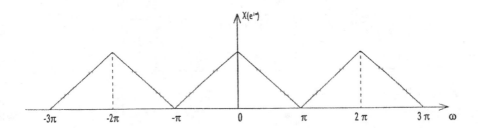

Figure 2A.7 The spectrum of the original signal.

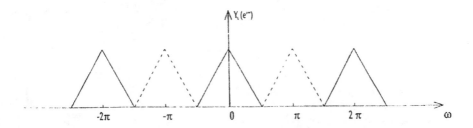

Figure 2A.8 The spectrum of the upsampled signal.

Chapter 3

APPLICATION OF WAVELETS IN THE SOLUTION OF OPERATOR EQUATIONS

In the previous chapter, we discussed the application of wavelet techniques for image and signal processing. We illustrated how these techniques can be used to approximate a function with adequate accuracy and also developed an efficient way to compute them. In this chapter, the objective is to study the suitability of the wavelet concept in the solution of operator equations, namely, their use as basis and testing functions in the solution of boundary value problems. The effect of this basis on the condition number of the system impedance matrix is also addressed. Numerical examples are presented to illustrate various concepts. This chapter provides the groundwork for the next three chapters where a wavelet methodology is used for efficient solutions of large real and sparse or complex matrix equations. These large matrices are generated when an operator equation is converted to a matrix equation, for example, via the method of moments using wavelets as expansion and testing functions. Here we also apply a multiresolution decomposition on the solution, even though it is an unknown as opposed to applying multiresolution techniques on the dense complex impedance matrix. In this chapter we outline the requirements of a basis in the solution of operator equations and how the application of a wavelet-like basis affects the numerical formulation.

3.1   INTRODUCTION

The main purpose of our discussion of wavelets in this chapter is primarily in the approximation of functions. However, there are two types of approximations. One is the approximation of a function in an efficient manner using the least number of parameters in a finite dimensional space. The other type deals with the approximation of a function also in a finite dimensional space satisfying an integrodifferential equation. In the former case, one can deal with the continuous representation, where the goal is to approximate a function $f(t)$, first in an infinite dimensional space, through

$$f(t) = \sum_{n=-\infty}^{\infty} \sum_{m=-\infty}^{\infty} a_{nm} \, \psi_{mn}(t) \tag{3.1}$$

Usually, one would rather deal with a finite dimensional space, where for the discrete case, the limits are not infinite but finite. Then one obtains

$$f(t) = \sum_{m=1}^{S} a_{mM} \, \phi_{mM}(t) + \sum_{n=0}^{M} \sum_{m=1}^{S} a_{nm} \, \psi_{mn}(t) \tag{3.2}$$

where the summations in (3.2) are finite. Observe, that for the finite summation case, one needs both the scaling functions ϕ and wavelets ψ in contrast to the continuous case in (3.1) where only the wavelets are needed. This has already been illustrated in the previous chapter. In (3.2), the scaling functions approximate the low-frequency components up to the dc limit (i.e., zero frequency). In signal and image processing, and in the numerical solution of operator equations as well as in various other applications, the approximation given by (3.2) has been used.

In this chapter, we are interested in the second problem of interest, namely, solution of operator equations of the form

$$A \, \Xi = \Psi \tag{3.3}$$

where, in general, A is a known integrodifferential operator, and the goal is to find the unknown function Ξ for the given excitation Ψ. The general approach of solving an operator equation is to expand the unknown $\Xi(t)$, which is a function of the variable t (t here represents a variable and not time) in terms of known basis functions $x_i(t)$ with some constant coefficients α_i. Therefore, we transfer the solution for an unknown function to the solution of some unknown coefficients α_i which are to be solved for from the known excitation Ψ. If we approximate the unknown function Ξ with a weighted sum of the known functions $x_i(t)$, then

$$\Xi(t) = \sum_{i=1}^{N} \alpha_i \, x_i(t) \tag{3.4}$$

Substitution of (3.4) into (3.3) results in

$$A\Xi(t) = \sum_{i=1}^{N} \alpha_i \, A \, x_i(t) = \Psi(t) \tag{3.5}$$

In a Galerkin procedure, the error in the functional representation $\Psi(t) - A\,\Xi(t)$ is weighted to zero with respect to the chosen basis functions $x_j(t)$, for $j = 1, ..., N$. This results in a matrix equation:

$$\begin{bmatrix} <Ax_1;x_1> & <Ax_2;x_1> & \cdots & <Ax_N;x_1> \\ <Ax_1;x_2> & <Ax_2;x_2> & \cdots & <Ax_N;x_2> \\ & & \vdots & \\ <Ax_1;x_N> & Ax_2;x_N> & \cdots & <Ax_N;x_N> \end{bmatrix} \begin{bmatrix} \alpha_1 \\ \alpha_2 \\ \vdots \\ \alpha_N \end{bmatrix} = \begin{bmatrix} <Y;x_1> \\ <Y;x_2> \\ \vdots \\ <Y;x_N> \end{bmatrix} \quad (3.6a)$$

or

$$[B][\alpha] = [Z] \quad (3.6b)$$

from which the matrix [α] is solved from

$$[\alpha] = [B]^{-1}[Z] \quad (3.7)$$

The inner products in (3.6a) are defined by

$$<c;d> \equiv \int_a^b c(x)\overline{d}(x)\,dx, \text{ and } \|C\|^2 = \langle C; C \rangle \quad (3.8)$$

where the overbar represents the complex conjugate and the limits a and b for the integral define the extension of the domain. The goal is to explore the suitability of the use of a wavelet type of basis for x_i and study the properties of the resulting matrix equations introduced in (3.6). In (3.8) the norm $\|\bullet\|$ is related to the inner product as defined.

It is important to observe that in the solution of operator equations, it is not necessary that the functions x_i be orthogonal or for that matter, that they even be linearly dependent [1–7]. All that is required in this solution procedure is that the matrix [B] be nonsingular, which implies that:

1. Ax_i must be linearly independent.
2. The functions Ax_i must span Y.

Property 1 guarantees that B is nonsingular and property 2 guarantees that a meaningful solution will be obtained. Let us illustrate this with an example. Consider the solution of

$$\frac{d^2X}{dt^2} = 2 + \sin t; \text{ for } 0 \le t \le 2\pi \quad (3.9)$$

with the boundary condition $X(0) = 0 = X(2\pi)$. The solution of this differential equation is given by

$$X = t(t - 2\pi) - \sin t \quad (3.10)$$

Now if we want to solve this problem using the principles outlined in the development of (3.3) to (3.6), we need to expand the unknown X in terms of a set of basis functions. An obvious choice (which will turn out to be incorrect) for the function X would be [4]

$$X(t) = a_o + \sum_{n=1}^{\infty} (a_n \sin nt + b_n \cos nt) \qquad (3.11)$$

the classical Fourier series expansion for X. In this case, application of (3.4) to (3.6) leads to the solution

$$X = \sin t \qquad (3.12a)$$

Observe that this does not provide the correct solution of (3.9). What is the problem? The problem is that we have made a mathematically incorrect choice for the basis x_i by selecting a complete orthogonal Fourier series expansion for the unknown. The important point to make is that even though the basis is complete, it does not satisfy condition 2. The choice of an orthogonal basis is irrelevant in the solution of operator equations. What is required is that the set of functions Ax_i be complete. When one chooses the Fourier expansion for the unknown as represented by (3.11), the set Ax_i in that case results in $\sum_n -(a_n n^2 \sin nt + b_n n^2 \cos nt)$ and has the constant term missing. Therefore, the choice of the basis (3.11) does not constitute a complete set for Ax_i. Hence, the correct choice for the basis will be

$$X(t) = a_o + c_1 t + c_2 t^2 + \sum_{n=1}^{\infty} (a_n \sin nt + b_n \cos nt) \qquad (3.12b)$$

Under these circumstances, the linearly dependent basis chosen in (3.12b) has the required completeness property for Ax_i. In this case, the functions Ax_i form a complete set even though the set x_i is linearly dependent. One obtains the correct solution using the procedure of (3.6). Therefore, to deal with the solution of operator equations, one may have to deal with linearly dependent sets. However, as outlined in Mikhlin [1], for the system matrix to be well conditioned, the basis needs to be "strongly minimal" or a Riesz basis. This implies that if any one of the elements of the basis is deleted then the set is no longer complete. The right mathematical framework to deal with these types of basis is the concept of frames, which has been widely used in the classification of the wavelet basis. The redundancy in the basis may be necessary to generate the complete set of functions for Ax_i. The redundancy in the basis can be quantified using the concept of frames.

Consider the two-dimensional $\Re^2$ space. An example of such a space may be the plane of this paper. As illustrated by Vetterli and Kovacevic [8], if one selects the coordinate axes e_0 and e_1 to be: $e_0 = [1,0]^T$ and $e_1 = [0,1]^T$, where T denotes the transpose of a matrix, then these two vectors (or column matrices) form an orthonormal set. If we now consider the vectors ϕ_0 and ϕ_1 as shown in Figure 3.1(a), then it is easy to see that $\phi_0 = \dfrac{[1,1]^T}{\sqrt{2}}$ and $\phi_1 = \dfrac{[1,-1]^T}{\sqrt{2}}$ form an orthonormal basis, that is,

$$< \phi_i ; \phi_j > = \delta_{i,j} = \begin{cases} 1, & \text{for } i = j \\ 0, & \text{for } i \neq j \end{cases} \tag{3.13}$$

where $\delta_{i,j}$ is the Dirac delta function. A vector x in the space can be approximated by using the basis ϕ_i, with $\alpha_i = < x; \phi_i >$.

We now introduce the dual basis defined by $\tilde{\phi}_i$ for the vectors ϕ_i so that

$$\tilde{\phi}_0 = [1,-1]^T ; \tilde{\phi}_1 = [0,1]^T \;\; and \;\; \phi_0 = [1,0]^T ; \phi_1 = [1,1]^T \tag{3.14}$$

Here the dual basis $\tilde{\phi}_i$ of the basis ϕ_i, as illustrated in Figure 3.1(b), satisfies

$$< \phi_i ; \tilde{\phi}_j > = \delta_{i,j} \tag{3.15}$$

A vector x of the space $\Re^2$ can be expanded either using the dual basis $\tilde{\phi}_i$ with β_i $= < x; \tilde{\phi}_i >$ or using the usual basis ϕ_i with α_i as the expansion coefficients.

We next consider the linearly dependent set

$$\phi_0 = [1,0]^T ; \phi_1 = [-1/2 ; \sqrt{3}/2]^T \;\; and \;\; \phi_2 = [-1/2 ; -\sqrt{3}/2]^T \tag{3.16}$$

as shown in Figure 3.1(c). For this case one needs to introduce the concept of frames. The theory of frames was originally developed by Duffin and Schaeaffer [9].

A family of square integrable functions ϕ_j is called a frame if two constants C and D exist such that $C > 0, D < \infty$ so that for all functions f

$$C \| f \|^2 \leq \sum_j \left| < f ; \phi_j > \right|^2 \leq D \| f \|^2 \tag{3.17}$$

where the norm $\| \bullet \|$ is defined in (3.8). Here C and D are called the frame bounds. If the two frame bounds C and D are equal, then it is called a tight frame. For that case, if $\| \phi_j \| = 1$, and $C = D$, then either C or D provides a quantitative measure as to how many additional functions ϕ_j have been provided than are necessary. The significance of this parameter will be explained later. However, it tells us quantitatively the number of redundant basis functions that exists in the expansion of the given function. When $C = D = 1$, then the ϕ_j form an orthonormal basis, with $\|\phi_j\| = 1$.

Therefore, for any vector $f = [f_1; f_2]^T$, we have

$$\sum_{j=1}^{3} \left| < f; \phi_j > \right|^2 = \|f_1\|^2 + \left| -\frac{1}{2} f_1 + \sqrt{3}/2 \, f_2 \right|^2 + \left| -\frac{1}{2} f_1 - \frac{\sqrt{3}}{2} f_2 \right|^2$$

$$= \frac{3}{2} \left[|f_1|^2 + |f_2|^2 \right] = \frac{3}{2} \|f\|^2$$

(3.18)

Therefore, here $C = D = 3/2$ and the three vectors ϕ_j form a tight frame. But they are not orthonormal as $C = D \neq 1$. Here the frame bound is given by $C = 3/2$ and this gives the "redundancy ratio" (three vectors in a two-dimensional space).

Orthonormal Basis	Biorthonormal Basis	Linearly Dependent Basis
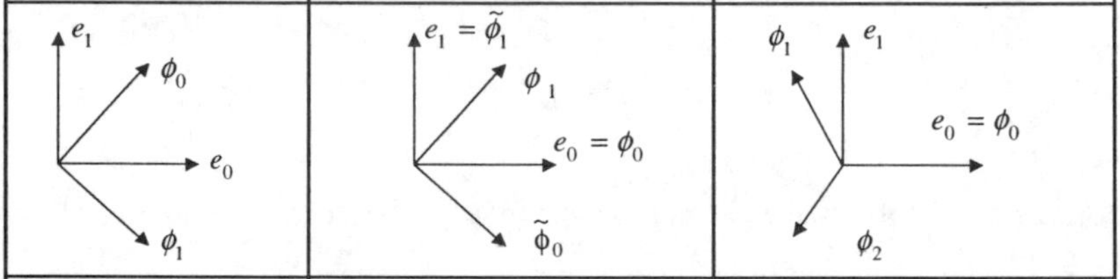		
$\phi_0 = \frac{1}{\sqrt{2}}\begin{bmatrix}1\\1\end{bmatrix}$ $\phi_1 = \frac{1}{\sqrt{2}}\begin{bmatrix}1\\-1\end{bmatrix}$ $x = \sum_i \alpha_i \phi_i$ $\alpha_i = \langle x; \phi_i \rangle$	$\phi_0 = \begin{bmatrix}1\\0\end{bmatrix}$ $\tilde{\phi}_0 = \begin{bmatrix}1\\-1\end{bmatrix}$ $\phi_1 = \begin{bmatrix}1\\1\end{bmatrix}$ $\tilde{\phi}_1 = \begin{bmatrix}0\\1\end{bmatrix}$ $\langle \phi_0, \tilde{\phi}_1 \rangle = \langle \phi_1, \tilde{\phi}_0 \rangle = 0$ $x = \sum_i \alpha_i \phi_i = \sum_j p_j \tilde{\phi}_j$ with $p_j = \langle x; \tilde{\phi}_j \rangle$	$\phi_0 = \begin{bmatrix}1\\0\end{bmatrix}$ $\phi_1 = \frac{1}{2}\begin{bmatrix}-1\\\sqrt{3}\end{bmatrix}$ $\phi_2 = \frac{1}{2}\begin{bmatrix}-1\\-\sqrt{3}\end{bmatrix}$ Use Concept of Frames
(a)	(b)	(c)

Figure 3.1 The concept of (a) the orthonormal basis, (b) the biorthonormal basis, and (c) the linearly dependent basis.

We now generate a frame operator F defined by

$$[Ff]_j = < f; \phi_j >$$

(3.19)

In addition, we define another set of vectors $\tilde{\phi}_j$ by

$$\tilde{\phi}_j = [F^H F]^{-1} \phi_j \tag{3.20}$$

where H denotes the conjugate transpose of a matrix, then the functions $\tilde{\phi}_j$ form the dual frame. It can be shown from [8] that

$$\sum_j < f; \phi_j > \tilde{\phi}_j = f = \sum_j < f; \tilde{\phi}_j > \phi_j \tag{3.21}$$

This provides an expansion formula for f given the inner products $< f; \phi_j >$. This also describes a recipe for writing f as a superposition of the functions ϕ_j. Therefore, for the example described in this section, we have

$$f = \frac{2}{3} \sum_{j=1}^{3} < f; \phi_j > \phi_j \tag{3.22}$$

since $[F^H F] = \frac{3}{2}[I]$, where $[I]$ is the identity matrix. However, this representation is not unique. As $\sum_{j=1}^{3} \phi_j = 0$ then, one can write

$f = \frac{2}{3} \sum_{j=1}^{3} \big[< f; \phi_j > + E \big] \phi_j$ for any arbitrary constant E.

The above cursory description of frames explains how this methodology is used to deal with a linearly dependent basis. Next, we show how to use the concept of frames in the solution of operator equations.

3.2 APPROXIMATION OF A FUNCTION BY WAVELETS

Approximation of a function by orthogonal wavelets is equivalent to combining portions of the functions, which have been filtered by a bank of filters. Typically, the filters have bandwidth, which progressively gets reduced by a factor of 2 as we go to the low-frequency region as illustrated earlier in Figure 2.6. The popularity of the wavelets is primarily due to the fact that in many biological systems (for example, the human ear or eyes), at least at the first step of processing, the input is decomposed by a bank of constant-Q filters. The Western music scale is also based on this type of dyadic decomposition of the spectrum. This appears to be a good fit for biological systems. So, by looking at a signal through a number of constant-Q filters, the observation window changes depending on the frequencies we are looking at. Here, the question is: Can the same type of low-complexity approximation be used to arrive at an efficient solution to electromagnetic field problems?

The answer to this problem is quite straightforward. If the excitation is impulse-like, then it has a broad spectrum, and whether one decomposes into a band of constant-Q responses or uses the uniform partition of the frequency band as with a Fourier technique, nothing is gained by choosing one form of the representation over the other because all bands are full. However, if the spectrum of the excitation has localized spectral components, then, depending on the frequency content of the waveform, one of the two expansions (i.e., using the Fourier or wavelet theory) may have a more efficient representation.

The other point to keep in mind is that for approximation of a function in the discrete case, we have to use the representation given by (3.2). However, there is an equivalent representation in terms of filtering the signal by a bank of constant-Q filters. In this alternate representation, one does not require any scaling functions or wavelets for the discrete case. The entire representation can be done in terms of low- and highpass filtering operations. The scaling functions and the wavelets can be related through the low- and highpass structures of the filters as seen earlier in Figure 2.1. The concept of dyadic decomposition of a wide bandwidth signal (i.e., filtering by constant-Q filters) is shown in Figures 2.5 and 2.6. The various expressions for the scaling and wavelet functions and how they relate to the filter representations are illustrated next for two of the popular wavelets [10].

(A) The Haar wavelet

In this case, the scaling function is defined by

$$\phi(t) = 1 \quad \text{for} \quad 0 \le t < 1 \tag{3.23}$$

and the wavelet is given by

$$\psi(t) = \begin{cases} 1, & \text{for} \quad 0 \le t < 1/2 \\ -1, & \text{for} \quad \dfrac{1}{2} \le t < 1 \\ 0, & \text{otherwise} \end{cases} \tag{3.24}$$

The synthesis filters (i.e., the filters on the receiver end as illustrated in Figure 2.1) corresponding to these scaling functions and wavelets are given by

$$h'[n] = [\delta(n) + \delta(n-1)]/\sqrt{2}$$
$$g'[n] = [\delta(n) - \delta(n-1)]/\sqrt{2}$$
$$H'(e^{j\omega}) = \sqrt{2}\, e^{-j\omega/2} \cos \omega/2 \tag{3.25}$$
$$G'(e^{j\omega}) = \sqrt{2}\, j\, e^{-j\omega/2} \sin \omega/2$$

The analysis filters of Figure 2.1 (i.e., the filters on the transmitter end) are simply the time-reversed versions of the synthesis filters, that is,

$$h[n] = h'[-n] \tag{3.26}$$

For a detailed explanation on the developments of the above expressions and what they represent, the reader is referred to Chapter 2. The Haar scaling function and wavelets are equivalent to the choice of expansion functions with finite support that form the conventional subdomain basis functions in the method of moments. The filter involved in the Haar wavelet is the only compactly supported conjugate mirror filter that has a linear phase.

(B) The Shannon or the sinc wavelet

The Haar basis has poor frequency resolution because it is a piecewise constant. The other extreme is the sinc function. In this case the scaling function is given by

$$\phi(t) = \frac{\sin \pi t}{\pi t} \tag{3.27}$$

So that its Fourier transform $\tilde{\phi}(\omega)$ is

$$\tilde{\phi}(\omega) = \begin{cases} 1, & -\pi \le \omega < \pi \\ 0, & \text{otherwise} \end{cases} \tag{3.28}$$

The wavelet function therefore is given by

$$\psi(t) = \frac{\sin \dfrac{\pi t}{2}}{\dfrac{\pi t}{2}} \cos \frac{3\pi t}{2} \tag{3.29}$$

The various synthesis filters are given by (see Figure 2.1)

$$h'(n) = \frac{1}{\sqrt{2}} \sin \frac{(\pi/2)n}{(\pi/2)n}$$

$$g'(n) = (-1)^n \, h'[-n+1] \tag{3.30}$$

$$H'(e^{j\omega}) = \begin{cases} \sqrt{2}, & \text{for} \quad -\pi/2 \le \omega < \pi/2 \\ 0, & \text{otherwise} \end{cases}$$

$$G'(e^{j\omega}) = -e^{-j\omega} \, H'(-e^{-j\omega})$$

The choice of this type of scaling function and wavelet is equivalent to choosing entire domain basis functions in the method of moments.

When one tries to approximate a function by some basis, the natural question that arises is does it give rise to Gibb's phenomenon? Gibb's phenomena are directly related to the approximation of a discontinuous function by a set of smooth basis functions. There seems to be some misunderstanding as to what constitutes Gibb's phenomenon. It deals not with the failure of the Fourier series to converge to a point of jump discontinuity but rather with the *overshoot* of the partial sums in the limit [11].

A Gibb's phenomenon does not occur if the partial sums are replaced by the Cesaro sums (i.e., average of the partial sums). Since the wavelet expansions have convergence properties similar to Cesaro summability, we might expect them not to exhibit Gibb's phenomena.

It is important to note that certain wavelet expansions do exhibit this overshoot phenomenon. In particular, many of the Daubechies wavelets, the Shannon (sinc wavelet), and hat (Franklin) wavelets, and some of the Meyer wavelets do display Gibb's phenomenon. (In the literatures different names have been used for the same wavelet family and therefore we have listed all the names here to avoid confusion.) In contrast, the use of Haar wavelets in approximating a discontinuous function does not produce Gibb's phenomenon, because they themselves are discontinuous functions. In many cases of electromagnetic characterization, therefore, it appears that the use of the Haar wavelet could approximate the component of the current parallel to an edge of a conducting structure much better because the current has a singularity. In that case, use of discontinuous basis functions will provide a better functional approximation for the unknown in the solution of operator equations than any of the smoother functions. Also, at a feed point, the current has a discontinuous second derivative and hence the basis functions cannot be very smooth near the feed structures.

3.3 SOLUTION OF OPERATOR EQUATIONS

Consider the approximation of a function in $\Re^4$ (a Euclidean space of dimension 4 [11]). A possible basis for the space is

$$\phi = \begin{bmatrix} 1 \\ 1 \\ 1 \\ 1 \end{bmatrix}; \quad \psi = \begin{bmatrix} 1 \\ 1 \\ -1 \\ -1 \end{bmatrix} \quad \psi_{01} = \begin{bmatrix} 1 \\ -1 \\ 0 \\ 0 \end{bmatrix} \quad \text{and} \quad \psi_{02} = \begin{bmatrix} 0 \\ 0 \\ 1 \\ -1 \end{bmatrix} \tag{3.31}$$

So, any vector Q in $\Re^4$ can be approximated by

$$Q = a_0\phi + a_1\psi + a_2\psi_{01} + a_3\psi_{01} \tag{3.32}$$

where a_i are some constants. The above four functions, ϕ, ψ, ψ_{01}, and ψ_{02}, form a complete set for $\Re^4$. One can relate these four functions to the scaling functions ϕ (also called the father of the wavelet) and the wavelet function ψ (also called the mother of the wavelet) for a Haar wavelet, followed by two daughters ψ_{01} and ψ_{02}. A daughter is considered to be a replica of the mother. This is a typical wavelet-type representation. Now, it is possible to have another orthogonal basis for $\Re^4$ which may be composed only of the four sons of the Haar wavelet as given by

$$\phi_{01} = \begin{bmatrix} 1 \\ 0 \\ 0 \\ 0 \end{bmatrix} ; \phi_{02} \begin{bmatrix} 0 \\ 1 \\ 0 \\ 0 \end{bmatrix} ; \phi_{03} = \begin{bmatrix} 0 \\ 0 \\ 1 \\ 0 \end{bmatrix} ; \phi_{04} \begin{bmatrix} 0 \\ 0 \\ 0 \\ 1 \end{bmatrix} \tag{3.33}$$

Here, the sons are scaled versions of the father, namely, the scaling functions. Hence, the vector Q can also be represented by the four sons (another orthogonal basis) through

$$Q = C_0\phi_{01} + C_1\phi_{02} + C_2\phi_{03} + C_3\phi_{04} \tag{3.34}$$

It is interesting to point out that (3.34) forms the usual subsectional basis often used in numerical electromagnetics. This subsectional basis provides a better conditioned impedance matrix than the entire domain basis. In addition, the integrations required in evaluating the elements of the impedance matrix are more time consuming and need to be carried out in a very careful fashion for an entire domain wavelet basis. Sometimes, the wavelet representation is advantageous, because it may

1. Provide a fast algorithm for generating the constant coefficients a_i in the approximation of Q in (3.32).
2. Lead to a sparse impedance matrix after applying a threshold when they are used as expansion and weighting functions in the approximation of an unknown in the solution of an operator equation.

In the next three chapters, various possibilities using wavelets are explored in the solution of operator equations. Namely, wavelets can be introduced in three different ways:

1. We use a subdomain basis and convert the operator equation to a matrix equation. Then we use a wavelet-like decomposition with a threshold to make the impedance matrix sparse. We then solve the sparse matrix equation in an efficient way using a sparse matrix solver or an iterative method like the conjugate gradient. This procedure will be discussed in Chapter 4.
2. We use the wavelets themselves as the basis function. In this case, they are entire domain basis or overlapping subsectional basis as the case may

be and are used directly in the solution of an operator equation. This is described in Chapter 5 where the wavelets are used as basis and testing functions in a finite element method.

3. A multiscale basis may be used to represent the solution of an operator equation even though it is unknown. By thresholding the coefficients of a multiscale basis at different scales based on the second derivative of the extrapolated solution from a lower scale, one can eliminate rows and columns in the impedance matrix. In this way, one can set up a modified impedance matrix of smaller dimension than the original matrix. However, this smaller dimensioned matrix is typically dense. This method is presented in Chapter 6.

We conclude this chapter, by illustrating one of the areas of concern in the utilization of wavelets as basis functions. We illustrate this problem by using the wavelets as expansion functions in the solution of a sample integral equation.

3.4 WAVELET BASIS IN THE SOLUTION OF INTEGRAL EQUATIONS

Consider the problem of the charge distribution on a thin wire (of length L and radius a) maintained at a constant voltage V. If the wire is assumed to lie along the z-axis, the integral equation that relates the charge distribution to the potential V is given by [12]

$$\frac{1}{4\pi\varepsilon_0} \int_0^L \frac{q(z')}{\sqrt{a^2 + (z-z')^2}}\, dz' = V(z) \qquad (3.35)$$

For the method of moments, when using a subdomain basis, we attempt to solve integral equation (3.35) by expanding the charge $q(z)$ in terms of a set of basis functions with unknown coefficients, that is,

$$q(z) \simeq \sum_{i=1}^M b_i \phi_i(z) \qquad (3.36)$$

Substituting (3.36) in (3.35) we get

$$\sum_{i=1}^M b_i \frac{1}{4\pi\varepsilon_0} \int_0^L \frac{\phi_i(z')}{\sqrt{a^2 + (z-z')^2}}\, dz' \cong V(z) \qquad (3.37)$$

By using the method of moments and by choosing the same functions $\phi_i(z)$ as weighting functions, one obtains Galerkin's method. This results in

$$\sum_{i=1}^{M} \frac{b_i}{4\pi\varepsilon_o} \int_0^L dz\, \phi_j(z) \int_0^L \frac{\phi_i(z')dz'}{\sqrt{a^2+(z-z')^2}} \simeq \int_0^L V(z)\phi_j(z)dz \qquad (3.38)$$

and can be written in compact matrix form as

$$[A]_{M\times M}\,[b]_{M\times 1} = [V]_{M\times 1} \qquad (3.39)$$

where

$$A_{ji} = \int_0^L \phi_j(z)dz \int_0^L \frac{\phi_i(z')dz'}{\sqrt{a^2+(z-z')^2}} \qquad (3.40)$$

$$V_j = \int_0^L V(z)\phi_j(z)\,dz \qquad (3.41)$$

To obtain the unknown coefficient b_i, we need to solve (3.39), resulting in

$$[b]_{M\times 1} = [A]_{M\times M}^{-1}\,[V]_{M\times 1} \qquad (3.42)$$

The solution of (3.42) necessitates that the condition number of the matrix $[A_{ji}]$ be finite and not too large. The condition number of a matrix $C[A]$ is defined by the ratio of the largest singular value to the smallest singular value. For a square matrix $[A]$, the singular values are related to the eigenvalues of the matrix $[A]^H[A]$ where the superscript H denotes the conjugate transpose of a matrix. Hence the condition number becomes the ratio of the square root of the maximum eigenvalue of $[A]^H[A]$ to that of the square root of the minimum eigenvalue. In other words, the quantity $\log_2\{C[A]\}$ tells us how many binary bits are necessary to represent the matrix. Therefore, if $\log_2\{C[A]\} > w$, then w is the number of bits needed in a computer that will be used for carrying out the computations. Therefore, if the quantity $\log_2\{C[A]\}$ is 20, then solution of (3.42) cannot be done on a computer where computations are carried out using 16 bits. Hence, the basis functions, which constitute the matrix A, must be chosen in such a way that if M increases, the condition number $C[A]$ will not increase. We solve (3.42) in two different ways. In the first procedure, we choose the basis functions to be the usual subdomain basis denoted by both solid and dotted lines in Figure 3.2(e) where $M = 9$ including the two half dotted triangles at the end. For the second case, we choose the basis by using the wavelet principles, namely, the dilated and shifted version of the same function as illustrated in Figure 3.2(a – d) and by the functions represented by the solid line of Figure 3.2(e). So there are two sets of bases. One, the usual subsectional basis represented by the nine functions of Figure 3.2(e). The second basis can be formed by the nine functions represented by the solid lines of Figure 3.2(a – e). For either choice of the basis, the solution would be approximated by nine functions and the final solution would be identical independent of the choice of the basis. When we use the subdomain basis in the conventional method of moments using the nine functions illustrated by Figure 3.2(e), then we observe that the condition number $C[A]$ of the impedance matrix does not increase with respect to the number of unknowns. This is indicated

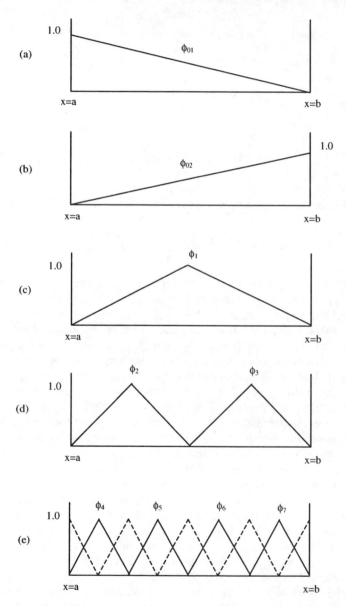

Figure 3.2 The wavelet-like basis (a – e) shown by solid lines and the subdomain basis in (e) shown by both solid and dashed lines.

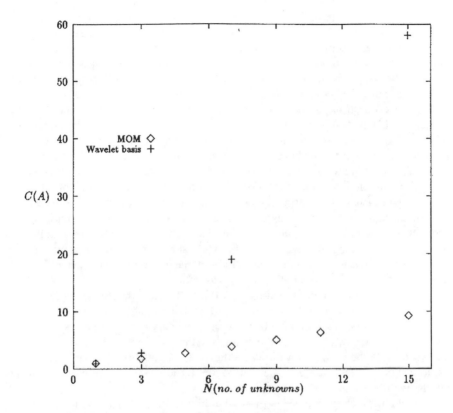

Figure 3.3 The growth of the condition number as a function of the number of unknowns.

by the diamonds in Figure 3.3. However, when we use the wavelet type of basis, the condition number increases very rapidly as shown by the + symbols in Figure 3.3. Therefore, unless the precision of the computing environment is simultaneously increased with the increase of the number of unknowns when using the wavelet-type basis, the increase in the condition number can be a problem in the solution of matrix equations.

For the type of integral equation defined by (3.35), the condition number of the system matrix is known to increase at a rate of $\theta\left(\dfrac{1}{h}\right)$ as a function of the number of unknowns. Here θ ($\bullet$) denotes of "the order of" and h is the subsection length defined by $L/(M-1)$. If one considers the numerical solution of the integral equation relating the charge distribution on a wire charged to a constant potential, utilizing both subdomain and wavelet-type bases, then a measure of the efficiency of the solution procedure is how the condition number of the system matrix changes with the number of basis functions. In Figure 3.3, the condition number of the system matrices for the two approaches is compared. We can see that the subdomain basis for the method of moments performs better than a wavelet-type basis. This is not surprising, because Mikhlin has pointed out [1, p. 43] that for the system matrix in a variational method to have a small condition number, the basis should be "strongly minimal in the corresponding energy space." The rise in the condition number due to the entire domain basis is known to occur in numerical computations. That is why in traditional method of moments calculations, one usually chooses subdomain basis functions. However, Mikhlin points out [1] that appropriately scaling the basis functions in some situations might not deteriorate the condition number of the impedance matrix that much. How to choose such a scaled basis for a general geometry is still an open question.

It is interesting to note that in computational electromagnetics, invariably we deal with convolutional operators. For convolution forms of the integral operator, a good approach would definitely be to look at the problem in the transform domain because a Fourier transform essentially converts the convolution to a product and therefore "diagonalizes" the operator. Hence, computationally the most efficient method to solve any operator equations arising in electromagnetics will be to use the iterative conjugate gradient method along with the FFT. Any other method will take more CPU time for execution [13].

However, the problem of dealing with the FFT is that the discretization has to be done on a uniform grid. In addition, it is not possible to solve a problem completely in the spectral domain because the boundary conditions are specified in the original domain. Moreover, the nature of the solution of the problem is not known outside the region of validity of the integral equation. Hence, we need to explore other semi-optimum methodologies. The wavelets provide such an alternative.

In the next chapters, we explore the application of the wavelets for the solution of matrix equations and the solution of the differential form of Maxwell's equations. We further explore an adaptive multiscale technique for solution of operator equations.

REFERENCES

[1] S. G. Mikhlin, *The Numerical Performance of a Variational Method*, Wolters Noordhoff, Groningen, The Netherlands, 1971.

[2] M. A. Krasnosel'skii et al., *Approximate Solution of Operator Equations*, Wolters Noordhoff, Groningen, The Netherlands, 1972.

[3] I. Stakgold, *Green's Function and Boundary Value Problems*, Wiley, New York, 1979.

[4] K. Rektorys, *Variational Methods in Mathematics, Science and Engineering*, D. Reidel Publishing Company, Dordrecht, Holland, 1975.

[5] T. K. Sarkar, "A note on the choice of weighting functions in the method of moments," *IEEE Trans. Antennas and Propag.*, Vol. AP-33, April 1985, pp. 436–441.

[6] T. K. Sarkar, A. R. Djordjevic, and E. Arvas, "On the choice of expansion and weighting functions in the numerical solution of thin wire integral equations," *IEEE Trans. Antennas and Propag.*, Vol. AP-33, Sept. 1985, pp. 988–996.

[7] M. Salazar-Palma et al., *Iterative and Self-Adaptive Finite-Elements in Electromagnetic Modeling*, Artech House, Norwood, MA, 1998.

[8] M. Vetterli and J. Kovacevic, *Wavelets and Subband Coding*, Prentice Hall, Upper Saddle River, NJ, 1995.

[9] R. J. Duffin and A. C. Schaeaffer, "A class of nonharmonic Fourier series," *Trans. Amer. Math. Soc.*, Vol. 72, 1952, pp. 341–366.

[10] G. Strang, "Wavelets from filter banks," in *Wavelets, Multilevel Methods and Elliptic PDE's,* M. Ainsworth et al., Eds., Oxford University Press, Oxford, U.K., 1997.

[11] G. G. Walter, *Wavelets and Other Orthogonal Systems with Applications*, CRC Press, Boca Raton, FL, 1994.

[12] R. F. Harrington, *Field Computation by Moment Methods*, Macmillan, New York, 1968.

[13] T. K. Sarkar, Ed., *Application of Conjugate Gradient Method in Electromagnetics and Signal Analysis*, PIER Vol. 5, Elsevier, New York, 1991.

Chapter 4

SOLVING MATRIX EQUATIONS
USING THE WAVELET TRANSFORM

The conventional wavelet transform used in one-dimensional image and signal processing is a dyadic decomposition of the signal. In this case, the signal is filtered by a bank of constant-Q filters, where the bandwidth of each filter gets reduced by a factor of 2, as we proceed from the high- to the low-frequency regimes. When one translates the wavelet transform to two dimensions, primarily for image analysis, one usually carries out a dyadic product of two one-dimensional decompositions. Since a matrix is a two-dimensional image, one can use wavelet techniques to compress the matrix, and then a solution technique based on sparse matrices is used for efficient solution of such systems. However, for the solution of large dense complex matrix equations, one requires a tensor product of the two one-dimensional decompositions of the matrix as opposed to the dyadic decomposition. This chapter illustrates through examples how to implement a tensor form of the wavelet transform in two dimensions. We call it a wavelet-like transform in order to differentiate it from the wavelet transform that we refer to as the dyadic decomposition. Both the wavelet and the wavelet-like transforms are carried out in the discrete case using filter banks. The use of a different type of filter bank might lead to sparser matrices than when using the conventional Daubechies wavelets based on finite impulse response (FIR) filters [1, 2]. In this methodology, the concept of the fast Fourier transform (FFT) technique can also be utilized for efficient numerical computation. However, this is achieved at the expense of robustness, which may lead to large reconstruction errors due to the presence of Gibb's phenomenon. In addition, when dealing with dense complex matrices arising in numerical electromagnetics, we show that FIR filters with a linear phase response may sometimes yield better compression than using filters with the same magnitude response but with a nonlinear phase response. Numerical examples are presented to illustrate these principles.

4.1 INTRODUCTION

Consider the solution of a matrix equation

$$[A][X] = [Y] \qquad (4.1)$$

where $[A]$ is a $Q \times Q$ complex dense matrix and $[Y]$ is a $Q \times 1$ given excitation vector. The goal is to find the $Q \times 1$ unknown vector $[X]$. Note that Q has to be an integer power of 2 for the wavelet techniques to be applicable. However, later in the chapter a modified method is presented where Q need not be an integer power of 2. In the solution of large dense complex matrix equations by the wavelet-like transform the goal is to transform a dense complex matrix into a very sparse matrix. One possible way to make matrix $[A]$ sparse is to make it diagonal. However, that requires θ (Q^3) operations to achieve it, where θ ($\bullet$) represents that the variable is of the order of the quantity between parenthesis. This takes many more operations than solving (4.1) by an LU decomposition, which takes θ $(Q^3/3)$ operations. An alternate procedure to make matrix $[A]$ sparse is uses methodologies of the wavelet transform. Wavelet techniques have been used very successfully for compressing images and since a matrix can be interpreted as a digitized image, it is hoped that this methodology can be applied too.

However, it is important to note that the conventional wavelet transform that is based on a dyadic decomposition [2–4] cannot be applied to this problem. So here we use the wavelet-like transform involving the tensor product [1, 5]. Figure 4.1 illustrates the wavelet-like transform in terms of the scaling functions and wavelets. The coefficients A_{ijk} in Figure 4.1 are the wavelet coefficients, and the products between the scaling function ϕ and the wavelet ψ are shown explicitly to illustrate the tensor product. The two-dimensional wavelet-like transform is implemented as a sequence of 1-D wavelet transforms applied first to the rows and then to the columns of the matrix. The key ingredient of the tensor product in implementing the wavelet-like transform is that the output from the highpass filters are further filtered by a bank of high- and lowpass filters. The transform domain representation of the wavelet-like transform is shown in Figure 4.2.

In contrast, the dyadic decomposition, which is commonly used in image and signal processing, is presented in Figure 4.3 in terms of the scaling functions and wavelets. The transform domain representation of the dyadic wavelet decomposition is shown in Figure 4.4. In Figures 4.1 through 4.4 the blocks H and G represent the low- and highpass FIR filters. It is very important to note that the dyadic decomposition is not applicable to the solution of matrix equations. The key differences between the wavelet transform based on the dyadic and the tensor representations are clearly seen in Figures 4.1 and 4.3. A few papers are available on the application of the wavelet-like transform for the solution of matrix equations [6–8].

In the wavelet-like transform based on the tensor product the original matrix equations in (4.1) are preprocessed by an orthogonal matrix $[W]$ so that

$$[W]^H [W][A][W]^H [W][X] = [Y] \tag{4.2}$$

where the superscript H denotes the conjugate transpose of a matrix. Equivalently,

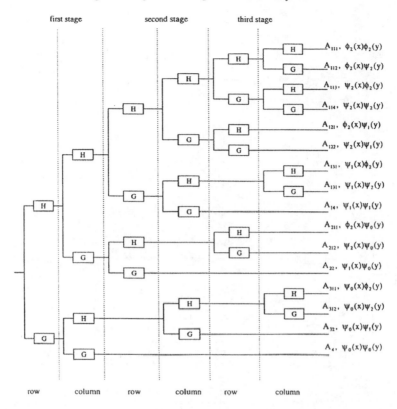

Figure 4.1 A wavelet-like decomposition of a matrix using the tensor product of the scaling function and wavelets.

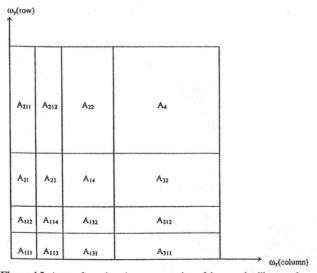

Figure 4.2 A transform domain representation of the wavelet-like transform.

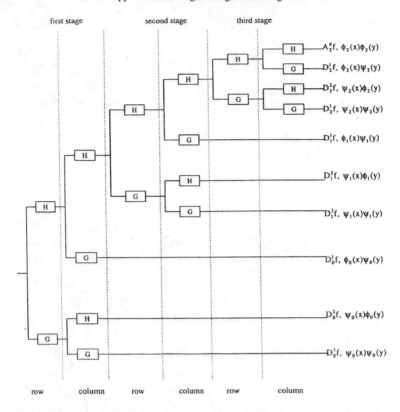

Figure 4.3 A dyadic wavelet decomposition of a matrix in terms of the scaling functions and wavelets.

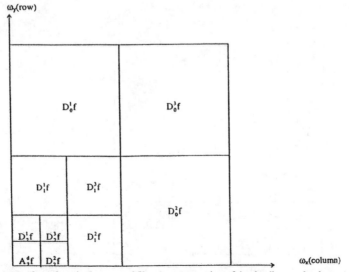

Figure 4.4 A transform domain (in terms of filters) representation of the dyadic wavelet decomposition.

$$[B] [Z] = [F] \qquad (4.3)$$

where

$$[B] = [W][A] [W]^H \qquad (4.4)$$

$$[Z] = [W] [X] \qquad (4.5)$$

$$[F] = [W][Y] \qquad (4.6)$$

and

$$[W] [W]^H = [W]^H [W] = [I] \quad \text{(identity matrix)} \qquad (4.7)$$

where $[I]$ is the identity matrix, which is a diagonal matrix with unity elements on the main diagonal and zero everywhere else. Preprocessing of $[A]$ by an orthogonal matrix $[W]$ leads to $[B]$, which is a sparse matrix. From a theoretical point of view it is claimed that for the type of convolutional kernels used in electromagnetics, the sparse matrix $[B]$ would have $\theta(N)$ elements instead of the $\theta(N^2)$ elements of the original matrix $[A]$ and that the process of making it sparse requires $\theta(N)$ operations. Since matrix $[B]$ contains only $\theta(N)$ elements, it can now be stored completely in the main memory of the computer. If one uses an iterative technique like the conjugate gradient method to solve the modified equation (4.3), then there may be some computational advantages. This is because at each iteration the conjugate gradient method requires two matrix-vector products. The computational load will now be $\theta(N)$ instead of $\theta(N^2)$ and will thereby reduce the number of page faults if N is large. From a purely theoretical point of view, both matrices $[A]$ and $[B]$ have the same condition number, which is defined as the absolute value of the ratio of the maximum to the minimum singular values of a matrix as explained in Chapter 3. This is because $[W]$ is an orthogonal matrix and from a purely theoretical point of view, the orthogonal transformation described by (4.7) should not change the condition number of matrices $[A]$ and $[B]$ in (4.4). Then a threshold is applied to the elements of $[B]$ to eliminate the small values below some number having a magnitude in order to make it sparse. The application of a threshold to the matrix elements changes the condition number of $[B]$ from that of $[A]$. However, it is important to point out from our experience that these theoretical conjectures never actually hold in real numerical experimentation and the actual results, even though quite encouraging, still fall short of the theoretical expectations [9]. We now illustrate how to obtain the wavelet transform first for a vector and then for a matrix.

4.2 IMPLEMENTION OF THE WAVELET-LIKE TRANSFORM BASED ON THE TENSOR PRODUCT

For the one-dimensional case, there is no difference between the dyadic wavelet transform or the tensorial form of the wavelet-like transform of a vector. The difference shows up for the two-dimensional case, namely, for a matrix. The wavelet-like transform of a matrix $[A]$ is then equivalent to taking two 1-D wavelet transforms as illustrated by Figure 4.1. First, we illustrate how to numerically implement the 1-D wavelet transform for a vector. We select the FIR filters $h(n)$ as illustrated in Chapter 2. Then we will use $h'(n)$ and $g'(n)$ to find the discrete wavelet coefficients utilizing (2.77a) and (2.77b). What is going to be different is that we are going to express (2.77a) and (2.77b) as circular correlations with respect to $h'(n)$ and $g'(n)$ or as circular convolutions with respect to $h(n)$ and $g(n)$, respectively. The convolutions can be computed in a very efficient way by utilizing the FFT. As an example, let us choose a FIR filter of order 5 (i.e., length is equal to 6). We then create the following 8×8 orthogonal matrix $[P]$ in terms of the FIR filter coefficients $h'(n)$ or $h(n)$ as

$$[P] = \begin{bmatrix} h'_0 & h'_1 & h'_2 & h'_3 & h'_4 & h'_5 & 0 & 0 \\ 0 & 0 & h'_0 & h'_1 & h'_2 & h'_3 & h'_4 & h'_5 \\ h'_4 & h'_5 & 0 & 0 & h'_0 & h'_1 & h'_2 & h'_3 \\ h'_2 & h'_3 & h'_4 & h'_5 & 0 & 0 & h'_0 & h'_1 \\ - & - & - & - & - & - & - & - \\ -h'_5 & +h'_4 & -h'_3 & +h'_2 & -h'_1 & +h'_0 & 0 & 0 \\ 0 & 0 & -h'_5 & +h'_4 & -h'_3 & +h'_2 & -h'_1 & +h'_0 \\ -h'_1 & +h'_0 & 0 & 0 & -h'_5 & +h'_4 & -h'_3 & +h'_2 \\ -h'_3 & +h'_2 & -h'_1 & +h'_0 & 0 & 0 & -h'_5 & +h'_4 \end{bmatrix}$$

$$= \begin{bmatrix} h_5 & h_4 & h_3 & h_2 & h_1 & h_0 & 0 & 0 \\ 0 & 0 & h_5 & h_4 & h_3 & h_2 & h_1 & h_0 \\ h_1 & h_0 & 0 & 0 & h_5 & h_4 & h_3 & h_2 \\ h_3 & h_2 & h_1 & h_0 & 0 & 0 & h_5 & h_4 \\ - & - & - & - & - & - & - & - \\ -h_0 & +h_1 & -h_2 & +h_3 & -h_4 & +h_5 & 0 & 0 \\ 0 & 0 & -h_0 & +h_1 & -h_2 & +h_3 & -h_4 & +h_5 \\ -h_4 & +h_5 & 0 & 0 & -h_0 & +h_1 & -h_2 & +h_3 \\ -h_2 & +h_3 & -h_4 & +h_5 & 0 & 0 & -h_0 & +h_1 \end{bmatrix}$$

$$(4.8)$$

We have chosen $[P]$ as an 8×8 matrix since we need the size of the row to be an integer power of 2. In the expression of $[P]$ in (4.8) we have used h_k to represent the quantities $h(k)$ as defined in Chapter 2, to simplify the notation. This equation is the matrix form of the discrete wavelet transform of (2.77a) and (2.77b). Inside $[P]$ we have the filter coefficients, which are 6 in number because we have chosen a fifth-order filter. The first four rows are due to the filters $h'(n)$ and the last four rows are due to $g'(n)$ [see (2.17) for the relation between $h'(n)$ and $g'(n)$]. Therefore, premultiplying a vector (say, $[Y]$) by $[P]$ is equivalent to implementing the high- and lowpass filtering of a signal followed by decimation of the data by a factor of 2 as illustrated in the transmitter portion of Figure 2.1. The matrix vector product of $[P]$ with $[Y]$ can be identified with the outputs $u(n)$ and $v(n)$ as illustrated in Figure 2.1. The first four elements of matrix $[P]$ are equivalent to

$u(n)$ and the last four are $v(n)$. The matrix vector product has already incorporated the subsampling by a factor of 2. The subsampling by a factor of 2 is accomplished by the shift between the elements of each row of the matrix $[P]$. Now for $[P]$ to be an orthogonal matrix, it is necessary for the following three equations, which are similar to (2.29) (where the normalization constant is set to unity) and (2.30) (i.e., the filter is orthogonal to its two shifted versions), to hold:

$$\sum_{i=0}^{N=5} h(i)^2 = 1 \tag{4.9}$$

$$h(0)\,h(2) + h(1)\,h(3) + h(2)\,h(4) + h(3)\,h(5) = 0 \tag{4.10}$$

$$h(0)\,h(4) + h(1)\,h(5) = 0 \tag{4.11}$$

Here, because the length of the FIR filters is six, we need orthogonality conditions (4.10) and (4.11) instead of one equation only as used in (2.30). Finally, from the boundary conditions for the filter $H(z = 1) = \sqrt{2}$ and $G(z = 1) = 0$ we have

$$h(0) + h(2) + h(4) = \frac{1}{\sqrt{2}} = h(1) + h(3) + h(5) \tag{4.12}$$

Equations (4.9)–(4.12) provide four independent equations. One needs two more equations to uniquely solve for the six values of $h(i)$, $i = 0, 1, \ldots, 5$. If one follows the Daubechies procedure of making the wavelets smooth, one would need to enforce the value of the first two derivatives of $\left. \dfrac{d^p G'}{dz^p} \right|_{z=1} = 0$ for $p = 1$ and 2, leading, [from (2.18)], to

$$-0 \times h(0) + 1 \times h(1) - 2 \times h(2) + 3 \times h(3) - 4 \times h(4) + 5 \times h(5) = 0 \tag{4.13}$$

$$-0 \times h(0) + 1 \times h(1) - 4 \times h(2) + 9 \times h(3) - 16 \times h(4) + 25 \times h(5) = 0 \tag{4.14}$$

This is equivalent to setting the higher order moments of $g'(n)$ to zero and this guarantees smoothness of the wavelets. This in turn provides a recipe so that the discrete wavelet coefficients of the transform drop off rapidly as one goes to the dilated scales from a fine scale. The number of zeros of $\left. \dfrac{d^p G'}{dz^p} \right|_{z=1}$ tells us how many basis functions are needed in the wavelet expansion for approximating any function. The smoother the function and the higher the order of zeros, the faster the expansion coefficients go to zero and the fewer coefficients we need to keep in the wavelet transform. For piecewise continuous functions that may have jump discontinuities, a wavelet basis consisting of discontinuous functions like Haar is better because it may not produce any Gibb's phenomenon. We keep more of the wavelet coefficients in the neighborhoods of the discontinuities of the function by going to a higher scale for the wavelet expansion. The finer discretization adapts to approximating a function in a way that Fourier methodology finds difficult.

If a function $x(t)$ has p derivatives, its wavelet coefficients $d_{k,n}$ decay like 2^{-np} and are approximately bounded by the values

$$\left| d_{k,n} \right| = \left| \int x(t)\, \psi_{n,k}(t)\, dt \right| \leq J\, 2^{-np} \left\| x^{(p)}(t) \right\| \tag{4.15}$$

where the nomenclature of (2.65) has been followed. In (4.15) J is a constant, $x^{(p)}(t)$ represents the pth derivative of $x(t)$, and n represents the value of the scale. However, for the discrete case, this equation does not hold because for discrete samples, it is difficult to define the existence of derivatives and hence the bounds given by (4.15) do not hold. In addition, since we are dealing with a finite number of terms in the expansion of a function by wavelets, the drop-off rate of the wavelet coefficients is of little significance because the wavelet coefficients will cease to exist after a finite number of scales anyway.

The solution of (4.9) to (4.14) can be obtained analytically and has been given by Daubechies [2] as

$$h(5) = h'(0) = \frac{1 + \sqrt{10} + \sqrt{5 + 2\sqrt{10}}}{16\sqrt{2}} \tag{4.16a}$$

$$h(4) = h'(1) = \frac{5 + \sqrt{10} + 3\sqrt{5 + 2\sqrt{10}}}{16\sqrt{2}} \tag{4.16b}$$

$$h(3) = h'(2) = \frac{10 - 2\sqrt{10} + 2\sqrt{5 + 2\sqrt{10}}}{16\sqrt{2}} \tag{4.16c}$$

$$h(2) = h'(3) = \frac{10 - 2\sqrt{10} - 2\sqrt{5 + 2\sqrt{10}}}{16\sqrt{2}} \tag{4.16d}$$

$$h(1) = h'(4) = \frac{5 + \sqrt{10} - 3\sqrt{5 + 2\sqrt{10}}}{16\sqrt{2}} \tag{4.16e}$$

$$h(0) = h'(5) = \frac{1 + \sqrt{10} - \sqrt{5 + 2\sqrt{10}}}{16\sqrt{2}} \tag{4.16f}$$

Once the coefficients of the FIR filters are available, one can form the orthogonal matrix $[P]$ by substituting the above values for the FIR filter coefficients $h(n)$.

Let us say we have a vector $[Y_1]$ of length $Q = 2^5 = 32$. We now know how to extend the wavelet transform to $[Y_1]$. We create a $[P_1]$ matrix, which is 32×32 and only 6 elements of any of its rows are populated by the elements of (4.16), and a matrix of size 32×32, similar to (4.8), is formed. Twenty-six elements per row of the matrix are zero. Note that the first 16 rows of $[P_1]$ are formed by $h'(m)$, or to be more precise by $h(m)$, and the last 16 rows by $g'(m)$, or to be more precise by $g(m)$, since we are dealing with the transmitter part of Figure 2.1. We premultiply $[Y_1]$ by $[P_1]$ and obtain a vector $[Y_2]$. The last 16 elements of $[Y_2]$ are $d_{k,0}$ and they are fixed. This is the result of filtering $[Y_1]$ by $g(n)$ as illustrated in Figure 2.5. We premultiply the first 16 elements of $[Y_2]$, that is, $[Y_2']$ as shown in Figure 4.5 by an orthogonal matrix $[P_2]$ which is 16×16. Again only 6 of the elements of any row of matrix $[P_2]$ are nonzero. The result will be a vector $[Y_3]$ of 16 elements. The last 8 elements are fixed, as they are $d_{k,1}$, and the first 8 elements of $[Y_3]$, namely, $[Y_3']$ are again premultiplied by the orthogonal matrix $[P]$ of (4.8) as shown in Figure 4.5 producing $c_{k,2}$ and $d_{k,2}$. The final result is the wavelet decomposition of the vector $[Y_1]$. The resultant composite vector of 32 elements includes the wavelet coefficients resulting from the discrete wavelet transform. It is interesting to note that when dealing with a finite length vector we have a wavelet decomposition series containing both the scaling functions and wavelets. Specifically, for the case we have just considered, we obtain a transformed vector of 32 elements, $[T]$, which is given by

$$[T] = \begin{bmatrix} Y_4' \\ Y_4'' \\ Y_3'' \\ Y_2'' \end{bmatrix} \tag{4.17}$$

where

$$T(k) = \begin{cases} c_{k,2} & \text{for} \quad k = 1, ..., 4 \\ d_{k-4,2} & \text{for} \quad k = 5, ..., 8 \\ d_{k-8,1} & \text{for} \quad k = 9, ..., 16 \\ d_{k-16,0} & \text{for} \quad k = 17, ..., 32 \end{cases} \tag{4.18}$$

where the coefficients $c_{k,2}$ are due to the scaling functions $\phi_{2,k}$ and the remainder of the coefficients, $d_{k,i}$ ($i = 0, 1, 2$) are due to wavelets $\psi_{i,k}$. Note that in this case even though we have carried out a discrete wavelet transform, we do not need to know anything about the scaling and the wavelet functions $\phi_{2,k}$ and $\psi_{i,k}$, respectively. The discrete wavelet transform of $[Y_1]$ is depicted in Figure 4.5. The choice of the filter $h'(m)$ completely defines the entire procedure. This very important feature unfortunately is often not spelled out explicitly in the literature. Here, the dimension of the filter defines how many scales one can consider. Because the filter length is 6, we cannot go beyond the second scale in this example, because the filtering process cannot be continued any further.

$$[P_1]_{32\times32} \; [Y_1]_{32\times1} \;\rightarrow\; [Y_2]_{32\times1} \rightarrow \begin{bmatrix} \begin{bmatrix} Y_2' \end{bmatrix} \\ \\ \begin{bmatrix} Y_2'' \end{bmatrix} \end{bmatrix} \;\rightarrow\; 16 \;\; \text{Wavelet} \qquad d_{k0}, \; k = 1,...,16$$

$$\text{Coefficients}$$

$$[P_2]_{16\times16} \; [Y_2']_{16\times1} \;\rightarrow\; [Y_3]_{16\times1} \rightarrow \begin{bmatrix} \begin{bmatrix} Y_3' \end{bmatrix} \\ \\ \begin{bmatrix} Y_3'' \end{bmatrix} \end{bmatrix} \;\rightarrow\; 8 \;\; \text{Wavelet} \qquad d_{k1}, \; k = 1,...,8$$

$$\text{Coefficients}$$

$$[P]_{8\times8} \; [Y_3']_{8\times1} \;\rightarrow\; [Y_4]_{8\times1} \rightarrow \begin{bmatrix} \begin{bmatrix} Y_4' \end{bmatrix} \\ \\ \begin{bmatrix} Y_4'' \end{bmatrix} \end{bmatrix} \begin{matrix} \rightarrow 4 \;\; \text{Scaling \; Function} \\ \text{Coefficients} \qquad c_{k2}, \; k = 1,...,4 \\ \rightarrow 4 \;\; \text{Wavelet Coefficients} \quad d_{k2}, \; k = 1,...,4 \end{matrix}$$

Figure 4.5 Principles of the wavelet transform applied to a column vector of 32 coefficients.

To compute the inverse of the discrete wavelet transform of the resultant vector, one simply reverses the procedure as illustrated by Figure 2.7. In this case we start with the smallest level of the hierarchy and work our way through. We will need the inverse of the $[P_i]$ matrices. The inverse of these matrices in this case are simply their transpose because these matrices are real orthogonal matrices. To compute the two-dimensional wavelet transform one follows the same procedure as in the computation of the FFT. One first deals with the transform of all the rows and then evaluates the transform of all the columns.

After taking the discrete wavelet transform, we set to zero all the elements whose magnitudes are below some prespecified threshold values. The value of the threshold is determined by the desired accuracy of the solution. In this way, we generate a sparse vector from a full vector without going through the computationally intensive similarity transforms.

4.3 NUMERICAL EVALUATION OF THE WAVELET-LIKE TRANSFORM

Next we consider the solution of the matrix equation $[A][X] = [Y]$ utilizing the discrete wavelet transform (DWT). The discrete wavelet transform actually does not solve any matrix equations. What the wavelet transform does is preprocess matrix A and make it sparse by applying some thresholds to a processed matrix obtained after the application of a sequence of similarity transforms [4]. The basic principles as outlined by Beylkin, Coifman, and Rokhlin [4] based on [10] is that if the matrix elements are generated from a kernel such that the magnitude of the elements of matrix $[A]$ decay from the diagonal as $\dfrac{1}{|i-j|^{\alpha}}$, where i and j may be the row or the column number, and α is a positive real number, then if the 2-D discrete wavelet transform is applied to system matrix $[A]$, the resulting system matrix would be sparse if all the elements below a threshold ε are set to zero. Typically one would have only $10Q \log_{10}\left(\dfrac{1}{\varepsilon}\right)$ nonzero elements in matrix $[A]$, which is of size Q with ε as the truncation level, that is, elements of the resultant matrix whose absolute value is less than ε will be discarded. So for a 2,048 × 2,048 matrix, the resultant system matrix would be sparse by a factor of about 30 if the threshold is set to $\varepsilon = 0.001$ [5, 9]. As an example, consider a real matrix $[A]$ of dimension $Q \times Q$ where Q is large. Let the elements of $[A]$ be defined by

$$A(i, j) = \begin{cases} 1, & \text{if } i = j \\ \dfrac{1}{\sqrt{|i-j|}}, & \text{for } i \neq j \end{cases} \tag{4.19}$$

We now apply a wavelet transform to matrix $[A]$. This is equivalent to pre- and postmultiplying $[A]$ by a number of orthogonal matrices. Let $[W]$ be the product of the orthogonal matrices $[P_i]$ as explained earlier for the 1-D discrete wavelet transform as given by Figure 4.5. Then

$$[W] = [P] \,.... \, [P_2] \, [P_1] \tag{4.20}$$

Even though the sizes of the various matrices $[P_i]$'s are not the same we make them the same by supplementing, say, $[P_2]$, by a diagonal identity matrix $[I]$ to make it the same size as $[P_1]$. So that

$$[P_2] = \begin{bmatrix} [P_2]_{k \times k} & 0 \\ 0 & [I]_{k \times k} \end{bmatrix}_{2k \times 2k} \tag{4.21}$$

Since the product of all orthogonal matrices is an orthogonal matrix, $[W]$ is an orthogonal matrix. When the wavelet transform is applied to the system of equations $[A][X] = [Y]$, one obtains the following expansion, which is identical to (4.2):

$$[W][A][W]^{H}[W][X] = [W][Y] \tag{4.22}$$

Since $[W]$ is an orthogonal matrix, we have

$$[W]^{H}[W] = [I] \tag{4.23}$$

Here, $[Z] = [W][X]$ is the 1-D wavelet transform of vector $[X]$. $[F]$ is the 1-*D* wavelet transform of vector $[Y]$ because $[F] = [W][Y]$, and $[B] = [W][A][W]^{H}$ is the 2-*D* wavelet-like transform of $[A]$. Hence (4.22) reduces to (4.3) from which the unknown $[X]$ is solved for using (4.5) from

$$[X] = [W]^H [Z] \tag{4.24}$$

The wavelet-like transform of $[A]$ which is $[B]$ has been computed by a series of 1-D transforms to its rows and columns as illustrated in Figure 4.1. The application of the wavelet-like transform does not make the matrix $[B]$ sparse. It is the application of the threshold operation, where elements below a certain prespecified value are set to zero, that produces a sparse matrix.

The computation time involved for the evaluation of the wavelet-like transforms in the processing of a matrix is now investigated. We consider a filter of length $L = N + 1$ and the data matrix of length Q. Then the 1-D wavelet transform of $[Y]$ may be done (we carry out the initial product at the highest stage of resolution and then downsample) by a number of mathematical operations given by

$$QL\left[1 + \frac{1}{2} + \frac{1}{4} + \ldots\right] = 2QL \tag{4.25}$$

To carry out the 2-D wavelet transform of $[A]$, we require $(2QL)^2$ operations. Therefore to produce the sparse system (4.3), we require $4Q^2L^2 + 2QL$ operations, resulting in a matrix $[B]$, which after the threshold is applied contains at most of the order of θ (Q) non-zero elements. If we now apply the conjugate gradient method to solve the sparse system, we will require θ $(2Q)$ multiplications per iteration to carry out the two matrix-vector products usually required in one cycle of the conjugate gradient method [11]. Even if the conjugate gradient method converges in at most Q steps (where Q is the number of unknowns), then we have solved (4.3) in Q θ $(2Q)$ operations in addition to $4Q^2L^2 + 2QL$. Observe that the total operation count θ $[\sim Q^2]$ is significantly lower than the conventional $\theta[Q^3/3]$ operations typically required for solving a matrix equation of size Q. This essentially is the contribution of Beylkin, Coifman, and Rokhlin [4].

However, it is interesting to note that if a function has a variation of the form $\dfrac{1}{|i - j|^\alpha}$, then this may be the result of a convolution. In that case, the FFT would be much faster than the discrete wavelet transform because the FFT essentially diagonalizes the operator/matrix. However, for other cases even when the variation is not due to a convolution the wavelet result still holds.

The most disturbing fact about the DWT is that the condition number of the matrix changes after the transform. We have shown earlier that the wavelet-like transform of the matrix $[A]$, which is $[B]$, has been obtained through a series of orthogonal transformations. Therefore, by definition, the condition number of the matrix should not change as one goes from $[A]$ to $[B]$. However, even though from a theoretical point of view, the condition number of matrices $[A]$ and $[B]$ should not change after we go through a series of orthogonal transformations, from a practical viewpoint, the slight change in the condition number is due to numerical roundoff and truncation errors during the computational process. Furthermore, when a threshold is applied by deleting the smaller elements of the matrix, the change in the condition number is definitely dramatic. As will become clear from the results to be presented, the change in the condition number of the transformed matrix depends on the order in which the FIR filters are used and on the threshold used to eliminate the small coefficients of the matrix.

In the first example, matrix [A] given by (4.19) is assumed to be real. We will consider various lengths $L = 4, 8, 16$, or 32 of the Daubechies FIR filters $h(n)$. Then, we compute $[B] = [W][A][W]^H$ with $[W]$ being given by (4.20). We apply a threshold to the elements of [B] to obtain the matrix $[B_a]$. In general, the matrix $[B_a]$ will be a sparse matrix. For example, if the threshold ε is set at 10^{-3}, the size of [A] is $Q = 512$, and if we apply a FIR filter of length $L = 4$, we find that only 7.58% of the elements of B_a are nonzero as seen from Table 4.1.

To estimate the error that we have incurred, we take the sparse matrix $[B_a]$ and try to reconstruct [A] by computing

$$[A_a] = [W]^H [B_a][W] \tag{4.26}$$

We define an average value of the error δ between the elements of $[A_a]$ and [A] through

$$\delta = \frac{1}{Q^2} \sum_i \sum_j |A_a(i, j) - A(i, j)| \tag{4.27}$$

From Tables 4.1 through 4.6, we can see that matrix [A] can be recovered from $[B_a]$ to provide $[A_a]$ with an average error, which is lower than the value of the threshold used to eliminate the wavelet coefficients of [A] (or equivalently the elements of matrix [B]). The number of [P] matrices [see (4.20)] used for obtaining the results appearing in Tables 4.1 through 4.6 are $INT\left[\log_2\left(\frac{Q}{N+1} \right) \right]$ where $INT[\bullet]$ stands for the integer part of the argument within the brackets. In addition, it is important to observe that the wavelet transform of a real function is always real and hence easy to compute.

The other interesting property to note is that increasing the length of the filters does not lead to a monotonic increase of the sparsity of the resulting matrix as shown in Tables 4.1 through 4.5. In addition, the condition number of $[B_a]$ does not seem to follow a systematic rule when the length of the filters is increased. The only conclusion that can be drawn from the tables is that there is very little consistency in the results to draw any general conclusions.

Table 4.1
Threshold = 0.001; $Q = 512$; Cond[A] = 4.45×10^6

Length of filter $L = N+1$	% of elements nonzero	Error in reconstruction δ	Cond[B_a]
4	7.58%	0.58×10^{-4}	4.66×10^6
8	6.28%	0.54×10^{-4}	5.56×10^6
16	6.41%	0.52×10^{-4}	5.74×10^7
32	6.54%	0.51×10^{-4}	1.57×10^7

Table 4.2
Threshold = 0.001; $Q = 1,024$; Cond[A] = 3.02×10^7

Length of filter $L = N+1$	% of elements nonzero	Error in reconstruction δ	Cond[B_a]
4	3.88%	0.38×10^{-4}	2.56×10^8
8	3.17%	0.33×10^{-4}	3.69×10^7
16	3.18%	0.33×10^{-4}	8.43×10^7
32	3.1%	0.32×10^{-4}	1.53×10^7

Table 4.3
Threshold = 0.001; $Q = 2,048$; Cond[A] = 6.80×10^7

Length of filter $L = N+1$	% of elements nonzero	Error in reconstruction δ	Cond[B_a]
4	1.97%	0.24×10^{-4}	5.01×10^7
8	1.6%	0.20×10^{-4}	5.48×10^7
16	1.58%	0.19×10^{-4}	3.29×10^7
32	1.51%	0.19×10^{-4}	6.61×10^7

Table 4.4
Threshold = 0.0001; $Q = 512$; Cond[A] = 4.45×10^6

Length of filter $L = N+1$	% of elements nonzero	Error in reconstruction δ	Cond[B_a]
4	11.9%	0.78×10^{-5}	5.03×10^6
8	9.9%	0.63×10^{-5}	5.84×10^6
16	10.1%	0.61×10^{-5}	4.59×10^6
32	9.9%	0.53×10^{-5}	1.12×10^7

Table 4.5
Threshold = 0.0001; $Q = 1,024$; Cond[A] = 3.02×10^7

Length of filter $L = N+1$	% of elements nonzero	Error in reconstruction δ	Cond[B_a]
4	6.25%	0.54×10^{-5}	2.77×10^7
8	5.06%	0.41×10^{-5}	2.55×10^7
16	5.09%	0.38×10^{-5}	2.98×10^7
32	4.84%	0.33×10^{-5}	2.27×10^7

Table 4.6
Threshold = 0.0001; $Q = 2,048$; Cond[A] = 6.80×10^7

Length of filter $L = N+1$	% of elements nonzero	Error in reconstruction δ	Cond[B_a]
4	3.2%	0.35×10^{-5}	5.98×10^7
8	2.56%	0.25×10^{-5}	5.27×10^7
16	2.55%	0.23×10^{-5}	6.64×10^7
32	2.36%	0.21×10^{-5}	4.42×10^7

As a second example, consider the electromagnetic scattering from an array of wires randomly spaced. We consider 56 thin wire antennas. Six are of length 2.7λ and radius 0.005λ. The remaining 50 wires are 3λ long and have the same radius. The 56 wires are randomly distributed inside a parallelepiped of dimensions $27\lambda \times 25\lambda \times 21\lambda$. We utilize the method of moments to solve for the radiation from such a structure. The impedance matrix [A] that results from the 56 wires is of size $Q = 2,048$ and as usual is a full complex matrix. Before we apply the wavelet transform, we rescale the matrix elements by utilizing the following scaling

$$A(i,j) \quad \Rightarrow \quad \frac{A(i,j)}{\sqrt{A(i,i)}\,\sqrt{A(j,j)}} \quad \text{for } i,j = 1,...., Q \qquad (4.28)$$

Thus, through (4.28) we have scaled all the elements of the matrix so that all the diagonal elements are unity in magnitude. This transformation is necessary; otherwise, the results obtained from the wavelet applications are not good. We apply the wavelet-like transform to the rescaled matrix [A] and the threshold operation is applied separately to the real and the imaginary parts of [A], because the wavelet-like transform of a real function is real. For filter $h(n)$ we choose different lengths of Daubechies filters, namely, 4, 8, 16, and 32. The goal for this problem is to determine for this $2,048 \times 2,048$ complex matrix [A] what length of filter provides a maximally sparse matrix. The results of this numerical experiment are summarized in Tables 4.7 through 4.12. For example, if we choose a filter of length $L = 8$ and set the threshold to 0.001, so that all the wavelet coefficients of magnitude below 0.001 are discarded, then we observe that approximately 88.5% of the elements of the real and imaginary parts of $[B_a]$ are zeros. The wavelet-like transform is applied separately to the real and imaginary parts of matrix $[A]$ separately. The choice of a higher value of threshold (say, 0.01; i.e., all the wavelet coefficients of magnitude less than 0.01 are discarded) leads to approximately 96% of the elements of the real and the imaginary parts of the matrix $[B_a]$ being zero. From an intuitive viewpoint, it would seem that increasing the length of the filter would provide a greater sparsity of the matrix if the threshold is kept the same. However, from this numerical experiment it appears that there is no such correlation. This clearly indicates that the bound of (4.15) developed for the conventional wavelet transform does not apply to the wavelet-like transforms.

We take the nonzero elements of the matrix $[B_a]$ and observe how accurately one can reconstruct the original matrix $[\tilde{A}]$ from $[B_a]$. The reconstruction errors for the real and the imaginary parts of the matrices are evaluated separately. The average value of the reconstruction error is defined by

$$\delta_{\text{real}} = \frac{1}{Q^2} \sum_i \sum_j | A_{\text{real}}(i, j) - \tilde{A}_{\text{real}}(i, j)| \qquad (4.29a)$$

$$\delta_{\text{imag}} = \frac{1}{Q^2} \sum_i \sum_j | A_{\text{imag}}(i, j) - \tilde{A}_{\text{imag}}(i, j)| \qquad (4.29b)$$

Table 4.7
Threshold = 0.1; Q = 2,048; Real Part

Length of filter $L = N+1$	% of elements nonzero	Error in reconstruction δ
4	1.03%	0.37×10^{-2}
8	1.06%	0.38×10^{-2}
16	1.16%	0.43×10^{-2}
32	1.53%	0.48×10^{-2}

Table 4.8
Threshold = 0.1; Q = 2,048; Imaginary Part

Length of filter $L = N+1$	% of elements nonzero	Error in reconstruction δ
4	1.07%	0.38×10^{-2}
8	1.11%	0.39×10^{-2}
16	1.25%	0.43×10^{-2}
32	1.68%	0.49×10^{-2}

Table 4.9
Threshold = 0.01; $Q = 2{,}048$; Real Part

Length of filter $L = N+1$	% of elements nonzero	Error in reconstruction δ
4	3.62%	0.95×10^{-3}
8	3.72%	0.87×10^{-3}
16	4.36%	0.88×10^{-3}
32	5.64%	0.94×10^{-3}

Table 4.10
Threshold = 0.01; $Q = 2{,}048$; Imaginary Part

Length of filter $L = N+1$	% of elements nonzero	Error in reconstruction δ
4	3.66%	0.95×10^{-3}
8	3.80%	0.87×10^{-3}
16	4.44%	0.88×10^{-3}
32	5.99%	0.94×10^{-3}

Table 4.11
Threshold = 0.001; $Q = 2{,}048$; Real Part

Length of filter $L = N+1$	% of elements nonzero	Error in reconstruction δ
4	13.89%	0.15×10^{-3}
8	11.45%	0.12×10^{-3}
16	12.65%	0.12×10^{-3}
32	15.51%	0.13×10^{-3}

Table 4.12
Threshold = 0.001; $Q = 2{,}048$; Imaginary Part

Length of filter $L = N+1$	% of elements nonzero	Error in reconstruction δ
4	13.92%	0.16×10^{-3}
8	11.50%	0.12×10^{-3}
16	12.70%	0.12×10^{-3}
32	16.65%	0.13×10^{-3}

A 2,048 × 2,048 matrix has 4,194,304 elements. To obtain a sparse matrix of $\theta(Q)$ nonzero elements, one should obtain a sparsity of the order of 99.95%. If the result is of the form θ (10Q) we still should obtain a sparsity of the order of 99.5%. However, for an extremely large threshold of 0.1 we observe that the number of nonzero elements is $\theta(22.5Q)$. For the threshold of 0.001, we get the best compression θ (235.2Q) when we use a FIR filter of length 8. The only consistency in the results is that the average reconstruction error δ for the reconstruction of the matrix elements is much smaller than the threshold value ε.

When we decrease the subsection length for the wires, the complex matrix [A] becomes that of dimension 4,096 × 4,096. The results of the transformation for this case are shown in Tables 4.13 through 4.18. As before, it is seen that the degree of sparseness depends not only on the order of the filter but also on the values of the threshold applied to the elements of the matrix and one cannot observe any definitive trends.

The results indicate that the reconstruction errors for both the real and the imaginary parts of both the matrices of size Q = 2,048 or 4,096 do not show any pattern with the increase in length of the FIR filters. However, the error is always below the threshold. For example, when the threshold is 0.001, the reconstruction errors for both the Q = 2,048 and 4,096 matrices decrease with the increase of the length of the filter and then increase when the length of the filter becomes 32. Here also there is no systematic rule as to what is the best possible strategy. It appears that for these two examples a FIR filter of length 8 is sufficient.

This is the result of a trade-off between the sparsity of the transformed matrix and accuracy on one hand and the number of operations counts on the other.

Table 4.13
Threshold = 0.1; Q = 4,096; Real Part

Length of filter $L = N+1$	% of elements nonzero	Error in reconstruction δ
4	0.51%	0.31×10^{-2}
8	0.50%	0.31×10^{-2}
16	0.45%	0.32×10^{-2}
32	0.60%	0.33×10^{-2}

Table 4.14
Threshold = 0.1, Q = 4,096; Imaginary Part

Length of filter $L = N+1$	% of elements nonzero	Error in reconstruction δ
4	0.52%	0.31×10^{-2}
8	0.52%	0.31×10^{-2}
16	0.54%	0.33×10^{-2}
32	0.66%	0.35×10^{-2}

Table 4.15
Threshold = 0.01; $Q = 4{,}096$; Real Part

Length of filter $L = N+1$	% of elements nonzero	Error in reconstruction δ
4	2.39%	0.86×10^{-3}
8	2.27%	0.74×10^{-3}
16	2.42%	0.72×10^{-3}
32	2.79%	0.75×10^{-3}

Table 4.16
Threshold = 0.01; $Q = 4{,}096$; Imaginary Part

Length of filter $L = N+1$	% of elements nonzero	Error in reconstruction δ
4	2.42%	0.86×10^{-3}
8	2.32%	0.74×10^{-3}
16	2.48%	0.71×10^{-3}
32	2.86%	0.74×10^{-3}

Table 4.17
Threshold = 0.001; $Q = 4{,}096$; Real Part

Length of filter $L = N+1$	% of elements nonzero	Error in reconstruction δ
4	11.18%	0.15×10^{-3}
8	7.95%	0.11×10^{-3}
16	7.78%	0.089×10^{-3}
32	9.04%	0.096×10^{-3}

Table 4.18
Threshold = 0.001; $Q = 4{,}096$; Imaginary Part

Length of filter $L = N+1$	% of elements nonzero	Error in reconstruction δ
4	11.23%	0.15×10^{-3}
8	7.99%	0.10×10^{-3}
16	7.83%	0.089×10^{-3}
32	9.13%	0.096×10^{-3}

One should also be aware that as the length of the filter increases, the operation count also increases. Hence a good trade-off appears to be a filter of length 8. However, for a different problem, the conclusion may be different.

Next we look at the condition number of the matrix for size $Q = 2,048$. It is true that from a purely theoretical point of view, we are performing a series of orthogonal transformations and hence the condition number of the matrix should not change. However, because numerical computations are not perfect, we observe a slight change in the condition number after the computations. We also compute the condition number of the sparse matrix after the threshold has been applied. The condition number of the original scaled matrix is 121.9. However, after the wavelet-like transform and the threshold operation have been applied with different lengths of the filters, there appears to be a significant change in the condition number, even though from a purely theoretical viewpoint this should not occur. It is rather consistent that increasing the length of the FIR filters, in general, increases the condition number of the transformed matrix. The results are shown in Table 4.19.

Table 4.19
Threshold = 0.001; $Q = 2,048$; Cond [A] = 121.91

Length of the filter	4	8	16	32
Condition number after applying wavelet-like transform	188.76	198.40	247.40	288.63
Condition number after threshold	200.99	197.20	242.81	288.14
Threshold = 0.01				
Length of the filter	4	8	16	32
Condition number after applying wavelet-like transform	188.76	198.40	247.40	288.63
Condition number after threshold	184.38	192.19	246.87	299.61
Threshold = 0.1				
Length of the filter	4	8	16	32
Condition number after applying wavelet-like transform	188.76	198.40	247.40	288.63
Condition number after threshold	129.18	133.04	155.10	159.90

The interesting part is that when the threshold is applied to the transformed matrix to make it sparse, the condition number in some cases may actually be slightly better for the sparse system. Also, when the threshold is large (i.e., $\varepsilon >$ 0.01), the condition number actually is better. The consistent observation that is remarkable for the wavelet-like transforms is that, even though we truncate the wavelet coefficients to ε when we reconstruct the approximate original matrix $[\tilde{A}]$ from the truncated wavelet coefficients of the sparse matrix $[B]$, the average error of the reconstruction δ is always invariably less than ε. In fact, in all numerical experiments that we have conducted, we never found a situation where the average reconstruction error was not significantly less than ε.

Finally, we present the results of solving the sparse thresholded wavelet-like transformed complex matrix obtained for this thin wire structure. The excitation for this example is an electromagnetic field incident from the broadside direction. The sparse matrix equation is now solved by the iterative conjugate gradient method to observe how efficient the solution procedure is. In our experimentation, we stopped the iteration for the conjugate gradient method when the normalized error between the successive residuals (the residual of a matrix equation is given by $[R_n] = [A][X_n] - [Y]$), δ_R, was of the same order as ε, that is,

$$\frac{\left\| [R_{n+1}] \right\| - \left\| [R_n] \right\|}{\left\| [Y] \right\|} \leq \delta_R \approx \varepsilon \tag{4.30}$$

where the numerator defines the difference in the norms of the residuals between the $n + 1$ and n stages of iteration.

We observed that for the case of $Q = 2{,}048$, the conjugate gradient method took 22 iterations to produce the same order of accuracy between the residuals obtained after successive iterations as ε. The error in the solution δ_x is defined by

$$\delta_x = \frac{1}{Q} \sum_{i=1}^{Q} \left\| x(i) - x_{22}(i) \right\| \tag{4.31}$$

where $x_{22}(i)$ represents the solution after $n = 22$ iterations. It is interesting to note that as the length of the filter increases, δ_x becomes smaller, as seen in Table 4.20. This is also seen for the case of $Q = 4{,}096$ in Table 4.21.

Table 4.20
Threshold = 0.001; $Q = 2{,}048$; Error in Residuals (δ_R)

Length of the filter	4	8	16	32
Error in solution δ_x	0.0080	0.0079	0.0077	0.0075
Number of iterations	22	22	22	22

Table 4.21
Threshold $= 0.001$; $Q = 4{,}096$; Error in Residuals (δ_R)

Length of the filter	4	8	16	32
Error in solution δ_x	0.018	0.0179	0.017	0.017
Number of iterations	22	22	22	22

In summary, in the solution of a matrix equation using the wavelet-like transform, we first transform the full complex matrix to a sparse matrix and then solve the sparse matrix equation iteratively using the conjugate gradient method. The number of iterations taken by the conjugate gradient method for the solution of the original dense matrix equations is identical to that of the sparse matrix problem. However, for the sparse matrix obtained by applying the wavelet-like transform problem, we are performing only $0.02Q^2$ multiplications per iteration as opposed to Q^2 for the original problem. In addition, for the sparse matrix since 98% of the elements of the matrix are zero, we can store the entire $2{,}048 \times 2{,}048$ matrix, or $4{,}096 \times 4{,}096$ matrix in the main memory of the computer. In doing so, we do not incur any page faults. This significantly reduces the computation time as far as the solution of the matrix equations is concerned.

4.4 SOLUTION OF LARGE DENSE COMPLEX MATRIX EQUATIONS USING A WAVELET-LIKE METHODOLOGY BASED ON THE FAST FOURIER TRANSFORM

An alternate methodology to that of the FIR Daubechies filters which may result in a greater sparsity at the expense of robustness [2, p. 98] is described in this section. The goal of this new implementation is to make the wavelet coefficients decay much faster as suggested in [1, p. 231]. However, no numerical recipes are prescribed regarding how this can be achieved. It is clear that the use of a very high order Daubechies filter would accomplish the goal (namely, reduction in redundancy and hence more sparsity and less robustness). But this is not feasible computationally, because too many equations need to be solved analytically. The question of efficient computational implementation is also a requirement. It is seen that the use of an ideal quadrature mirror filter would definitely satisfy the goal of achieving maximum sparsity for a given threshold as the dimension of the problem approaches infinity [12]. However, then the question is how to carry out the numerical computations efficiently. We propose to use the FFT with a symmetric extension of the data as presented in the next sections.

4.4.1 Implementation of the DWT Through the FFT

The conventional DWT is implemented by means of the highpass and lowpass filters with transfer functions $G(z)$ and $H(z)$, respectively, as illustrated in Figure 2.2. Note that the filters are overlapping for realizability conditions. This may also give rise to redundancy and, hence, leads to robustness. (Namely, very few bits of accuracy in the wavelet coefficients are necessary to reproduce the original function with great accuracy.) It is important to note that the decay of the wavelet coefficients depends on the order of the zero at the origin of the filters shown in Figure 2.2. The greater the order of the zero at the origin, the faster the decay of the wavelet coefficients [1, p. 231]. Hence, one needs to have a very high order filter. However, it is extremely difficult to construct FIR filters with the required quadrature symmetry for lengths greater than 32 for computational reasons. Hence, the question is what to do when the lengths of the FIR filter need to be very large. We propose the use of ideal FIR filters as shown in Figure 2.3 as opposed to the low-order FIR filters described in [1]. However, if we utilize the FFT, because the responses of the filters we propose are quite sharp, this procedure is going to introduce Gibb's phenomenon in the response and this will limit the degree of sparseness. Or equivalently, it will be less robust for a prespecified error in the reconstruction of the original vector from its wavelet coefficients. In this section, a symmetric extension of the function is used to minimize the discontinuity as illustrated in [1] in order to reduce the Gibb's phenomenon. The contribution of the section is a novel implementation of a very high order FIR filter (in this case, an ideal one as shown in Figure 2.3) without significant display of Gibb's phenomenon.

Consider an impulse-like signal $x(n)$ as shown in Figure 4.6. It is finite and discrete so that it is represented by the sequence $x(n)$, $n = 1, 2, ..., NN$. (Here $NN = 2^{p+1}$, where p is an integer.) The signal of length NN is extended in a symmetric fashion·as shown in Figure 4.7. The symmetric extension $x(n)$ is carried out by adjoining the following two sequences to form:

$$\bar{x}(n) = [\{ x(n) \} \{ x(NN + 1 - n) \}]; \quad \text{for} \quad n = 1, 2, ..., NN \qquad (4.32)$$

The DWT of the signal is carried out by first taking the FFT of the symmetrically extended signal of Figure 4.7. The FFT is relevant here because the highpass and lowpass filters are ideal. After taking the FFT, the signal is split into two parts (as seen in Figure 2.1), the lowpass parts and the highpass parts, using the nonoverlapping filters of Figure 2.3. Now, if we choose the highpass part described by all the coeffcients from locations $\dfrac{NN}{2} + 1$ to $\dfrac{3NN}{2}$, then in Matlab notation, one has

$$H_1(z) = \left[\bar{X} \left(\frac{NN}{2} + 1 : \frac{3NN}{2} \right) \right] * 0.5 \qquad (4.33)$$

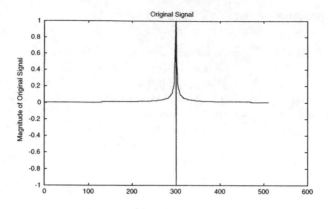

Figure 4.6 An impulse-like signal.

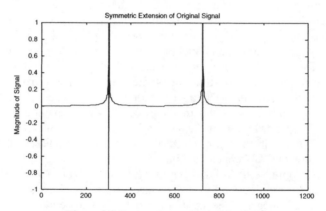

Figure 4.7 Symmetric extension of the signal.

where $H_1(z)$ is defined to be the filtered output of the highpass filter of the first stage, where the normalization factor of 0.5 has been taken into account. The length of this highpass part is the same as that of the original signal ($NN = 2^{p+1}$). Next we take the inverse Fourier transform of this highpass part $H_1(z)$ and the result will be the vector $h_1(n)$ as shown in Figure 4.8. The first half of this signal $h_1(n)$ will then be the discrete wavelet coefficients obtained after stage 1 (i.e., at the first level).

The length of the signal, which has been passed through the lowpass part, is now NN and it is again passed through a highpass and a lowpass half-band filter. This corresponds to the second stage of the wavelet decomposition as shown in Figure 2.5. Again we choose, the signal to be passed through a highpass filter to be

$$ H_2(z) = \left[\overline{X}\left(\frac{NN}{4}+1 : \frac{NN}{2} \right) \overline{X}\left(\frac{3NN}{2}+1 : \frac{3NN}{2} + \frac{NN}{4} \right) \right] * 0.25 \quad (4.34) $$

where again we have a normalization factor of $1/2^2$ and the length is $NN/2$. Then taking the inverse FFT of this highpass part will yield the wavelet coefficients at the end of the second stage and the result will be a vector $h_2(n)$. This will also be a symmetrically extended version of the DWT coefficients as shown in Figure 4.9. The first half of this signal corresponds to the discrete wavelet coefficients of the second stage and its length is $NN/4$.

This procedure is continued until the discrete wavelet coefficients are computed at level $p - 1$ given by $2^{p+1} = NN$. Hence if $NN = 512$, then $p = 8$. For this example, we have seven levels of DWT coefficients, as shown in Figures 4.8 through 4.15. Note that at the final stage, we apply the DWT to both the lowpass part and the highpass part because at the final step, the scaling functions and wavelet coefficients are obtained using the output from both stages of the filter. The resultant composite vectors of NN elements are the discrete wavelet coefficients as shown in Figure 4.16.

When we carry out the DWT using the FFT, the frequencies that are most dominant in the original signal will appear in the form of large amplitudes in the discrete wavelet coefficients. However, the low-amplitude discrete wavelet coefficients, which contain information regarding the spectrum of the signal, which are small in power, will be deleted through thresholding. Hence, the low-amplitude signal can be discarded by applying a threshold without much loss of information. Even though some effects of the Gibb's phenomenon are visible in the discrete wavelet coefficients of the second, third, and fourth stages of the discrete wavelet coefficients, by making a symmetric extension of the signal, they are drastically reduced at all of the higher stages. The reconstruction is carried out exactly in the same reverse order after a symmetric extension is made at every level of the signal. Without a threshold procedure, this reverse procedure should result in a perfect reconstruction.

4.4.2   Numerical Implementation of the DWT Through the FFT for a Two-Dimensional System (i.e., a Matrix)

Next we apply the discrete wavelet-like transform to make a complex large dense matrix [A] of size $Q \times Q$ sparse. Here, Q has to be an integer power of 2 for the discrete wavelet-like transform (DW_LT) to be applicable. This is generated from the tensor product of two 1-D bases as illustrated in [13] and in Section 4.3. If matrix [A] is not of size 2^m, where m is an integer, then either matrix [A] can be augmented by a matrix with unity on the diagonal and zero everywhere else or the procedure described in Section 4.3 can be applied. Before we apply the DW_LT using the FFT, we first rescale the matrix elements utilizing the following diagonal scaling:

$$A\,(i,j) \leftarrow \frac{A\,(i,j)}{\sqrt{A\,(i,i)\,A\,(j,j)}} \text{ with } i,\,j = 1, ..., Q \qquad (4.35)$$

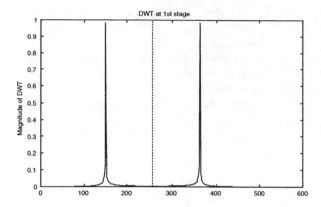

Figure 4.8 Discrete wavelet coefficients at the first stage (highpass): 256 coefficients.

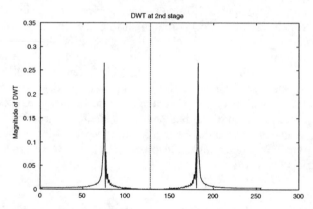

Figure 4.9 Discrete wavelet coefficients at the second stage (highpass): 128 coefficients.

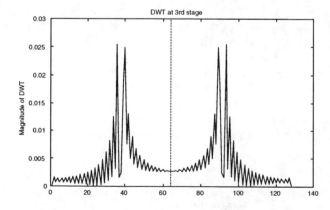

Figure 4.10 Discrete wavelet coefficients at the third stage (highpass): 64 coefficients.

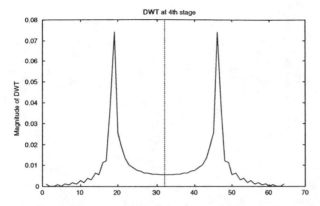

Figure 4.11 Discrete wavelet coefficients at the fourth stage (highpass): 32 coefficients.

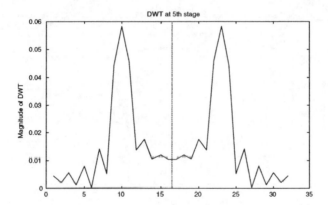

Figure 4.12 Discrete wavelet coefficients at the fifth stage (highpass): 16 coefficients.

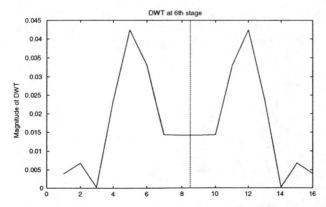

Figure 4.13 Discrete wavelet coefficients at the sixth stage (highpass): 8 coefficients.

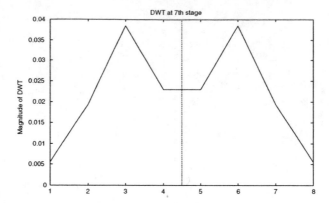

Figure 4.14 Discrete wavelet coefficients at the seventh stage (highpass): 4 coefficients.

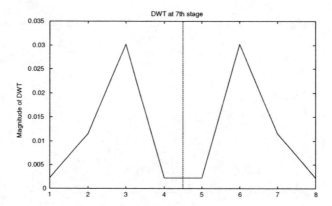

Figure 4.15 Discrete coefficients of the scaling function at the seventh stage (lowpass): 4 coefficients.

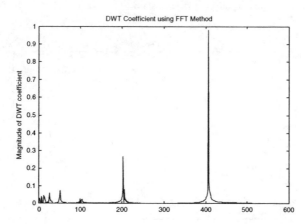

Figure 4.16 Discrete scaling function wavelet coefficients of the impulse-like signal: 512 coefficients.

This diagonal scaling seems necessary when we deal with large impedance matrices resulting from the discretization of integral equations utilizing subsectional bases in the conventional method of moments. This is because the dynamic range of its values is quite large. In addition, the real part of the impedance matrix is several orders of magnitude lower in value than the imaginary part. Otherwise, the degree of compression using the wavelet principles is not very good.

For application to a 2-D structure, like a matrix, a sequence of 1-D DWT using FFT is applied to every row and column of the matrix as shown in Figure 4.1. In addition, this transform is applied separately to the real and the imaginary parts of the matrices because the ratio between the real and the imaginary parts of the matrices is quite large. Hence the transform is applied separately to each real and imaginary part of the matrices rather than to the whole complex matrix at once.

4.4.3 Numerical Results

Consider the analysis of electromagnetic scattering from a conducting sphere and a plate illuminated by an incident plane wave from the broadside direction. We consider the electric field integral equation in the context of MM utilizing a triangular patch subsectional basis and testing functions. The triangular patch subsections and the operating frequency are chosen in such a way that it gives rise to a large dense complex matrix of size $4,096 \times 4,096$ for both the examples. In addition, the excitation on the righthand side is also a matrix of size $4,096 \times 1$. We next try to make these large dense complex impedance matrices arising in MM sparse through the use of DW_LT implementing the FIR Daubechies filter of length 8 following the procedure of [9] and using the technique presented in Section 4.4.1. After the $4,096 \times 4,096$ matrix is appropriately scaled diagonally as outlined in Section 4.4.2, we apply the wavelet-like transform to make the matrix sparse. We use two different methods, namely, the use of a low-order Daubechies FIR filter and an ideal filter implemented through the DW_LT, to make the impedance matrix sparse. We use the following abbreviations: D-DW_LT represents the results for the discrete wavelet-like transform implementing the Daubechies FIR filters, whereas FFT represents the results obtained by the current methodology. The threshold value implies that the elements of the real and imaginary parts of the transformed matrix are discarded when in absolute value the elements are below ε. Hence, an a priori threshold can be set for these problems.

As a first example consider the electromagnetic scattering from a sphere. For a matrix of size $4,096 \times 4,096$, there are $16,777,216$ elements. The degree of sparseness, that is, the number of nonzero elements, depends on the threshold values applied to the matrix elements after they are processed either by the Daubechies filters or by the FFT. For both threshold values of $\varepsilon = 0.01$ and 0.001, the proposed method achieves a sparser matrix than when filtering using the Daubechies filters in Table 4.22. However, the average error in the reconstruction is larger as can be seen in Table 4.22. Here, the reconstruction error is defined as the square root of the sum of the mean squared values between the original matrix elements and the elements of the reconstructed matrix using the nonzero values of the elements of the impedance matrix left after the threshold has been applied. This is equivalent to less robustness in the reconstruction and this occurs as predicted.

Table 4.22
Percentage of Elements That Are Nonzero (Real Part)

Threshold value	$\varepsilon = 0.01$	$\varepsilon = 0.001$	$\varepsilon = 0.0001$
D-DW_LT	1.34	8.13	35.05
FFT method	0.51	4.76	43.96

Percentage of Elements That Are Nonzero (Imaginary Part)

Threshold value	$\varepsilon = 0.01$	$\varepsilon = 0.001$	$\varepsilon = 0.0001$
D-DW_LT	0.21	6.05	33.40
FFT method	0.03	0.30	4.14

Reconstruction Error (Real Part)

Threshold value	$\varepsilon = 0.01$	$\varepsilon = 0.001$	$\varepsilon = 0.0001$
D-DW_LT	4.9130×10^{-4}	1.4924×10^{-4}	2.3131×10^{-5}
FT method	3.4181×10^{-3}	1.4131×10^{-3}	3.5381×10^{-4}

Reconstruction Error (Imaginary Part)

Threshold value	$\varepsilon = 0.01$	$\varepsilon = 0.001$	$\varepsilon = 0.0001$
D-DW_LT	4.3875×10^{-4}	1.5038×10^{-4}	2.2308×10^{-5}
FFT method	1.0906×10^{-3}	5.8295×10^{-4}	2.0124×10^{-4}

However, the reconstruction error is still quite low for most practical applications. Thus the proposed method may be used to generate a greater degree of sparseness. If we now use an iterative method to solve this sparse matrix equation, like the conjugate gradient, then the CPU time required per iteration would be reduced because fewer elements in the matrix will be involved in carrying out the matrix-vector product. For a low threshold value of 0.0001, the

Daubechies FIR filters produce a greater degree of sparsity only for the real part of the impedance matrix, because we know that this method has more redundancy and hence it will be more robust. Therefore, for a low value of the truncation threshold the Daubechies FIR filters will lead to fewer elements. The presence of the Gibb's phenomenon increases the oscillations in the coefficients. It has been our experience that the larger the matrix, the smaller the effect of the Gibb's phenomenon introduced by the FFT method presented, because the magnitude of the elements of the matrix changes more slowly when it is symmetrically extended. Due to the presence of the Gibb's phenomenon, the average reconstruction error in the FFT procedure is sometimes larger than the truncation level ε. It is conjectured that for larger values of NN exceeding tens of thousands the proposed method would be much better.

Next, we consider the electromagnetic scattering from a flat plate, which has been so discretized as to yield an impedance matrix of size $4,096 \times 4,096$. Again we apply the two different approaches as described earlier. The results are shown in Table 4.23.

For threshold values of $\varepsilon = 0.01$ and 0.001, the proposed method achieves a more sparse matrix than when using the Daubechies FIR filters, except for the real part where the degree of sparseness produced by both methods is approximately the same for $\varepsilon = 0.001$. For the imaginary part of the matrix, except for $\varepsilon = 0.01$, the FFT method achieves greater sparsity. The average error in the reconstruction is larger for the FFT method for both the real and the imaginary parts of the matrix. Here, the reconstruction error is defined as the square root of the sum of the mean squared values between the original matrix elements and the elements of the reconstructed matrix from the thresholded wavelet coefficients. This is equivalent to less robustness in the reconstruction and this occurs as predicted.

No mathematical proofs are available for the finite wavelet-like transform, so one can only conjecture the trends and this trend is not at all guaranteed all the time.

Even though the FFT implementation of the DW_LT displays Gibb's phenomenon, it can still achieve sparsity greater than 99% for a $4,096 \times 4,096$ complex dense matrix that arises in a typical electromagnetic scattering problem. It has been our experience that the results of the FFT become better as the size of the matrix equations increases. Hence, when solving large complex dense matrix equations, where these matrices need to be stored on a disk for solution purposes, a highly sparse large matrix can now be stored entirely in the main memory of the computer without resulting in a significant loss in accuracy. However, even though the degree of sparsity is often better than the Daubechies method for relatively moderate threshold values, the average reconstruction error is larger. This method will make it possible to solve large dense complex matrix equations utilizing modest computer resources.

Table 4.23
Percentage of Elements That Are Nonzero (Real Part)

Threshold value	$\varepsilon = 0.01$	$\varepsilon = 0.001$	$\varepsilon = 0.0001$
D-DW_LT	1.70	6.44	24.93
FFT method	0.79	6.61	44.17

Percentage of Elements That Are Nonzero (Imaginary Part)

Threshold value	$\varepsilon = 0.01$	$\varepsilon = 0.001$	$\varepsilon = 0.0001$
D-DW_LT	0.01	1.11	17.89
FFT method	0.02	0.04	6.04

Reconstruction Error (Real Part)

Threshold value	$\varepsilon = 0.01$	$\varepsilon = 0.001$	$\varepsilon = 0.0001$
D-DW_LT	1.21×10^{-3}	2.3×10^{-4}	3.27×10^{-5}
FFT method	8.8×10^{-3}	4.24×10^{-3}	3.75×10^{-4}

Reconstruction Error (Imaginary Part)

Threshold value	$\varepsilon = 0.01$	$\varepsilon = 0.001$	$\varepsilon = 0.0001$
D-DW_LT	3.3×10^{-4}	2.02×10^{-4}	3.16×10^{-5}
FFT method	3.26×10^{-4}	3.24×10^{-4}	1.58×10^{-4}

4.5 UTILIZATION OF CUSTOM FIR FILTERS

In the previous sections, conventional filters like Daubechies and perfect reconstruction filters were implemented by using the Z-transform technique. The assumption has been that the signal/vectors are of finite duration. However, if we want to use a very high order filter, then it is computationally more efficient to employ the FFT rather than the discrete Fourier transform (DFT). It is important to recollect that when we employ the FFT we are periodically extending the finite length signal/vector as illustrated in Section 4.4. Therefore, in order to produce perfect reconstruction we need to introduce boundary filters. The boundary filters deal with the wrapping of the data into a circular matrix when we carry out an

FFT. This introduces some error at the ends of the data and filters need to be designed to handle that. The question is how to do that for the general case and what type of filters not only will be computationally efficient but also will provide better reconstruction error. These issues are addressed in this section.

We investigate various filter configurations, which have the potential to provide greater sparsity and minimum reconstruction errors than the conventional FIR filters. Specifically we delve into the design of filters with a linear phase response. We think filters with a linear phase may have better properties than the Daubechies filters. From natural phenomena in electromagnetics, we know that systems with a linear phase response have less dispersion, because they tend to provide a perfect delay if the magnitude response is flat. So the philosophy here is to employ lowpass and highpass filters that have flat amplitude responses, together with a linear phase response. Then the filters will not produce dispersion of the signal. Therefore, from intuition it appears that they will perform better when generating sparse matrices, which case arises when an integrodifferential equation is discretized using the conventional method of moments. We also illustrate how to employ linear phase filters in the perfect reconstruction methodology introduced in Chapter 2.

Typically, in a wavelet methodology, we apply an iterative technique to compute the filter coefficients for computational efficiency as described by (2.77). We also employ orthogonal filter banks. However, this assumption/restriction is not necessary, as we will demonstrate. The goal here is to find the set of high- and lowpass filters that is going to provide the best sparsity. It is important to note that different design criteria for the filters are possible, where the criteria may be to minimize the truncation error when applying the threshold as this will try to maximize the signal-to-noise ratio of the data after the threshold operation is applied to the processed matrix. For practical data transmission these criteria may be useful and a neural network-based approach utilizing optimization may be used to carry out the wavelet transform in such cases [14–16]. Here, we use filter banks with a finite transition band of width TR as shown in Figure 4.17. In contrast, in Section 4.4, the filters used were perfect and had zero transition band.

In Chapter 2, we considered signals/vectors that were of finite duration and we used the Z transform to carry out the filter design and the various computations. However, in our quest to design very high order filter banks and to implement them efficiently we later took recourse in the FFT. (Actually, we are computing the Fourier series and not the transform. The Fourier transform is actually a misnomer as we are just computing the Fourier series.) We use the FFT to carry out the filtering operations. However, inadvertently when employing the FFT technique, one is essentially making the data vector repeat itself because one is constructing a Fourier series. Hence, unless we use ideal filter banks, aliasing can present a big problem. Therefore, in this section we want to employ filters with a finite transition band designated as TR as shown in Figure 4.17 and still employ the FFT for computational efficiency. To this end, we use the principles outlined by Strang and Nguyen [1] in Section 8.5 of their book.

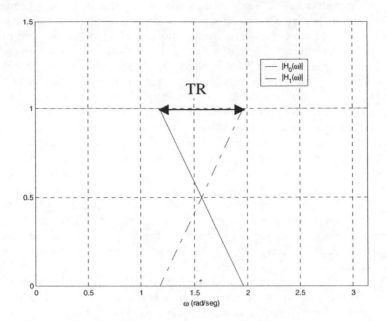

Figure 4.17 Response of a filter with a finite transition band.

If we want to minimize the discontinuities of the signal at the boundaries where in an FFT implementation they are allowed to repeat themselves, we need to symmetrically extend the signal first and then make it periodically repeat itself. In this way, one may reduce the discontinuities in the data at the boundary and yet be able to use the computationally efficient FFT by implementing it through the discrete cosine transform (DCT) and the discrete sine transform (DST) to carry out filtering operations. In this way, we can extend the signal symmetrically first and then periodically. This makes it possible to minimize the discontinuity in the extension of the signal. We also develop filters that are nonorthogonal.

4.5.1 Perfect Reconstruction Conditions Using Complex Filters with a Finite Transition Band

We illustrate how to design the filters for the general case. If we have a data vector of N points, and we want to employ the DFT instead of the Z transform presented in Chapter 2, then we observe that the DFT of a vector/signal of N points is another vector/signal of N points, so that:

$$x[n] \xrightarrow{\quad DFT_N \quad} X[k] \qquad (4.36)$$

As in the Z transform, the DFT of a finite convolution, denoted by $\overset{N}{*}$, becomes:

$$x[n] \xrightarrow{\quad DFT_N \quad} X[k]$$

$$h[n] \xrightarrow{\quad DFT_N \quad} H[k] \qquad (4.37)$$

$$x[n] \overset{N}{*} h[n] \xrightarrow{\quad DFT_N \quad} X[k]H[k]$$

The DFT of downsampling the vector/signal $x[n]$ of N (N is assumed to be even) points by a factor of 2, is a vector/signal of $N/2$ points given by:

$$x[n] \xrightarrow{\quad DFT_N \quad} X[k]$$

$$(\downarrow 2)x[n] \xrightarrow{\quad DFT_{N/2} \quad} X[k] + X\left[k + \frac{N}{2}\right] \qquad (4.38)$$

Observe that if $x[n]$ and $X[k]$ are considered to be periodic of period N, then $(\downarrow 2) x[n]$ and $X[k] + X[k + N/2]$ are periodic signals of period $N/2$. After the signal is filtered, it is subsampled to produce vectors/signals of half-length duration.

The upsampling by a factor of 2 results in a vector/signal also of length $N/2$ and is given by:

$$x[n] \xrightarrow{\quad DFT_{N/2} \quad} X[k]$$

$$(\uparrow 2)x[n] \xrightarrow{\quad DFT_N \quad} X\left[k \bmod \frac{N}{2}\right] \qquad (4.39)$$

Observe that if we consider $X[k]$ to be periodic of period $N/2$, then upsampling does not result in a DFT, which is different from the original, but the value of index k is changed. Before upsampling, each index k is related to a frequency $\omega = 2\pi \, k/(N/2)$ rad/sec and the range of k is from 0 to $(N/2) - 1$. After upsampling, each index k is now identified with the frequency $\omega = 2\pi \, k/N$ rad/sec and the range of k is from 0 to $N - 1$. With the above definitions and the model defined by Figure 4.18, the perfect reconstruction condition for developing such a filter pair is given by

$$H_0[k]F_0[k] + H_1[k]F_1[k] = 2e^{j\frac{2\pi}{N}mk} \quad \text{for} \quad m \in \{0, 1, ..., N-1\} \qquad (4.40)$$

where m is related to a circular displacement of the data. In addition, for no aliasing, we obtain the condition

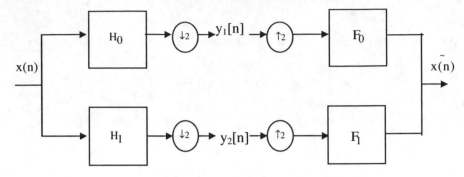

Figure 4.18 A multiresolution filtering scheme.

$$H_0\left[k + N/2\right]F_0[k] + H_1\left[k + N/2\right]F_1[k] = 0 \qquad (4.41)$$

Thus, we choose the four analysis and synthesis filters described in Chapter 2 as

$$F_0[k] = H_1\left[k + N/2\right]$$
$$F_1[k] = -H_0\left[k + N/2\right] \qquad (4.42)$$

for removing aliasing. In addition, for no distortion, we require

$$P[k] - P\left[k + N/2\right] = 2e^{j\frac{2\pi}{N}mk} \qquad \text{for} \qquad m \in \{0,1,\ldots,N-1\} \qquad (4.43)$$

by utilizing (4.39), (4.40), and (4.42). Here $P[k] = H_0[k]F_0[k]$. We define $R[k]$ as

$$R[k] = P[k]e^{-j\frac{2\pi}{N}mk} \qquad (4.44)$$

We further assume that m is odd. Because we need to enforce the no-distortion condition and require perfect reconstruction of the data, we need to satisfy

$$R[k] + R\left[k + N/2\right] = 2 \qquad (4.45)$$

where $R[k] = H_0[k]F_0[k]e^{-j\frac{2\pi}{N}mk}$. If, in addition, we require orthogonality of the filter banks, then we must have

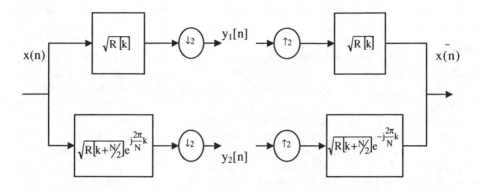

Figure 4.19 A multiresolution filtering scheme using an orthonormal filter bank.

$$\left| H[k]^2 \right| + \left| H\left[k + \frac{N}{2}\right] \right|^2 = 2 \tag{4.46}$$

If we choose $m = 1$, then an orthonormal filter bank is obtained as per Figure 4.19. Here, $R[k]$ is a vector of N coefficients which satisfies

$$R[k] + R\left[k + \frac{N}{2}\right] = 2 \tag{4.47}$$

Much more freedom in the design of the filter banks can be achieved if the filters are not orthogonal. There may be many choices for such filter banks. A possible realization of the process is shown in Figure 4.20, where $R[k] = S[k]T[k]$ and condition (4.47) must be satisfied.

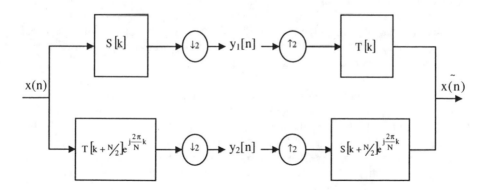

Figure 4.20 A multiresolution scheme using a nonorthogonal filter bank.

4.5.2 Application of the Discrete Cosine Transform to a Symmetric Extension of the Data

The periodic extension of the data in the computation of the FFT produces discontinuities along the borders of the signal of finite duration. We can avoid this problem by making a symmetric extension of the signal before extending it periodically. The even symmetry considered is of type II (i.e., odd antisymmetry) as illustrated by Strang and Nguyen [1]. This implies that the data are now two times the original number of samples. Filters that will operate on this signal must also be symmetric in order to retain the symmetry in the outputs of the filters. The components of the system are shown in Figure 4.21. Both $y_1[n]$ and $y_2[n]$ are also symmetric, but $y_1[n]$ has even symmetry and $y_2[n]$ has odd symmetry. Here $R[k]$ is the DFT of a symmetric filter consisting of $2N$ points. In addition, if the following condition $R[k] + R[k+N] = 2$ is satisfied, the filter bank satisfies the conditions of no aliasing (4.41) and no distortion (4.40). Using a terminology similar to that of Strang and Nguyen [1, p.131], we define $E^{II}\{x[n]\}$ to be the symmetric extension of type II as originally done by Vaidyanathan [12] of the original data/signal $x[n]$ consisting of N points. Therefore in this symmetric extension of type II we have,

$$x[-n] = x[n-1] \qquad (4.48)$$

In Figure 4.21, $H_I[k]$ is purely imaginary and of unity magnitude for $0 \le k \le N-1$ and equal to an imaginary quantity whose magnitude is unity, but with a negative sign, for $-N \le k \le -1$. The reason for this design is that, using the present flow diagram, a symmetric extension of the data/signal need not be carried out in an explicit fashion. Instead, with this algorithm we can use the DCT for even symmetric signals, and the DST for odd symmetric signals, to carry out all operations in this procedure. This is illustrated by Figure 4.22, where the prefix I in front of either the DCT or DST represents an inverse operation. It is important to observe that all computations for this algorithm are carried out using real numbers if it is assumed that the vector $R[k]$ is real. Furthermore, the filter banks constructed in this procedure are also orthogonal.

It is interesting to observe that in this design nonorthogonal filter banks for the analysis and the synthesis filters as described in Chapter 2 for perfect reconstruction can also be considered if $R[k]$ is factored in the following way. We replace the block $(R[k])^{1/2}$ by the analysis filter $S[k]$. Also, $T[k]$ is defined to be the synthesis filter in the top part of Figure 4.20. In addition, we must have $R[k] = S[k]$ $T[k]$. Therefore, the block $(R[k+N])^{1/2}$ is replaced in the lower part of Figure 4.20 by the analysis filter $T[k+N]$ and then $S[k+N]$ becomes the synthesis filter.

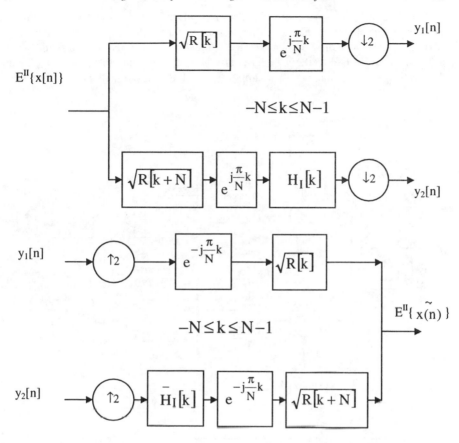

Figure 4.21 Perfect reconstruction procedure for a periodic signal.

4.5.3 Numerical Results

We now apply the above principles to the numerical solution of a dense large complex matrix equation whose size is not related to a power of 2. Furthermore, the following steps are carried out in the solution procedure.

Step 1: Generate an impedance matrix using any analysis procedure to convert an operator equation to a matrix equation.

Step 2: Carry out a diagonal scaling of the matrix so that all the elements on the diagonal are unity. This operation is defined by (4.28) and is applied to the original complex dense matrix $[A_{or}]$ to form the processed matrix $[A]$.

Step 3: Separate matrix $[A]$ into real and imaginary parts because the wavelet transforms of a real matrix are real and hence the computational load is significantly reduced if the wavelet transform is applied to real matrices. Furthermore, invariably in most electromagnetic problems the magnitudes of the real and the imaginary parts of the matrix [A] differ by at least two or three orders of magnitude. Hence it makes more sense to separate the impedance matrix into real and imaginary parts. Otherwise, the truncation error may be significant when a threshold is applied to the wavelet coefficients to eliminate the smaller elements from the processed matrix. The importance of this issue will be clear in later steps.

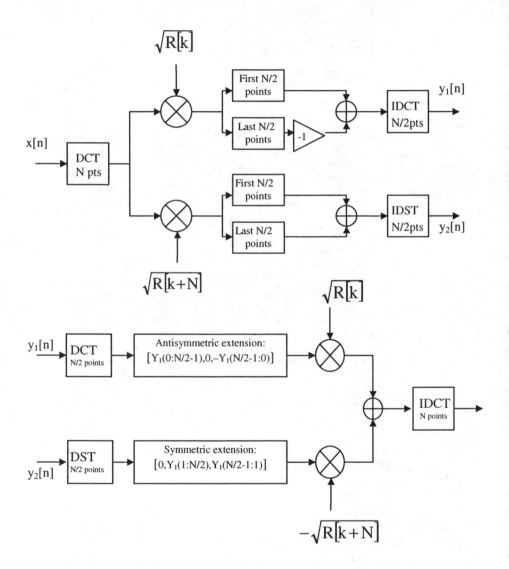

Figure 4.22 Numerical implementation of the perfect reconstruction procedure for a periodic signal.

Step 4: Apply the wavelet-like transform separately to the rows and the columns of matrix [A] as outlined in Section 4.3. This is carried out by using the FIR filters described in the present section and implementing them numerically using the DCT and the DST.

Step 5: Set to zero the matrix elements in the real or the imaginary parts of the postprocessed matrix [A], whose values are lower than a prespecified thereshold value of ε. Typically, the threshold value is chosen to be of the order of 0.001. Now if the wavelet-like transform is carried out simultaneously for the entire complex matrix, without separating it into real and imaginary parts, then the information about the majority of the elements will be lost in making the complex dense matrix sparse for the reasons specified in Step 3. This completes the wavelet transform of the matrix.

To observe how accurately we can recover the original matrix using the thresholded coefficients, we carry out the following operation to get an estimate of the original matrix.

$$[\tilde{A}] = IDWT\{Threshold_\varepsilon[DWT(A)]\} \tag{4.49}$$

where *IDWT* stands for the inverse DWT and *Threshold*$_\varepsilon$[*DWT(A)*] represents applying the threshold operation to the discrete wavelet coefficients of matrix [A]. The error in the reconstructed matrix can be calculated by using any one of the following norms:

$$\delta_1 = \frac{1}{N^2} \sum_{i,j} \left| A(i,j) - \tilde{A}(i,j) \right| \tag{4.50}$$

$$\delta_2 = \sqrt{\frac{1}{N^2} \sum_{i,j} \left| A(i,j) - \tilde{A}(i,j) \right|^2} \quad (\mathscr{L}_2 \text{ Norm}) \tag{4.51}$$

$$\delta_3 = \underset{i,j}{\text{Max}} \left\{ \left| A(i,j) - \tilde{A}(i,j) \right| \right\} \quad (\mathscr{L} \text{ Max-Norm}) \tag{4.52}$$

$$\delta_4 = \frac{1}{N} \underset{j}{\text{Max}} \left\{ \sum_i \left| A(i,j) - \tilde{A}(i,j) \right| \right\} \quad (\mathscr{L}_1 \text{ Norm}) \tag{4.53}$$

$$\delta_5 = \frac{1}{N} \underset{i}{\text{Max}} \left\{ \sum_j \left| A(i,j) - \tilde{A}(i,j) \right| \right\} \quad (\mathscr{L}_\infty \text{ Norm}) \tag{4.54}$$

Any of the five error criteria can be used to quantify the reconstruction error associated with the recovery of the original matrix from the compressed matrix.

4.5.3.1 Square Root Transition Band Filter

As a first example, we consider a filter whose finite transition band TR has a square root shape such that the shape of the magnitude response $(R[k])^{1/2}$ is shown in Figure 4.23. The response of the filters $H_0(z)$ and $H_1(z)$ of Figure 4.18 are shown in the same plot. The phase response of the filters is depicted in Figure 4.24. Hence, $H_1(z)$ has a linear phase response, whereas the phase response of $H_0(z)$ is zero. However, when implemented using the DCT and DST, the phase response of the two filters becomes different as shown in Figure 4.25.

In the simulations to be presented several values for TR and ε have been used. The matrix generated in this case arises from the electromagnetic scattering from a perfectly conducting sphere. The size of the matrix is $4,140 \times 4,140$, so it is not a composite power of 2. To apply the wavelet-like transform, we first apply one step of the wavelet transforms to the matrix and separate the lowpass and the highpass response as described in Section 4.3. Then we select a portion of the processed matrix which is of dimension 2^n, so that it is $4,096 \times 4,096$. We now apply the wavelet-like transform to this submatrix using various order FIR filters, that is, various widths of the transition band TR. A threshold is applied to the rest of the matrix elements and they are not involved in the rest of the computations. In Tables 4.24(a) and (b), the five rows of numbers correspond to the five different error norms associated with the reconstruction error δ_i. They correspond to the five different errors defined by (4.50) to (4.54).

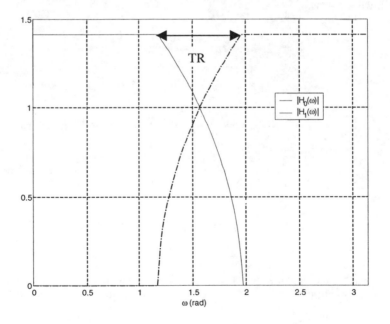

Figure 4.23 Magnitude response of the filters.

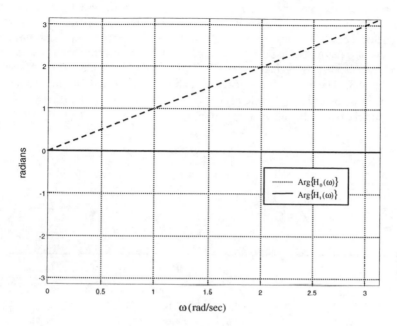

Figure 4.24 Phase response of the filters.

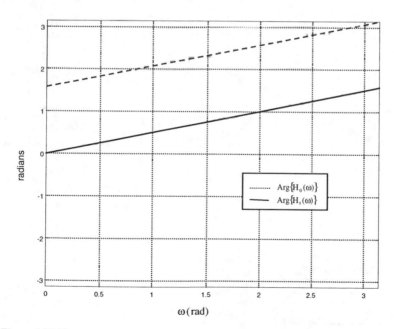

Figure 4.25 Phase response of the filters when numerically implemented with DCT and DST.

Table 4.24(a) presents the sparsity achieved for the real part and Table 4.24(b) corresponds to the imaginary part of the matrix for different values of the threshold parameter. The tables also provide the five reconstruction errors in creating the real and imaginary parts of the original matrix from its compressed state using a square root type filter response in the transition band, but with a linear phase response. A value of $TR = 0$ corresponds to the ideal filter described in Section 4.3. The last column in both Tables 4.24(a) and (b) represents the various values for δ_i when the elements of $\tilde{A}_{(ij)}$ are assumed not to exist (i.e., for only the matrix $[A]$ in the relevant equations).

Table 4.24(a)
Real Part

Type of Transition Band (TR)	Length TR	Threshold (ε) 0.1 — % Nonzero Elements	Threshold (ε) 0.1 — Error (δ_i)	0.01 — % Nonzero Elements	0.01 — Error (δ_i)	0.001 — % Nonzero Elements	0.001 — Error (δ_i)	0.0001 — % Nonzero Elements	0.0001 — Error (δ_i)	Matrix Norm for the Original Matrix
Square Root	0.75		0.15×10^{-02}		0.88×10^{-03}		0.28×10^{-03}		0.53×10^{-04}	
			0.10×10^{-01}		0.24×10^{-02}		0.47×10^{-03}		0.71×10^{-04}	
		0.05%	0.59	0.90%	0.94×10^{-01}	5.09%	0.79×10^{-02}	17.92%	0.64×10^{-03}	
			0.25×10^{-02}		0.13×10^{-02}		0.39×10^{-03}		0.69×10^{-04}	
			0.25×10^{-02}		0.13×0^{-02}		0.39×10^{-03}		0.69×10^{-04}	
	0.50		0.15×10^{-02}		0.98×10^{-03}		0.29×10^{-03}		0.53×10^{-04}	
			0.99×10^{-02}		0.26×10^{-02}		0.49×10^{-03}		0.72×10^{-04}	
		0.05%	0.68	0.84%	0.96×10^{-01}	5.17%	0.85×10^{-02}	17.36%	0.68×10^{-03}	
			0.25×10^{-02}		0.13×10^{-02}		0.41×10^{-03}		0.74×10^{-04}	
			0.26×10^{-02}		0.13×10^{-02}		0.41×10^{-03}		0.74×10^{-04}	
	0.33		0.15×10^{-02}		0.10×10^{-02}		0.29×10^{-03}		0.54×10^{-04}	0.13×10^{-02}
			0.10×10^{-01}		0.25×10^{-02}		0.48×10^{-03}		0.73×10^{-04}	0.21×10^{-01}
		0.04%	0.75	0.82%	0.10	4.90%	0.10×10^{-01}	16.40%	0.66×10^{-03}	1
			0.27×10^{-02}		0.14×10^{-02}		0.44×10^{-03}		0.79×10^{-04}	0.22×10^{-02}
			0.27×10^{-02}		0.14×10^{-02}		0.44×10^{-03}		0.79×10^{-04}	0.22×10^{-02}
	0.25		0.16×10^{-02}		0.10×10^{-02}		0.30×10^{-03}		0.57×10^{-04}	
			0.10×10^{-01}		0.26×10^{-02}		0.49×10^{-03}		0.77×10^{-04}	
		0.04%	0.77	0.86%	0.11	5.21%	0.87×10^{-02}	17.09%	0.66×10^{-03}	
			0.28×10^{-02}		0.15×10^{-02}		0.45×10^{-03}		0.82×10^{-04}	
			0.28×10^{-02}		0.15×10^{-02}		0.46×10^{-03}		0.83×10^{-04}	
Ideal Filter	0		0.22×10^{-02}		0.18×10^{-02}		0.66×10^{-03}		0.79×10^{-04}	
			0.10×10^{-01}		0.36×10^{-02}		0.90×10^{-03}		0.10×10^{-03}	
		0.04%	0.84	1.13%	0.18	12.02%	0.13×10^{-01}	65.22%	0.70×10^{-03}	
			0.38×10^{-02}		0.26×10^{-02}		0.10×10^{-02}		0.11×10^{-03}	
			0.38×10^{-02}		0.26×10^{-02}		0.10×10^{-02}		0.11×10^{-03}	

Table 4.24(b)

Imaginary Part

Type of Transition Band (TR)	Length TR	Threshold (ε)								Matrix Norm for the Original Matrix
		0.1		0.01		0.001		0.0001		
		% Nonzero Elements	Error	% Nonzero Elements	Error	% Nonzero Elements	Error	% Nonzero Elements	Error	
Square Root			0.22×10^{-3}		0.83×10^{-04}		0.16×10^{-04}		0.23×10^{-05}	
			0.33×10^{-3}		0.11×10^{-3}		0.22×10^{-04}		0.30×10^{-05}	
	0.75	0.01%	0.37×10^{-3}	0.89%	0.20×10^{-02}	7.26%	0.34×10^{-03}	26.15%	0.26×10^{-04}	
			0.38×10^{-3}		0.15×10^{-03}		0.29×10^{-04}		0.36×10^{-05}	
			0.38×10^{-3}		0.15×10^{-03}		0.29×10^{-04}		0.36×10^{-05}	
			0.20×10^{-3}		0.68×10^{-04}		0.12×10^{-04}		0.20×10^{-05}	
			0.30×10^{-03}		0.94×10^{-04}		0.17×10^{-04}		0.26×10^{-05}	
	0.50	0.01%	0.34×10^{-02}	0.67%	0.19×10^{-02}	4.82%	0.30×10^{-03}	17.24%	0.24×10^{-04}	
			0.36×10^{-03}		0.14×10^{-03}		0.29×10^{-04}		0.37×10^{-05}	
			0.36×10^{-03}		0.15×10^{-03}		0.29×10^{-04}		0.37×10^{-05}	
			0.20×10^{-03}		0.61×10^{-04}		0.12×10^{-04}		0.19×10^{-05}	0.23×10^{-03}
			0.29×10^{-03}		0.87×10^{-04}		0.17×10^{-04}		0.26×10^{-05}	0.37×10^{-03}
	0.33	0.01%	0.35×10^{-02}	0.53%	0.22×10^{-02}	3.98%	0.34×10^{-03}	14.77%	0.26×10^{-04}	0.34×10^{-02}
			0.36×10^{-03}		0.16×10^{-03}		0.30×10^{-04}		0.42×10^{-05}	0.35×10^{-03}
			0.36×10^{-03}		0.16×10^{-03}		0.30×10^{-04}		0.42×10^{-05}	0.35×10^{-03}
			0.20×10^{-03}		0.65×10^{-04}		0.12×10^{-04}		0.20×10^{-05}	
			0.30×10^{-03}		0.92×10^{-04}		0.18×10^{-04}		0.27×10^{-05}	
	0.25	0.01%	0.34×10^{-02}	0.50%	0.23×10^{-02}	3.91%	0.32×10^{-03}	14.22%	0.32×10^{-04}	
			0.36×10^{-03}		0.16×10^{-03}		0.31×10^{-04}		0.40×10^{-05}	
			0.36×10^{-03}		0.16×10^{-03}		0.31×10^{-04}		0.40×10^{-05}	
Ideal Filter			0.20×10^{-03}		0.65×10^{-04}		0.15×10^{-04}		0.28×10^{-05}	
			0.30×10^{-03}		0.92×10^{-04}		0.22×10^{-04}		0.37×10^{-05}	
	0	0.01%	0.36×10^{-02}	0.53%	0.25×10^{-02}	4.15%	0.50×10^{-03}	21.99%	0.45×10^{-04}	
			0.36×10^{-03}		0.17×10^{-03}		0.41×10^{-04}		0.56×10^{-05}	
			0.36×10^{-03}		0.17×10^{-03}		0.41×10^{-04}		0.56×10^{-05}	

4.5.3.2 Other Types of Filters

To improve the degree of sparsity with the reconstruction error, a filter with a smoother transition band is used next.

It is possible to have an infinite variety of filters with different shapes in the transition band. For example, the shape of the transition band may have a cosine shape (as shown in Figure 4.26), or could be approximated by a polynomial (as illustrated in Figure 4.27) or for that matter they may have other shapes. Here we illustrate filters that have different types of transition bands, which influences not only the degree of sparsity but also the reconstruction error.

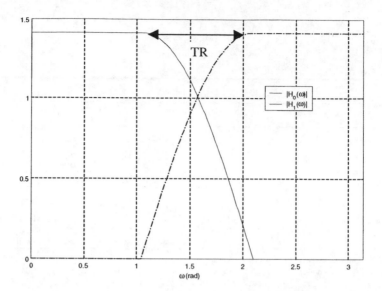

Figure 4.26 Filters with a transition band that has a cosine shape.

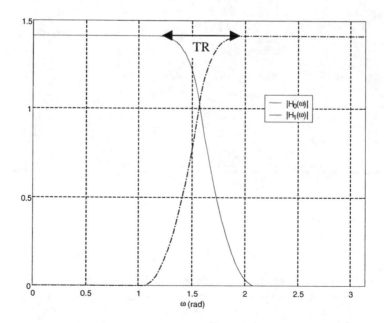

Figure 4.27 Filters with a transition band that has a polynomial shape.

4.5.3.3 Daubechies-Like Filters with a Linear Phase

After many numerical experiments with the same matrix, the best sparsity for the matrix and the reconstruction error are achieved by a filter whose magnitude response is similar to that of a FIR Daubechies filter (as shown in Figure 4.28), but has a linear phase response. The FIR Daubechies filter, however, has a nonlinear phase response, as seen from Figure 4.29.

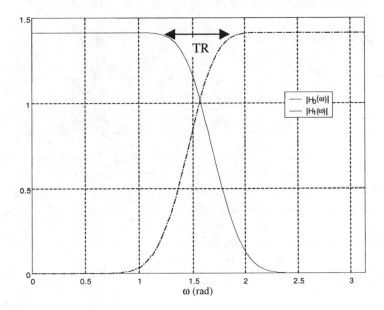

Figure 4.28 Filters with a transition band that has the same magnitude response as a Daubechies filter of order p.

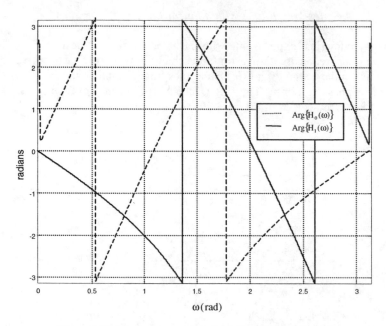

Figure 4.29 Phase response of the Daubechies filter of length 8.

Next, the Daubechies-like filter is now used to carry out the wavelet transform to compress a matrix equation as presented earlier of size 4,140 × 4,140. Table 4.25(a) provides the degree of sparsity and the reconstruction error achieved by using the Daubechies-like FIR filters on the real part of the impedance matrix. The five rows in each of the columns are associated with the reconstruction error corresponding to the five norms for the error that have been defined by (4.50) to (4.54). Different errors in the reconstruction and the compression ratio depend on the order of the filters chosen. Table 4.25(b) provides the compression and the reconstruction error achieved using the Daubechies-like filters for the imaginary part of the matrix.

Table 4.26(a) provides the compression and the reconstruction error achieved when using the actual Daubechies filters for the real part of the matrix. The five rows in each of the columns are associated with the reconstruction error corresponding to the five norms for the error that have been defined by (4.50) to (4.54). Table 4.26(b) provides the compression ratio and the reconstruction error achieved using the actual Daubechies filters for the imaginary part of the matrix. For all four cases in Tables 4.25 and 4.26, the filter banks have been chosen in such a way that the function $[R(k)]^{1/2}$ is equal in magnitude to that of a Daubechies filter. Thus the only difference between the two examples is in the phase response of the filters. In these tables ideal filter represents a filter with zero transition band (i.e., $TR = 0$).

Table 4.25(a)
Real Part

Type of Transition Band (TR)	Order	Threshold (ε)								∞
		0.1		0.01		0.001		0.0001		
		% Nonzero Elements	Error	% Nonzero Elements	Error	% Nonzero Elements	Error	% Nonzero Elements	Error	Matrix Norm for the Original Matrix
Daubechies-Like Filter	2	0.05%	0.14×10^{-02} 0.98×10^{-02} 0.61 0.23×10^{-02} 0.24×10^{-02}	0.84%	0.79×10^{-03} 0.22×10^{-02} 0.89×10^{-01} 0.12×10^{-02} 0.12×10^{-02}	3.66%	0.21×10^{-03} 0.36×10^{-03} 0.89×10^{-02} 0.32×10^{-03} 0.33×10^{-03}	10.34%	0.36×10^{-04} 0.51×10^{-04} 0.63×10^{-03} 0.63×10^{-04} 0.63×10^{-04}	
	4	0.04%	0.14×10^{-02} 0.99×10^{-02} 0.71 0.24×10^{-02} 0.24×10^{-02}	0.81%	0.83×10^{-03} 0.23×10^{-02} 0.96×10^{-01} 0.13×10^{-02} 0.13×10^{-02}	3.59%	0.20×10^{-03} 0.34×10^{-03} 0.86×10^{-02} 0.35×10^{-03} 0.34×10^{-03}	9.61%	0.32×10^{-04} 0.47×10^{-04} 0.67×10^{-03} 0.61×10^{-04} 0.61×10^{-04}	
	8	0.04%	0.15×10^{-02} 0.10×10^{-01} 0.76 0.25×10^{-02} 0.26×10^{-02}	0.81%	0.87×10^{-03} 0.23×10^{-02} 0.101 0.13×10^{-02} 0.13×10^{-02}	3.71%	0.21×10^{-03} 0.35×10^{-03} 0.93×10^{-02} 0.36×10^{-03} 0.35×10^{-03}	9.66%	0.32×10^{-04} 0.46×10^{-04} 0.63×10^{-03} 0.65×10^{-04} 0.65×10^{-04}	0.13×10^{-02} 0.20×10^{-01}
	16	0.04%	0.15×10^{-02} 0.10×10^{-01} 0.78 0.27×10^{-02} 0.27×10^{-02}	0.86%	0.94×10^{-03} 0.24×10^{-02} 0.11 0.14×10^{-02} 0.14×10^{-02}	4.23%	0.22×10^{-03} 0.37×10^{-03} 0.86×10^{-02} 0.38×10^{-03} 0.38×10^{-03}	10.86%	0.33×10^{-04} 0.48×10^{-04} 0.68×10^{-03} 0.67×10^{-04} 0.67×10^{-04}	1 0.21×10^{-02} 0.22×10^{-02}
	32	0.04%	0.16×10^{-02} 0.10×10^{-01} 0.80 0.28×10^{-02} 0.29×10^{-02}	0.96%	0.10×10^{-02} 0.25×10^{-02} 0.11 0.15×10^{-02} 0.15×10^{-02}	4.97%	0.24×10^{-03} 0.41×10^{-03} 0.89×10^{-02} 0.41×10^{-03} 0.41×10^{-03}	13.00%	0.35×10^{-04} 0.50×10^{-04} 0.78×10^{-03} 0.69×10^{-04} 0.70×10^{-04}	
Ideal Filter	∞	0.04%	0.22×10^{-02} 0.10×10^{-01} 0.83 0.38×10^{-02} 0.38×10^{-02}	1.13%	0.17×10^{-02} 0.36×10^{-02} 0.18 0.25×10^{-02} 0.25×10^{-02}	12.02%	0.66×10^{-03} 0.89×10^{-03} 0.12×10^{-01} 0.10×10^{-02} 0.10×10^{-02}	65.22%	0.78×10^{-04} 0.10×10^{-03} 0.70×10^{-03} 0.10×10^{-03} 0.10×10^{-03}	

Table 4.25(b)
Imaginary Part

Type of Transition Band (TR)	Order	0.1 % Nonzero Elements	0.1 Error	0.01 % Nonzero Elements	0.01 Error	0.001 % Nonzero Elements	0.001 Error	0.0001 % Nonzero Elements	0.0001 Error	∞ Matrix Norm for the Original Matrix
Daubechies-Like Filter	2		0.20×10^{-03}		0.71×10^{-04}		0.14×10^{-04}		0.22×10^{-05}	
			0.31×10^{-03}		0.10×10^{-03}		0.19×10^{-04}		0.28×10^{-05}	
		0.01%	0.35×10^{-02}	0.696%	0.20×10^{-02}	5.20%	0.32×10^{-03}	19.17%	0.26×10^{-04}	
			0.37×10^{-03}		0.14×10^{-03}		0.27×10^{-04}		0.38×10^{-05}	
			0.37×10^{-03}		0.14×10^{-03}		0.27×10^{-04}		0.37×10^{-05}	
	4		0.20×10^{-03}		0.63×10^{-04}		0.11×10^{-04}		0.15×10^{-05}	
			0.30×10^{-03}		0.89×10^{-04}		0.15×10^{-04}		0.21×10^{-05}	
		0.01%	0.34×10^{-02}	0.571%	0.21×10^{-02}	4.00%	0.36×10^{-03}	12.35%	0.26×10^{-04}	
			0.36×10^{-03}		0.15×10^{-03}		0.28×10^{-04}		0.31×10^{-05}	
			0.36×10^{-03}		0.15×10^{-03}		0.28×10^{-04}		0.31×10^{-05}	
	8		0.20×10^{-03}		0.61×10^{-04}		0.11×10^{-04}		0.14×10^{-05}	
			0.29×10^{-03}		0.86×10^{-04}		0.15×10^{-04}		0.20×10^{-05}	
		0.01%	0.34×10^{-02}	0.524%	0.22×10^{-02}	3.67%	0.33×10^{-03}	11.36%	0.26×10^{-04}	0.23×10^{-03}
			0.36×10^{-03}		0.16×10^{-03}		0.29×10^{-04}		0.31×10^{-05}	0.37×10^{-03}
			0.36×10^{-03}		0.16×10^{-03}		0.29×10^{-04}		0.31×10^{-05}	0.34×10^{-02}
	16		0.20×10^{-03}		0.62×10^{-04}		0.11×10^{-04}		0.16×10^{-05}	0.35×10^{-03}
			0.29×10^{-03}		0.87×10^{-04}		0.16×10^{-04}		0.22×10^{-05}	0.35×10^{-03}
		0.01%	0.33×10^{-02}	0.513%	0.23×10^{-02}	3.67%	0.35×10^{-03}	12.40%	0.28×10^{-04}	
			0.36×10^{-03}		0.16×10^{-03}		0.30×10^{-04}		0.34×10^{-05}	
			0.36×10^{-03}		0.16×10^{-03}		0.30×10^{-04}		0.34×10^{-05}	
	32		0.20×10^{-03}		0.63×10^{-04}		0.12×10^{-04}		0.18×10^{-05}	
			0.30×10^{-03}		0.89×10^{-04}		0.18×10^{-04}		0.25×10^{-05}	
		0.01%	0.34×10^{-02}	0.516%	0.23×10^{-02}	3.79%	0.37×10^{-03}	13.88%	0.31×10^{-04}	
			0.36×10^{-03}		0.16×10^{-03}		0.32×10^{-04}		0.38×10^{-05}	
			0.36×10^{-03}		0.16×10^{-03}		0.32×10^{-04}		0.38×10^{-05}	
Ideal Filter	∞		0.20×10^{-03}		0.65×10^{-04}		0.15×10^{-04}		0.28×10^{-05}	
			0.30×10^{-03}		0.92×10^{-04}		0.22×10^{-04}		0.37×10^{-05}	
		0.01	0.36×10^{-02}	0.54%	0.25×10^{-02}	4.15%	0.50×10^{-03}	21.991%	0.45×10^{-04}	
			0.36×10^{-03}		0.17×10^{-03}		0.41×10^{-04}		0.56×10^{-05}	
			0.36×10^{-03}		0.17×10^{-03}		0.41×10^{-04}		0.56×10^{-05}	

Table 4.26(a)
Real Part (Normalized Daubechies Filters)

Type of Transition Band (TR)	Order	Threshold (ε)								Matrix Norm
		0.1		0.01		0.001		0.0001		
		% Nonzero Elements	Error	% Nonzero Elements	Error	% Nonzero Elements	Error	% Nonzero Elements	Error	
Daubechies	2		0.13×10^{-02}		0.68×10^{-03}		0.19×10^{-03}		0.43×10^{-04}	
			0.10×10^{-01}		0.20×10^{-02}		0.33×10^{-03}		0.60×10^{-04}	
		0.05%	0.54	0.92%	0.90×10^{-01}	3.15%	0.71×10^{-02}	10.29%	0.66×10^{-03}	
			0.24×10^{-02}		0.11×10^{-02}		0.29×10^{-03}		0.78×10^{-04}	
			0.24×10^{-02}		0.11×10^{-02}		0.28×10^{-03}		0.78×10^{-04}	
	4		0.14×10^{-02}		0.83×10^{-03}		0.20×10^{-03}		0.33×10^{-04}	
			0.10×10^{-01}		0.22×10^{-02}		0.33×10^{-03}		0.47×10^{-04}	
		0.04%	0.70	0.93%	0.98×10^{-01}	3.83%	0.70×10^{-02}	10.18%	0.60×10^{-03}	
			0.26×10^{-02}		0.12×10^{-02}		0.29×10^{-03}		0.58×10^{-04}	
			0.27×10^{-02}		0.12×10^{-02}		0.29×10^{-03}		0.58×10^{-04}	
	8		0.15×10^{-02}		0.91×10^{-03}		0.21×10^{-03}		0.33×10^{-04}	0.13×10^{-02}
			0.10×10^{-01}		0.23×10^{-02}		0.36×10^{-03}		0.47×10^{-04}	0.20×10^{-01}
		0.04%	0.68	0.91%	0.10	3.98%	0.99×10^{-02}	10.38%	0.81×10^{-03}	1
			0.28×10^{-02}		0.14×10^{-02}		0.35×10^{-03}		0.59×10^{-04}	0.21×10^{-02}
			0.28×10^{-02}		0.13×10^{-02}		0.35×10^{-03}		0.59×10^{-04}	0.22×10^{-02}
	16		0.16×10^{-02}		0.10×10^{-02}		0.22×10^{-03}		0.32×10^{-04}	
			0.10×10^{-01}		0.24×10^{-02}		0.36×10^{-03}		0.47×10^{-04}	
		0.04%	0.70	1.01%	0.15	4.58%	0.17×10^{-01}	11.46%	0.10×10^{-02}	
			0.30×10^{-02}		0.15×10^{-02}		0.38×10^{-03}		0.74×10^{-04}	
			0.31×10^{-02}		0.15×10^{-02}		0.38×10^{-03}		0.74×10^{-04}	
	32		0.18×10^{-02}		0.11×10^{-02}		0.22×10^{-03}		0.31×10^{-04}	
			0.10×10^{-01}		0.26×10^{-02}		0.36×10^{-03}		0.44×10^{-04}	
		0.04%	0.78	1.40%	0.19	6.50%	0.12×10^{-01}	14.70%	0.11×10^{-02}	
			0.34×10^{-02}		0.19×10^{-02}		0.43×10^{-03}		0.75×10^{-04}	
			0.34×10^{-02}		0.18×10^{-02}		0.42×10^{-03}		0.75×10^{-04}	

Table 4.26(b)
Imaginary Part (Normalized Daubechies Filters)

Type of Transition Band (TR)	Order	Threshold (ε)								∞
		0.1		0.01		0.001		0.0001		
		% Nonzero Elements	Error	% Nonzero Elements	Error	% Nonzero Eleemnts	Error	% Nonzero Elements	Error	Matrix Norm
Daubechies	2		0.22×10^{-03}		0.90×10^{-04}		0.20×10^{-04}		0.27×10^{-05}	
			0.33×10^{-03}		0.12×10^{-03}		0.26×10^{-04}		0.34×10^{-05}	
		0.01%	0.37×10^{-02}	0.983%	0.18×10^{-02}	10.26%	0.28×10^{-03}	39.39%	0.21×10^{-04}	
			0.37×10^{-03}		0.20×10^{-03}		0.36×10^{-04}		0.42×10^{-05}	
			0.37×10^{-03}		0.20×10^{-03}		0.36×10^{-04}		0.42×10^{-05}	
	4		0.20×10^{-03}		0.70×10^{-04}		0.13×10^{-04}		0.20×10^{-05}	
			0.30×10^{-03}		0.98×10^{-04}		0.18×10^{-04}		0.26×10^{-05}	
		0.01%	0.48×10^{-02}	0.74%	0.19×10^{-02}	5.72%	0.29×10^{-03}	19.93%	0.26×10^{-04}	
			0.37×10^{-03}		0.14×10^{-03}		0.28×10^{-04}		0.34×10^{-05}	
			0.37×10^{-03}		0.14×10^{-03}		0.28×10^{-04}		0.34×10^{-05}	0.23×10^{-03}
	8		0.19×10^{-03}		0.62×10^{-04}		0.11×10^{-04}		0.14×10^{-05}	0.37×10^{-03}
			0.29×10^{-03}		0.87×10^{-04}		0.15×10^{-04}		0.19×10^{-05}	
		0.02%	0.36×10^{-02}	0.63%	0.22×10^{-02}	4.52%	0.35×10^{-03}	13.78%	0.26×10^{-04}	0.34×10^{-02}
			0.35×10^{-03}		0.15×10^{-03}		0.24×10^{-04}		0.29×10^{-05}	
			0.35×10^{-03}		0.15×10^{-03}		0.24×10^{-04}		0.29×10^{-05}	0.35×10^{-03}
	16		0.19×10^{-03}		0.61×10^{-04}		0.11×10^{-04}		0.14×10^{-05}	0.35×10^{-03}
			0.29×10^{-03}		0.86×10^{-04}		0.16×10^{-04}		0.20×10^{-05}	
		0.02%	0.36×10^{-02}	0.59%	0.21×10^{-02}	4.47%	0.38×10^{-03}	14.55%	0.41×10^{-04}	
			0.35×10^{-03}		0.16×10^{-03}		0.34×10^{-04}		0.37×10^{-05}	
			0.35×10^{-03}		0.16×10^{-03}		0.34×10^{-04}		0.37×10^{-05}	
	32		0.20×10^{-03}		0.61×10^{-04}		0.12×10^{-04}		0.16×10^{-05}	
			0.29×10^{-03}		0.88×10^{-04}		0.18×10^{-04}		0.23×10^{-05}	
		0.02%	0.32×10^{-02}	0.61%	0.24×10^{-02}	5.14%	0.48×10^{-03}	18.62%	0.54×10^{-04}	
			0.37×10^{-03}		0.17×10^{-03}		0.41×10^{-04}		0.45×10^{-05}	
			0.37×10^{-03}		0.17×10^{-03}		0.41×10^{-04}		0.45×10^{-05}	

From the four tables the following observations can be made:

- The smoother the behavior of the filters in the transition band, the better the degree of sparsity achieved when the wavelet-like transform is applied to the dense matrix. Also, the reconstruction error is less.
- We can see that for all the cases the Daubechies-like filters provide better sparsity and have lower reconstruction errors. It is important to note that the Daubechies filters have a nonlinear phase response as shown in Figure 4.29.
- The linear phase filters having the same magnitude response as the Daubechies filter (i.e., the Daubechies-like filters) do better in providing a sparser matrix for both the real and the imaginary parts of the matrix. Better sparsity in these cases is synonymous with fewer reconstruction errors for the impedance matrix.

In addition, several remarks can be made regarding the nonexistence of strict mathematical proofs to verify these results. For these reasons:

- The formula used to assess the decay of the wavelet coefficients defined by

$$d_{jk} \leq C 2^{-jp} \left\| f^{(p)}(t) \right\|$$ (4.55)

does not hold. Here d_{jk} are the wavelet coefficients of a vector/signal $f(t)$ and p is the number of zeros of the lowpass filter at $\omega = \pi$. The term $f^{(p)}(t)$ represents the pth derivative of the function $f(t)$. Expression (4.55) is deduced for infinite-duration signals with finite energy. If the signal is finite and is extended periodically, like we do in this method, the energy of the signal is not finite. Since we are dealing with the discrete case, theoretically $f(t)$ cannot be considered to be a continuous function with up to p derivatives. Although through a symmetrical extension we guarantee continuity, the existence of p derivatives at the boundary is not guaranteed.

- In addition, perhaps we may be committing a wavelet crime as stated by Strang and Nguyen [1], which has to do with representing a discrete function by a continuous wavelet series. The problem is that the dilation equation representing the wavelets has no solution in the discrete case.

4.5.4   A Note on the Characteristics of the Solution

With the filter banks having a magnitude response of the form $[R(k)]^{1/2}$ that is equal to the magnitude of the Daubechies filter, we now solve the original linear system of equations:

$$[A][X] = [Y]$$ (4.56)

with the stipulation that vector [Y] be equal to one for the first element while the other 4,139 components of the matrix are exactly zero. We use the same matrix [A] defined for the previous example.

We are going to solve this problem in two different ways:

- First, we solve the problem using an LU decomposition (the subroutine XGETRF from the LAPACK Library was used). This is useful for estimating the condition number of the original matrix (the subroutine XGECON from the LAPACK Library, which estimates the reciprocal condition number associated with a matrix, has also been used). The solution obtained with the LU decomposition is considered to be the reference solution.

- Second, at the first stage we apply the discrete wavelet transform to compress the matrix using a prespecified threshold. Then we use the conjugate gradient method to solve the sparse linear system. Finally, we apply an inverse discrete wavelet transform to the computed solution to obtain the final result.

The condition number of the matrix is very important because it controls the rate of convergence for the conjugate gradient method. Moreover, for a direct method, the accuracy of the solution is directly related to it. Thus a large condition number will lead to failure of direct methods [11] if there are not enough bits available for the computation. We also compute the condition number before and after the application of the threshold. The condition number for the original system is given by Table 4.27. Here we define the normalized residual $[\tilde{R}]$ of the solution by $[\tilde{R}] = \dfrac{\|[Y] - [A][X]\|}{\|[Y]\|}$. The error in the solution $[X]$ is defined with respect to the solution obtained from the LU decomposition procedure and is defined by $[E]$ through $[E] = \dfrac{\|[X_{LU}] - [X]\|}{\|[X_{LU}]\|}$. The following results were obtained, which are described in Tables 4.28 through 4.32 for different lengths of the filter.

Table 4.27
Condition Number of the Original Matrix Equation

Reciprocal condition number ($\mathcal{L}_1$ norm)	1.53×10^{-04}
Reciprocal condition number ($\mathcal{L}_\infty$ norm)	1.55×10^{-04}

In Tables 4.17 to 4.32 the reciprocal condition number provides an estimate of the inverse of the actual condition number. We would like this number to be as large as possible. The optimum performance is achieved by the DCT/DST bank filter with $[R(k)]^{1/2}$ equal in magnitude to that of a Daubechies filter of length 8.

Table 4.28
Transformed Matrix with a DCT/DST Bank Filter
(Magnitude Same as That of a Daubechies Filter of Length 4)

Before threshold

Reciprocal condition number ($\mathcal{L}_1$ norm)	3.10×10^{-05}
Reciprocal condition number ($\mathcal{L}_\infty$ norm)	3.12×10^{-05}

After the elements below the threshold $\varepsilon = 0.001$ were set to zero

---

Nonzero elements	3.68%
Reciprocal condition number ($\mathcal{L}_1$ norm)	3.61×10^{-07}
Reciprocal condition number ($\mathcal{L}_\infty$ norm)	5.97×10^{-07}
Residue for the transformed solution ($\mathcal{L}_2$ norm)	2.44
Residue for the final solution ($\mathcal{L}_2$ norm)	2.44
Error in the final solution ($\mathcal{L}_2$ norm)	1.77

After the elements below the threshold $\varepsilon = 0.0001$ were set to zero

---

Nonzero elements	10.67%
Reciprocal condition number ($\mathcal{L}_1$ norm)	1.54×10^{-05}
Reciprocal condition number ($\mathcal{L}_\infty$ norm)	1.55×10^{-05}
Residue for the transformed solution ($\mathcal{L}_2$ norm)	0.28
Residue for the final solution ($\mathcal{L}_2$ norm)	0.28
Error in the final solution ($\mathcal{L}_2$ norm)	0.17

Table 4.29
Transformed Matrix with a DCT/DST Bank Filter
(Magnitude Same as That of a Daubechies Filter of Length 8)

Before threshold	
Reciprocal condition number ($\mathcal{L}_1$ norm)	3.28×10^{-05}
Reciprocal condition number ($\mathcal{L}_\infty$ norm)	3.31×10^{-05}

After the elements below the threshold $\varepsilon = 0.001$ were set to zero

Nonzero elements	3.80%
Reciprocal condition number ($\mathcal{L}_1$ norm)	1.74×10^{-07}
Reciprocal condition number ($\mathcal{L}_\infty$ norm)	1.91×10^{-07}
Residue for the transformed solution ($\mathcal{L}_2$ norm)	2.62
Residue for the final solution ($\mathcal{L}_2$ norm)	2.62
Error in the final solution ($\mathcal{L}_2$ norm)	1.97

After the elements below the threshold $\varepsilon = 0.0001$ were set to zero

Nonzero elements	10.15%
Reciprocal condition number ($\mathcal{L}_1$ norm)	1.63×10^{-05}
Reciprocal condition number ($\mathcal{L}_\infty$ norm)	1.60×10^{-05}
Residue for the transformed solution ($\mathcal{L}_2$ norm)	0.26
Residue for the final solution ($\mathcal{L}_2$ norm)	0.26
Error in the final solution ($\mathcal{L}_2$ norm)	0.17

Table 4.30
Transformed Matrix with a DCT/DST Bank Filter
(Magnitude Same as That of a Daubechies Filter of Length 16)

Before threshold

Reciprocal condition number ($\mathcal{L}_1$ norm)	3.09×10^{-05}
Reciprocal condition number ($\mathcal{L}_\infty$ norm)	3.12×10^{-05}

After the elements below the threshold $\varepsilon = 0.001$ were set to zero

---

Nonzero elements	4.32%
Reciprocal condition number ($\mathcal{L}_1$ norm)	1.95×10^{-07}
Reciprocal condition number ($\mathcal{L}_\infty$ norm)	2.36×10^{-07}
Residue for the transformed solution ($\mathcal{L}_2$ norm)	3.05
Residue for the final solution ($\mathcal{L}_2$ norm)	3.05
Error in the final solution ($\mathcal{L}_2$ norm)	1.96

After the elements below the threshold $\varepsilon = 0.0001$ were set to zero

---

Nonzero elements	11.34%
Reciprocal condition number ($\mathcal{L}_1$ norm)	1.43×10^{-05}
Reciprocal condition number ($\mathcal{L}_\infty$ norm)	1.42×10^{-05}
Residue for the transformed solution ($\mathcal{L}_2$ norm)	0.30
Residue for the final solution ($\mathcal{L}_2$ norm)	0.30
Error in the final solution ($\mathcal{L}_2$ norm)	0.20

Table 4.31
Transformed Matrix with a DCT/DST Bank Filter ($TR = 0$; Ideal Filter)

Before threshold	
Reciprocal condition number ($\mathcal{L}_1$ norm)	1.89×10^{-05}
Reciprocal condition number ($\mathcal{L}_\infty$ norm)	1.90×10^{-05}

After the elements below the thereshold $\varepsilon = 0.001$ was set to zero	
Nonzero elements	12.10%
Reciprocal condition number ($\mathcal{L}_1$ norm)	1.23×10^{-07}
Reciprocal condition number ($\mathcal{L}_\infty$ norm)	2.73×10^{-07}
Residue for the transformed solution ($\mathcal{L}_2$ norm)	3.15
Residue for the final solution ($\mathcal{L}_2$ norm)	3.15
Error in the final solution ($\mathcal{L}_2$ norm)	1.47

After the elements below the threshold $\varepsilon = 0.0001$ were set to zero	
Nonzero elements	> 25%
Reciprocal condition number ($\mathcal{L}_1$ norm)	1.38×10^{-06}
Reciprocal condition number ($\mathcal{L}_\infty$ norm)	1.42×10^{-06}
Residue for the transformed solution ($\mathcal{L}_2$ norm)	—
Residue for the final solution ($\mathcal{L}_2$ norm)	—
Error in the final solution ($\mathcal{L}_2$ norm)	—

Table 4.32
Transformed Matrix with a Daubechies Filter of Length 8

Before threshold	
Reciprocal condition number ($\mathcal{L}_1$ norm)	2.30×10^{-05}
Reciprocal condition number ($\mathcal{L}_\infty$ norm)	2.30×10^{-05}

After the elements below the threshold $\varepsilon = 0.001$ were set to zero

Nonzero elements	4.07%
Reciprocal condition number ($\mathcal{L}_1$ norm)	3.48×10^{-07}
Reciprocal condition number ($\mathcal{L}_\infty$ norm)	2.00×10^{-07}
Residue for the transformed solution ($\mathcal{L}_2$ norm)	2.36
Residue for the final solution ($\mathcal{L}_2$ norm)	2.36
Error in the final solution ($\mathcal{L}_2$ norm)	1.7453

After the elements below the threshold $\varepsilon = 0.0001$ was set to zero

Nonzero elements	10.98%
Reciprocal condition number ($\mathcal{L}_1$ norm)	1.42×10^{-05}
Reciprocal condition number ($\mathcal{L}_\infty$ norm)	1.37×10^{-05}
Residue for the transformed solution ($\mathcal{L}_2$ norm)	0.28
Residue for the final solution ($\mathcal{L}_2$ norm)	0.28
Error in the final solution ($\mathcal{L}_2$ norm)	0.18

It is interesting to note that the condition number of the matrix changes by one order of magnitude before and after the application of the discrete wavelet-like transform even though the transform applied is orthogonal (since the filter banks are orthonormal). The only explanation possible is that this change must be due to the numerical errors in the computations. Next, the thresholding operation is applied to the transformed matrix where the elements below a certain value are set equal to zero. The condition number still changes. It is interesting to note that for the smaller thresholds of 0.001, the error in both the residuals and the solution is quite large. These results indicate that a moderate error in the final solution can be obtained for this example with a threshold equal to 0.0001. In that case, only 10% of the elements of the matrix are nonzero.

Figure 4.30 presents the first 250 components of the solution obtained using the LU decomposition. Figure 4.31 plots the first 250 components of the solution obtained with the application of the discrete wavelet-like transform [DCT/DST bank filter with $[R(k)]^{1/2}$ equal in magnitude to that of the Daubechies filter of length 8] with a threshold of $\varepsilon = 0.0001$. Figure 4.32 shows the same results but for a threshold value of $\varepsilon = 0.001$. It is seen that in this case even a 0.1% threshold may be inadequate (the matrix has been diagonally scaled to produce an element equal to one on the diagonal, and therefore the threshold is defined with respect to unity) to accurately compute the solution.

Hence, it appears that this discrete wavelet-like methodology of making a matrix sparse to improve the efficiency of the solution process depends on the type of problems that one may have at hand and it is not possible, in general, to know a priori the approximate quality of the solution. However, this technique does produce a compressed matrix and in most cases it does increase the efficiency of the solution procedure. The quality of the solution indeed depends on the threshold value ε.

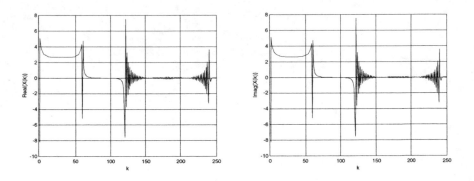

Figure 4.30 Solution obtained with the LU decomposition.

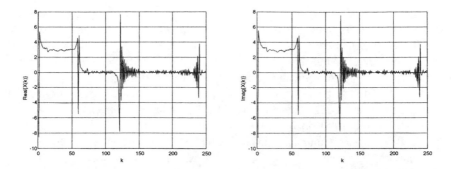

Figure 4.31 Solution obtained using the wavelet-like transform with a threshold of $\varepsilon = 0.0001$.

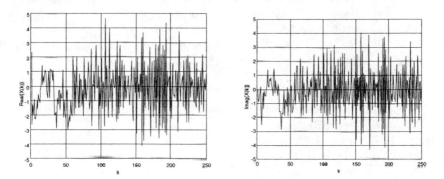

Figure 4.32 Solution obtained using the discrete wavelet-like transform with a threshold of $\varepsilon = 0.001$.

4.6 CONCLUSION

In this chapter, the discrete wavelet-like transform has been used for the solution of matrix equations. Namely, the discrete wavelet transform along with a threshold operation has been utilized to convert large dense complex matrices into sparse matrices. Once the matrices are made sparse so that the number of nonzero elements reduces from N^2 to $\theta(N)$ elements, this large sparse complex matrix can now be stored entirely into a computer's main memory resulting in a significant reduction in the number of page faults when carrying out matrix vector products. In this way the computational efficiency of the solution of large systems of equations can be increased.

REFERENCES

[1] G. Strang and T. Nguyen, *Wavelet and Filter Banks*, Wellesley Cambridge Press, Wellesley, MA, 1996.

[2] I. Daubechies, *Ten Lectures on Wavelets*, CBMS-NSF Regional Conference Series in Applied Mathematics, Philadelphia, SIAM, 1992.

[3] S. G. Mallat, "A theory for multiresolution signal decomposition: The wavelet representation," *IEEE Trans. Pattern Analysis and Machine Intelligence*, Vol. 11, No. 7, July 1990, pp. 674–693.

[4] G. Beylkin, R. Coifman, and V. Rokhlin, "Fast wavelet transforms and numerical algorithms," *Comm. Pure and Applied Math*, Vol. 44, pp. 141–183.

[5] B. Alpert et al., "Wavelet-like base for the fast solution of second-kind integral equations," *SIAM J. Sci. Comput.*, Vol. 14, No. 1, 1993, pp. 159–184.

[6] H. Kim and H. Ling, "On the application of fast wavelet transform to the integral equation solution of electromagnetic scattering problems," *Microwave and Optical Technology Letters*, Vol. 6, No. 3, March 1993, pp. 168–173.

[7] R. L. Wagner and W. C. Chew, "A study of wavelets for the solution of electromagnetic integral equations," *IEEE Trans. Antennas and Propagat.*, Vol. AP-43, No. 8, Aug. 1995, pp. 802–810.

[8] B. Z. Steinberg and Y. Leviatan, "On the use of wavelet expansions in the method of moments," *IEEE Trans. Antennas and Propagat.*, Vol. AP-41, No. 3, March 1993, pp. 610–619.

[9] W. H. Press et al., *Numerical Recipes—The Art of Scientific Computing*, Cambridge University Press, Cambridge, MA, 1992.

[10] A. P. Calderon, "Intermediate spaces and interpolation, the complex method," *Stud. Math.*, Vol. 24, 1964, pp. 113–190.

[11] T. K. Sarkar, *Application of the Conjugate Gradient Method to Electromagnetics and Signal Analysis*, PIER, Vol. 5, Elsevier, New York, 1991.

[12] P. P. Vaidyanathan, *Multirate Systems and Filters Banks*, Prentice Hall, Englewood Cliffs, NJ, 1992.

[13] T. K. Sarkar et al., "A tutorial on wavelets from an electrical engineering perspective, Part I: discrete wavelet techniques," *IEEE Antennas and Propagat. Mag.*, Vol. 40, No. 5, Oct. 1998, pp. 49–70.

[14] S. Schweid and T. K. Sarkar, "Iterative calculating and factorization of the autocorrelation function of orthogonal wavelets with maximal vanishing moments," *IEEE Trans. Circuits and Systems*, Vol. 42, No. 11, Nov. 1995, pp. 694–701.

[15] S. Schweid and T. K. Sarkar, "A sufficiency criteria for orthogonal QMF filters to ensure smooth wavelet decomposition," *Applied and Computational Harmonic Analysis*, Vol. 2, 1995, pp. 61–67.

[16] S. Schweid, "Projection Method in Minimization Techniques for Smooth Multistage QMF Filter Decomposition," Ph.D. thesis, Syracuse University, Syracuse, NY, 1994.

Chapter 5

SOLVING THE DIFFERENTIAL FORM OF MAXWELL'S EQUATIONS

The principles of dilation and shift are two important properties that are attributed to wavelets. In this chapter, we show that inclusion of such properties in the choice of a basis in Galerkin's method can lead to a slow growth of the condition number of the system matrix obtained from the discretization of the differential form of Maxwell's equations. This is because a large portion of the system matrix can be made diagonal. It is shown that for one-dimensional problems the system matrix can be perfectly diagonalized. For two-dimensional problems, however, the system matrix can be made mostly diagonal. The application of this new type of "dilated" basis (which, in principle, is similar to a Fourier basis; the wavelets have shifts in addition) as used in a Galerkin implementation of a finite element method may lead to efficient solution of waveguide problems. Typical numerical results are presented to illustrate the concepts.

5.1 INTRODUCTION

The finite difference [1] and the finite element method [2] have been developed during the last few years in the microwave area for efficient solution of the differential form of Maxwell's equations. Researchers have primarily focused their attention on the development of basis functions for treating boundaries with edges and open region problems, spurious-free solutions of eigenvalue problems and efficient solution of sparse matrix equations.

However, one of the problems with finite difference and finite element methods lies in the solution of a large matrix equation (either a direct solution with several right-hand sides or an eigenvalue problem). The problem here is that as the number of basis functions increases (and hence the dimension and size of the matrix), the condition number of the matrix also increases. The increase of the condition number of the matrix creates various types of solution problems. Typically the condition number of the system matrix grows as $\theta(1/h^2)$ [where $\theta(\bullet)$ denotes "of the order of $1/h^2$," where h is the discretization step]. This is in contrast to the electric field integral equation utilized in the method of moments where the growth of the condition number of the system matrix is $\theta(1/h)$ and for the magnetic field integral equation the growth of the condition number can be made independent of h. The above statements hold as long as the integral equations have a unique solution (that is, the problem is not solved at a frequency corresponding to an internal resonance of the closed structure) [3].

The condition number of a matrix directly dictates the solution procedure because a highly ill-conditioned matrix prohibits application of a direct matrix solver like Gaussian elimination [4, 5], and more sophisticated techniques like singular value decomposition may have to be introduced [6]. There are various ways to halt the increase in the condition number as the dimension of the matrix increases. One such procedure has been outlined by Mikhlin [5]. In [5], the basis functions are chosen in such a way that the growth of the condition number can be controlled. In this chapter, we utilize a particular set of basis functions primarily tailored to rectangular regions for an efficient solution of the resulting matrix equation. This particular choice of the basis is related to the "wavelet" concepts [7–9].

The basic philosophy of this chapter then lies in the choice of a particular set of basis functions (which of course is dependent on the nature of the problem, e.g., TM or TE, and on the particular shape of the domain) which attempts to diagonalize the system matrix that arises when Galerkin's method is applied to the solution of the differential form of Maxwell's equations. The ideal situation will of course be to make the large sparse "Galerkin system matrix" diagonal. Then the solution of such a matrix problem (either solution of the matrix equation due to different right-hand sides or solution of an eigenvalue problem) would be trivial. However, because of various boundary conditions, this goal cannot be achieved. Therefore, the next best procedure is an attempt to make, say, 80% of the system matrix diagonal. For example, in order to obtain the cut-off frequencies of a waveguide, one needs to solve for the eigenvalues of a large matrix equation. However, if the matrix is sparse and "almost" diagonal, then an iterative technique like the conjugate gradient [10–12] can be utilized to converge on the first few eigenvalues in a relatively few iterations to yield the cut-off frequencies. Some numerical examples are presented to illustrate this procedure [13, 14].

5.2   SOLUTION OF ONE-DIMENSIONAL PROBLEMS UTILIZING A WAVELET-LIKE BASIS

In the solution of operator equations, particularly differential equations, the above concepts of dilation and shift in the choice of the hybrid basis functions (a combination of scaling functions and wavelets) could provide some computational advantages. As an example, consider the solution of the one-dimensional differential form of Maxwell's equations, that is,

$$\nabla^2 u(x) = F(x) \quad \text{with} \quad a \leq x \leq b \tag{5.1}$$

where ∇^2 (called the Laplacian operator) $= d^2 / dx^2$ in one dimension and u is the unknown to be solved for the given excitation F. The boundary conditions are left undefined at this point because they can be either of Dirichlet type [either homogeneous, i.e., $u(a)$ or $u(b) = 0$, or inhomogeneous, $u(a) = A$ and $u(b) = B$], Neumann type [either homogeneous, i.e., du/dx (at $x = a$ and $b) = 0$ or inhomogeneous du/dx $(x = a) = C$ and du/dx $(x = b) = D$], or mixed type (i.e., Dirichlet on one part of the boundary and Neumann on the other). The development is independent of the nature of the boundary conditions. However, the boundary conditions are needed for the complete solution of the problem.

Galerkin's method is now used to solve (5.1), which gives the fundamental equations of the finite element method. Hence consider the weighting function $v(x)$, which multiplies both sides of (5.1). Then the product is integrated by parts from a to b to yield

$$-\int_a^b \frac{du}{dx}\frac{dv}{dx}dx - \frac{du(x)}{dx}\bigg|_{x=a} v(a) + \frac{du(x)}{dx}\bigg|_{x=b} v(b) = \int_a^b v(x)\,F(x)\,dx \qquad (5.2)$$

Next, it is assumed that the unknown $u(x)$ can be represented by a complete set of basis functions $\phi_i(x)$, which has first-order differentiability. Then

$$u(x) \approx u_N(x) = \sum_{i=1}^N a_i\,\phi_i(x) + a_{01}\,\phi_{01}(x) + a_{02}\phi_{02}(x) \qquad (5.3)$$

where a_i and a_{0j} are the unknowns to be solved. Basically, the functions $\phi_i(x)$ satisfy homogeneous boundary conditions and the functions ϕ_{0j}, for $j = 1$ and 2, take care of the inhomogeneous Dirichlet conditions on the boundaries.

In a Galerkin's procedure the weighting functions $v(x)$ are of the form

$$v(x) \equiv \phi_j(x)\,;\,\phi_{01}(x)\,;\,\phi_{02}(x) \qquad (5.4)$$

Substitution of (5.3) and (5.4) into (5.2) results in a system of equations that can be written in the following matrix form:

$$
\begin{bmatrix}
<\phi_1';\phi_1'> & \cdots & <\phi_1';\phi_n'> & <\phi_1';\phi_{01}'> & <\phi_1';\phi_{02}'> \\
<\phi_2';\phi_1'> & \cdots & <\phi_2';\phi_n'> & <\phi_2';\phi_{01}'> & <\phi_2';\phi_{02}'> \\
\vdots & & \vdots & \vdots & \vdots \\
<\phi_n';\phi_1'> & \cdots & <\phi_n';\phi_n'> & <\phi_n';\phi_{01}'> & <\phi_n';\phi_{02}'> \\
\cdots & \cdots & \cdots\cdots & \cdots\cdots & \cdots\cdots \\
<\phi_{01}';\phi_1'> & \cdots & <\phi_{01}';\phi_n'> & <\phi_{01}';\phi_{01}'> & <\phi_{01}';\phi_{02}'> \\
<\phi_{02}';\phi_1'> & \cdots & <\phi_{02}';\phi_n'> & <\phi_{02}';\phi_{01}'> & <\phi_{02}';\phi_{02}'>
\end{bmatrix}
\begin{bmatrix}
a_1 \\ a_2 \\ \vdots \\ a_n \\ \cdots \\ a_{01} \\ a_{02}
\end{bmatrix}
=
$$

$$
= -\begin{bmatrix} <F;\phi_1> \\ <F;\phi_2> \\ \vdots \\ <F;\phi_n> \\ \cdots\cdots\cdots \\ <F;\phi_{01}> \\ <F;\phi_{02}> \end{bmatrix} + \begin{bmatrix} 0 \\ \vdots \\ 0 \\ \cdots \quad \cdots \quad \cdots \quad \cdots \\ -\dfrac{du(x)}{dx}\bigg|_{x=a} v(a) \\ \dfrac{du(x)}{dx}\bigg|_{x=b} v(b) \end{bmatrix} \tag{5.5}
$$

or equivalently

$$
Z A = Y \tag{5.6}
$$

where the superscript $'$ denotes the first derivative of the function and $<c;d>$ denotes the classical Hilbert inner product and the norm, defined by

$$
<c;d> \equiv \int_a^b c(x)\overline{d}(x)\, dx
$$

$$
\|c\|^2 \equiv <c;c> \equiv \int_a^b c(x)\overline{c}(x)\, dx \tag{5.7}
$$

where the overbar above a variable represents the complex conjugate. The solution of (5.5) then provides the unknowns a_i, a_{01}, and a_{02}. The crux of the problem therefore lies in the solution of large matrix equations. The stability of the solution of large systems is dictated by the condition number of the matrix and by the number of effective bits t with which the solution is carried out on the computer. Specifically, in the solution of (5.6) if ΔZ is the error in the representation of Z, the error in the solution ΔX is bounded by [11]

$$
\frac{\|\Delta X\|}{\|X\|} \le \frac{cond\,(Z)}{1 - \sqrt{N}\,cond(Z)2^{-t}} \left[\frac{\|\Delta Z\|}{\|Z\|}\right] \tag{5.8}
$$

where N is the dimension of the matrix Z and the norm $\|\bullet\|$ is defined as the Euclidian norm in (5.7). Also $2^t > \sqrt{N} \times cond\,(Z)$. It is therefore clear that the choice of the basis functions, which determine the condition number of the matrix Z, has a tremendous influence on the efficiency and accuracy of the solution of (5.6). The problem with the finite element method lies in the solution of a large matrix equation. Also, as the number of basis functions increases, the condition number of the matrix also increases. An increase in the condition number of the matrix creates various types of numerical problems associated with the numerical solution of these matrix equations. For example, the condition number directly dictates the solution procedure, because a highly ill-conditioned matrix prohibits application of a direct matrix solver like Gaussian elimination and a more sophisticated technique like singular value decomposition may have to be introduced. There are various ways to eliminate the increase of the condition number as the dimension of the matrix increases. One way is to choose the basis functions such that the growth of the condition number is controlled.

A good way to choose the basis functions is shown in Figure 5.1. It is interesting to point out that these basis functions are similar to the classical triangular functions used in the method of moments. However, unlike in the method of moments, these basis functions are not subdomain basis functions. In the classical subdomain basis functions the choice would be the seven piecewise triangle functions as shown in Figure 5.1(e) and the two half triangles at the end. The nine basis functions would consist of the four solid line triangular functions and also the five dotted lines triangular functions shown in Figure 5.1(e). In the new basis, which we call the hybrid wavelet basis, the expansion functions are shown in Figure 5.1 marked by ϕ_{01}, ϕ_{02}, and ϕ_{1-7}. The difference is that instead of the three dotted triangular basis functions we have three different nested basis functions ϕ_1, ϕ_2, and ϕ_3, as shown in Figures 5.1(c) and (d), which in the finite element literature are called hierarchical basis functions. The functions ϕ_{01}, ϕ_{02} shown in Figures 5.1(a) and (b) are there to treat arbitrary boundary conditions. The basis functions shown by the solid lines in Figure 5.1 are termed the "wavelet" basis, as they are the dilated and shifted version of the same function. These basis functions are derived from the Battle-Lamarie type of wavelets. Here the basis is chosen using the scaling functions ϕ and not the wavelet functions ψ. The natural question that arises is, what is the advantage of this type of wavelet basis over the conventional subsectional basis functions? The disadvantage of the wavelet basis is clear, for example, for ϕ_1, ϕ_2, and ϕ_3, more calculations need to be carried out over the domain of interest as opposed to the three dotted triangular basis functions shown for the classical subsectional basis. This is because the inner products need to be carried out over the entire domain. However, for the final solution, both the subsectional and the wavelet-type basis provide the same information content about the approximation.

The distinct feature of the present approach is that a different choice of basis functions has been made for the unknown. One choice involves using the entire domain basis while the other one uses a subsectional basis. In either case, the span of the basis function is complete in the solution space. We now proceed with the solution procedure to illustrate the other salient features of using an entire domain basis.

Under the new basis functions we have

$$< \phi_i'; \phi_j' > = 0; \quad i \neq j \tag{5.9}$$

$$< \phi_i'; \phi_{01}' > = 0 \tag{5.10}$$

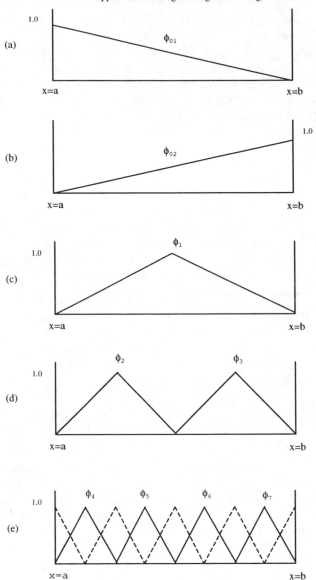

Figure 5.1 The basis functions: (a) ϕ_{01}, (b) ϕ_{02}, (c) ϕ_1, (d) ϕ_2 and ϕ_3, and (e) nine piecewise subsectional triangle functions.

$$< \phi_i' ; \phi_{02}' > = 0 \qquad (5.11)$$

Because of the proper choice of wavelet basis, we have (5.9)–(5.11) and therefore (5.5) reduces to

$$
\begin{bmatrix}
<\phi_1', \phi_1'> & \cdots & 0 & \vdots & & & \\
& \ddots & & \vdots & & [0] & \\
0 & \cdots & <\phi_n', \phi_n'> & \vdots & & & \\
\cdots & \cdots & \cdots & \vdots & \cdots & \cdots & \\
& [0] & & \vdots & <\phi_{01}', \phi_{01}'> & <\phi_{01}', \phi_{02}'> \\
& & & \vdots & <\phi_{02}', \phi_{01}'> & <\phi_{02}', \phi_{02}'>
\end{bmatrix}
\begin{bmatrix}
a_1 \\
\vdots \\
a_n \\
\cdots \\
a_{01} \\
a_{02}
\end{bmatrix}
$$

$$
= -
\begin{bmatrix}
<F; \phi_1> \\
\vdots \\
<F; \phi_n> \\
\cdots \\
<F; \phi_{01}> \\
<F; \phi_{02}>
\end{bmatrix}
+
\begin{bmatrix}
0 \\
\vdots \\
0 \\
\cdots \\
-\left.\dfrac{du(x)}{dx}\right|_{x=a} v(a) \\
\left.\dfrac{du(x)}{dx}\right|_{x=b} v(b)
\end{bmatrix}
\tag{5.12}
$$

In the solution of (5.12) the boundary conditions of the problem are implicitly provided. For example, if the boundary conditions are purely Dirichlet type, then a_{01} and a_{02} are known. For the Neumann condition, the derivatives of u at $x = a$ and $x = b$ are known, and the values for a_{0i} need to be solved. For mixed boundary conditions a combination of the above are required and a_{0j} needs to be solved at the jth boundary, which has a specified Neumann condition. Therefore, from (5.12) we have

$$
a_i = -\frac{<F; \phi_i>}{<\phi_i'; \phi_i'>}
\tag{5.13}
$$

$$
\begin{bmatrix}
<\phi_{01}'; \phi_{01}'> & <\phi_{01}'; \phi_{02}'> \\
<\phi_{02}'; \phi_{01}'> & <\phi_{02}'; \phi_{01}'>
\end{bmatrix}
\begin{bmatrix}
a_{01} \\
a_{02}
\end{bmatrix}
= -
\begin{bmatrix}
<F; \phi_{01}> \\
<F; \phi_{02}>
\end{bmatrix}
+
\begin{bmatrix}
-\left.\dfrac{du(x)}{dx}\right|_{x=a} v(a) \\
\left.\dfrac{du(x)}{dx}\right|_{x=b} v(b)
\end{bmatrix}
\tag{5.14}
$$

where two of the four parameters a_{01}, a_{02}, $\left.\dfrac{du(x)}{dx}\right|_{x=a}$, and $\left.\dfrac{du(x)}{dx}\right|_{x=b}$ have been fixed by the boundary conditions (Dirichlet, Neumann, or mixed) of the problem. The application of the wavelet-like basis is now clear. For one-dimensional problems, the system matrix can be made almost diagonal, except the terms containing the information on the boundary and hence its solution is nearly trivial.

In summary, if the dilation and shift principles of wavelets are utilized in the choice of basis functions such that their first derivatives are orthogonal to each other, then the system matrix corresponding to the unknowns can be diagonalized. Therefore, the choice of a wavelet-type basis makes the system matrix almost diagonal, simplifying the computational complexity. The basis functions in this case are neither the complete wavelet basis consisting of the father, mother, and the daughters, nor of the sons completely, as outlined in Chapter 3, but are instead hybrid in nature.

We now extend this procedure to the solution of two-dimensional structures and apply it to some waveguide problems.

5.3   SOLUTION OF $\nabla^2 U + k^2 U = F$ FOR TWO-DIMENSIONAL PROBLEMS UTILIZING A WAVELET-LIKE BASIS

Consider the solution of

$$\nabla^2 u(x, y) + k^2 u(x, y) = F(x, y) \qquad (5.15)$$

restricted to the two-dimensional region R defined by the contour Γ. In this section we have focused our attention only on rectangular regions and have assumed that any arbitrarily shaped region R is approximated by rectangular regions R of the type: $R : 0 \leq x \leq a;\ 0 \leq y \leq b$. To apply Galerkin's method, one integrates (5.15) in the domain R to obtain

$$\int_R v(x, y) \nabla^2 u(x, y)\ dx\,dy + k^2 \int_R u(x, y) v(x, y)\ dx\,dy = \int_R F(x, y) v(x, y)\ dx\,dy$$

$$(5.16)$$

which, after integration by parts, yields

$$- \int_R (\nabla u)(\nabla v)\ dr + \int_\Gamma v \frac{du}{dn}\ d\Gamma + k^2 \int_R uv\ dr = \int_R Fv\,dr \qquad (5.17)$$

where the incremental region dr stands for $dx\,dy$.

Generation of a wavelet-type basis in two dimensions can be done by utilizing the multiresolution analysis and essentially following the dyadic product of two one-dimensional constructions. Another method consists of obtaining the wavelets by utilizing the tensor product of two one-dimensional wavelet-like basis functions.

The development of a two-dimensional wavelet-type basis is carried out using the usual tensor product of the one-dimensional basis. Therefore, the three wavelets (ψ^1, ψ^2, and ψ^3) used in two dimensions can be generated from the following tensor products:

$$\psi^1(x, y) = \psi(x) \ \phi(y) \tag{5.18a}$$

$$\psi^2(x, y) = \phi(x) \ \psi(y) \tag{5.18b}$$

$$\psi^3(x, y) = \psi(x) \ \psi(y) \tag{5.18c}$$

where the significance of the scaling ϕ and wavelet ψ functions are explained in Chapter 2 [8, 9].

Let us approximate $u(x,y)$ on a quadrangle by

$$u(x, y) \approx \hat{u} = \sum_{i=1}^{M} \sum_{j=1}^{N} A_{ij}\phi_{ij}(x, y) + \sum_{i=1}^{4} \sum_{j=1}^{K} B_{ij}N_{ij}(x, y) + \sum_{i=1}^{4} C_i T_i(x, y) \tag{5.19}$$

where N_{ij} are the basis functions related to the four edges (N_{ij}, *for* $i = 1, 2, 3, 4$); the wavelet basis functions T_i ($i = 1, 2, 3, 4$) provide the matching conditions needed at the four vertices V_1, V_2, V_3, and V_4; and A_{ij}, B_{ij}, and C_i are the unknown constants to be solved for. In this expansion, the $\phi_{ij}(x,y)$ are zero on the rectangular boundary, and they are explicitly chosen as

$$\phi_{ij}(x, y) = \eta_i(x) \ \eta_j(y) = \sin\left(\frac{\pi i x}{a}\right)\sin\left(\frac{\pi j y}{b}\right) \tag{5.20}$$

These functions are not only orthogonal to themselves, but their partial derivatives are also orthogonal, that is,

$$< \phi_{ij} ; \phi_{pq} > = 0 \quad \text{for } i \neq p; \ j \neq q \tag{5.21}$$

$$< \nabla\phi_{ij} ; \nabla\phi_{pq} > = 0 \quad \text{for } i \neq p; \ j \neq q \tag{5.22}$$

$$\int_{\Gamma} \phi_{ij} \ \frac{\partial}{\partial n} \ \phi_{pq} \ d\Gamma = 0 \tag{5.23}$$

where the inner product in the two-dimensional rectangular region is defined as

$$< c ; d > \equiv \int_{0}^{a} dx \int_{0}^{b} dy \ c(x, y) \overline{d}(x, y) \tag{5.24}$$

The overbar denotes the complex conjugate. In addition, the edge basis functions N_{ij} are zero everywhere on the boundary (i.e., on all edges) except on edge E_i (refer to Figure 5.2). The two-dimensional basis has the representation

$$N_{1j}(x,y) = G(y)\eta_j(x)$$

$$N_{2j}(x,y) = H(x)\eta_j(y)$$

$$N_{3j}(x,y) = H(y)\eta_j(x) \tag{5.25}$$

$$N_{4j}(x,y) = G(x)\eta_j(y)$$

where the polynomials $G(x)$, $G(y)$, $H(x)$, and $H(y)$ are chosen based on the differentiability conditions. In this case,

$$G(x) = 1 - \frac{x}{a}$$

$$G(y) = 1 - \frac{y}{b}$$

$$H(x) = \frac{x}{a} \tag{5.26}$$

$$H(y) = \frac{y}{b}$$

Therefore, for example, N_{2j} only participates in providing the match corresponding to edge E_2.

In an analogous fashion, one can illustrate that the wavelet-like basis T_i in (5.19) provides the matching conditions needed for the four vertices V_1, V_2, V_3, and V_4 as shown in Figure 5.2. Specifically, the wavelet-like basis associated with the four vertices can be written as

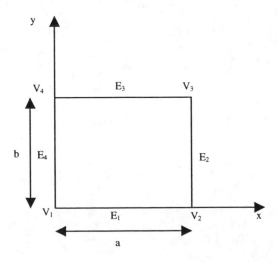

Figure 5.2 Geometry for the two-dimensional basis functions.

$$T_1(x, y) = G(x) G(y)$$
$$T_2(x, y) = H(x) G(y)$$
$$T_3(x, y) = H(x) H(y) \qquad (5.27)$$
$$T_4(x, y) = G(x) H(y)$$

Substituting (5.20), (5.25), and (5.27) in (5.19) and (5.17) and using the functions ϕ_{ij}, N_{ij}, and T_i as weighting functions in (5.17) results in a system matrix given by

$$[P][A] + k^2 [Q][A] = [V_F] + [V_B] \qquad (5.28)$$

where the system matrices $[P]$ and $[Q]$ have the form

$$[P]; [Q] = \begin{bmatrix} [D]_{N \times N} & [R]^T_{N \times M} \\ [R]_{M \times N} & [B]_{M \times M} \end{bmatrix} \qquad (5.29)$$

where $[D]$ is a diagonal matrix and $[R]$ and $[B]$ are sparse matrices. Here the superscript T stands for the transpose of a matrix. The system matrices again would be mostly diagonal. What percentage of the matrix is diagonal depends on how many rectangular regions the original region has been divided into and on the nature of the boundary conditions.

Case A: Dirichlet Boundary Conditions

If the boundary condition over Γ is purely Dirichlet, then the maximum dimension of the system matrix will be $L(N^2 + 4N + 4)$ where the original domain has been subdivided into L regions and the highest orders of approximation M, N, and K in (5.19) are assumed to be the same, let us say, equal to N; that is, they are considered to be the same in all L regions for comparison purposes.

Because of the choice of the wavelet-type basis, out of the maximum dimension of $L(N^2 + 4N + 4)$, the rank of the diagonal submatrix $[D]$ in (5.29) will be LN^2. This clearly demonstrates that as the number of unknowns N increases, the majority of the system matrix becomes diagonal. This is because the row size increase of matrices $[P]$ and $[Q]$ is dominated by the term LN^2 and so is the row size of the diagonal matrix $[D]$. The rectangular submatrix $[R]$ has the row dimension as M (the number of internal edges + number of internal corners) and the column dimension is LN^2. The square matrix $[B]$ has a row and column dimension of M (the number of internal edges + number of internal corners). Hence the size of $[B]$ goes up essentially as $\theta[(L + 1)N]$. Therefore the computational complexity goes up as $\theta[\{(L + 1)N\}^3]$ when the number of unknowns goes up by LN^2. This amounts to a significant decrease in the reduction of computational complexity as opposed to $N^3 M^3$.

Case B: Neumann Boundary Conditions

For this case, the diagonal submatrix is of the same size as that for the previous case of
Dirichlet boundary conditions. However, now the coefficients of all the matching
functions are unknowns. Hence, the size of the system matrix $[P]$ is $LN^2 + M$. The size of
the diagonal matrix $[D]$ is the same as before, that is, LN^2.

Case C: Mixed Boundary Conditions

It is easy to extrapolate the results to a mixed Dirichlet and Neumann condition. The
important point is that, due to the choice of the wavelet basis, a major portion of the
system matrix $[P]$ and $[Q]$ has been made diagonal.

5.4   APPLICATION TO SOME WAVEGUIDE PROBLEMS

As examples, consider the solution of cut-off frequencies of the transverse electric (TE)
and transverse magnetic (TM) modes of various conducting waveguide structures.
Therefore, in this case the objective is to solve an eigenvalue problem, $F = 0$. Therefore,
we are solving for

$$\nabla^2 u + k_c^2 u = 0 \qquad (5.30)$$

and the system matrix equation is

$$[P][A] + k_c^2[Q][A] = 0 \qquad (5.31)$$

The objective is to solve this generalized eigenvalue problem for the eigenvalue k_c^2 and
the eigenvectors $[A]$. Since $[P]$ and $[Q]$ are sparse, and mostly diagonally banded
matrices, the computational complexity has been greatly reduced and the conjugate
gradient method can be used efficiently to find a few of the generalized eigenvalues k_c^2
and the eigenvector associated with it.

As a first example consider the solution of the cut-off frequencies of the various
TE and TM modes of a rectangular waveguide of dimension 2 cm × 1 cm. Here we
could have used one region ($L = 1$) to solve the problem and increase N. But to
illustrate the flexibility and accuracy of the procedure we divide the rectangular region
into two regions $A^{(1)} = 1.3 \times 1$ and $A^{(2)} = 0.7 \times 1$ as shown in Figure 5.3. The basis
chosen are the same as in Section 5.2. Table 5.1 presents the cut-off wavenumbers of
the TM_{mn} mode as computed with this procedure and compared with the exact solution.
The * indicates that the order was not sufficient to perform reliable computation for the
modes for a rectangular waveguide with better than 1% accuracy. The exact solution is
obtained from [15]. Table 5.2 indicates the percent of the matrix that is diagonal. The
computational efficiency of the new basis now becomes clear. For this case, 95% of
the matrix is diagonal when a large number of unknowns are taken and as the number of

unknown increases, so does the size of the diagonal matrix. Table 5.3 presents the cutoff wave numbers for the TE_{mn} modes in a rectangular waveguide as computed with this procedure and compared with the exact solution. Table 5.4 shows that as the number of unknowns increases, so does the size of the diagonal block maintaining the computational efficiency. For the same value of N, the size of the matrix is different for the TE case as opposed to the TM case because many of the boundary terms go to zero for the TM case but not for the TE case.

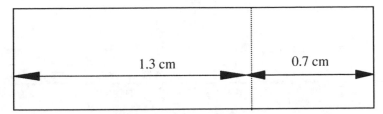

Figure 5.3 A rectangular waveguide.

Table 5.1
Cut-Off Wavenumbers of the TM_{mn} Modes

	TM_{11}	TM_{21}	TM_{31}	TM_{12}	TM_{41} & TM_{22}	TM_{32}	TM_{51}	TM_{42}
N=1	3.52	4.45	*	*	*	*	*	*
N=2	3.51	4.45	5.66	6.48	*	*	*	*
N=3	3.51	4.44	5.66	6.48	7.05	7.85	*	*
N=4	3.51	4.44	5.66	6.48	7.04	7.85	8.51	8.90
N=5	3.51	4.44	5.66	6.48	7.03	7.85	8.48	8.89
N=6	3.51	4.44	5.66	6.48	7.03	7.85	8.47	8.89
N=7	3.51	4.44	5.66	6.48	7.03	7.85	8.47	8.89
N=8	3.51	4.44	5.66	6.48	7.03	7.85	8.46	8.89
N=9	3.51	4.44	5.66	6.48	7.03	7.85	8.46	8.89
N=10	3.51	4.44	5.66	6.48	7.03	7.85	8.46	8.89
EXACT	3.51	4.44	5.66	6.48	7.02	7.85	8.46	8.89

*Order was not sufficient to obtain reliable computations.

Table 5.2
Percentage of the Matrix That Is Diagonal for the TM Modes

	Matrix size	Size of the diagonal block	% Diagonal
N=1	3	2	66.7%
N=2	10	8	80%
N=3	21	18	85.7%
N=4	36	32	88.9%
N=5	55	50	90.9%
N=6	78	72	92.3%
N=7	105	98	93.3%
N=8	136	128	94.1%
N=9	171	162	94.7%
N=10	210	200	95.2%

Table 5.3
Cut-Off Wavenumbers of the TE_{mn} Modes

	TE_{10}	TE_{20} & TE_{01}	TE_{11}	TE_{21}	TE_{30}	TE_{31}	TE_{40} & TE_{02}	TE_{12}
N=1	1.59	*	*	*	*	*	*	*
N=2	1.57	3.16	3.53	4.46	*	*	*	*
N=3	1.57	3.15	3.53	4.46	4.75	5.70	*	*
N=4	1.57	3.14	3.52	4.45	4.74	5.69	6.32	6.53
N=5	1.57	3.14	3.51	4.44	4.72	5.67	6.30	6.49
N=6	1.57	3.14	3.51	4.44	4.72	5.67	6.30	6.49
N=7	1.57	3.14	3.51	4.44	4.72	5.67	6.29	6.48
N=8	1.57	3.14	3.51	4.44	4.72	5.67	6.29	6.48
N=9	1.57	3.14	3.51	4.44	4.71	5.66	6.29	6.48
N=10	1.57	3.14	3.51	4.44	4.71	5.66	6.29	6.48
EXACT	1.57	3.14	3.51	4.44	4.71	5.66	6.28	6.48

*Order was not sufficient to obtain reliable computations.

Table 5.4
Percentage of the Matrix That Is Diagonal for the TE Modes

	Matrix size	Size of the diagonal block	% Diagonal
N=1	15	2	13.3%
N=2	28	8	28.6%
N=3	45	18	40%
N=4	66	32	48.5%
N=5	91	50	54.9%
N=6	120	72	60%
N=7	153	98	64%
N=8	190	128	67.4%
N=9	231	162	70.1%
N=10	276	200	72.5%

As a second example we consider an L-shaped waveguide as shown in Figure 5.4. The largest dimensions are all 1 cm. The structure has been subdivided into three subregions. This problem has been solved using a finite difference approximation [16] and an integral equation approach [17]. The results produced by this new approach are accurate and the convergence is very fast. Table 5.5 provides the cut-off wavenumbers for the first few dominant TM modes for the L-shaped waveguide. Table 5.6 shows that, as usual, as the number of unknowns increases, so does the size of the diagonal block. Table 5.7 provides the cut-off wavenumbers for the first few TE modes of the L-shaped waveguide. Again the percentage of the matrix that is diagonal increases consistently with the number of unknowns as shown in Table 5.8.

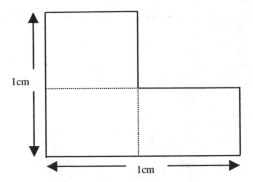

Figure 5.4 An L-shaped waveguide.

Table 5.5
Cut-Off Wavenumbers for the TM Modes of an L-Shaped Waveguide

	Mode 1	Mode 2	Mode 3	Mode 4	Mode 5	Mode 6	Mode 7
N=1	4.95	6.18	*	*	*	*	*
N=2	4.92	6.15	7.00	8.63	*	*	*
N=3	4.91	6.14	7.00	8.58	8.94	10.2	10.6
N=4	4.90	6.14	7.00	8.57	8.92	10.2	10.6
N=5	4.90	6.14	7.00	8.56	8.91	10.1	10.6
N=6	4.90	6.14	7.00	8.56	8.91	10.1	10.6
N=7	4.89	6.14	7.00	8.56	8.90	10.1	10.6
N=8	4.89	6.14	7.00	8.56	8.90	10.1	10.6
Ref. [14]	4.80	6.07	6.92	8.61			
Ref. [15]	4.87	6.13	6.99	8.55			

*Order was not sufficient to obtain reliable computations.

Table 5.6
Percentage of the Matrix That Is Diagonal for the TM Case

	Dimension of system matrix	Dimension of the diagonal block	% Diagonal
N=1	5	32	60%
N=2	16	12	75%
N=3	33	27	81.8%
N=4	56	48	85.7%
N=5	85	75	88.2%
N=6	120	108	90%
N=7	161	147	91.3%
N=8	208	192	92.3%

Table 5.7
Cut-Off Wavenumbers for the TE Modes of an L-Shaped Waveguide

	Mode 1	Mode 2	Mode 3	Mode 4	Mode 5	Mode 6
N=1	1.89	2.92	*	*	*	*
N=2	1.88	2.90	4.85	5.21	5.49	6.87
N=3	1.88	2.89	4.85	5.21	5.48	6.87
N=4	1.87	2.89	4.84	5.20	5.47	6.84
N=5	1.87	2.89	4.84	5.20	5.46	6.84
N=6	1.87	2.89	4.83	5.20	5.46	6.84
N=7	1.87	2.89	4.83	5.20	5.46	6.84
N=8	1.87	2.89	4.83	5.20	5.46	6.84
N=9	1.87	2.89	4.83	5.20	5.46	6.48
Ref. [14]	1.88	2.95	4.89	5.26	5.49	6.91
Ref. [15]	1.89	2.91	4.87	5.24		

*Order was not sufficient to obtain reliable computations.

Table 5.8
Percentage of the Matrix That Is Diagonal for the TE Case

	Dimension of system matrix	Dimension of the diagonal block	% Diagonal
N=1	21	3	14.3%
N=2	40	12	30%
N=3	65	27	41.5%
N=4	96	48	50%
N=5	133	75	56.4%
N=6	176	108	61.4%
N=7	225	147	65.3%
N=8	280	192	68.6%
N=9	341	243	71.3%

As a third example, consider the vaned rectangular waveguide shown in Figure 5.5. Table 5.9 provides the cut-off wavenumbers of the first few dominant TE modes. The results have been compared to that of finite difference solution technique [16] and an integral equation technique [17]. Again, as the number of unknowns increases, the majority of the system matrix is diagonal as shown in Table 5.10 and hence the computational efficiency increases. Finally, Table 5.11 provides the cut-off wavenumbers of TM modes of the vaned rectangular waveguide and Table 5.12 shows the size of the system matrix that is diagonal.

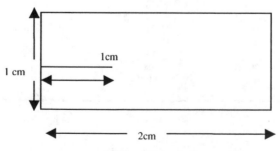

Figure 5.5 A vaned waveguide.

Table 5.9
Cut-Off Wavenumbers of the TE Modes of a Vaned Rectangular Waveguide

	Mode 1	Mode 2	Mode 3	Mode 4	Mode 5	Mode 6
N=1	1.60	2.21	3.32	3.55	4.58	*
N=2	1.57	2.17	3.17	3.31	4.26	5.41
N=3	1.57	2.15	3.15	3.30	4.26	4.75
N=4	1.57	2.14	3.15	3.30	4.25	4.74
N=5	1.57	2.13	3.14	3.30	4.25	4.72
N=6	1.57	2.12	3.14	3.30	4.25	4.72
N=7	1.57	2.12	3.14	3.30	4.25	4.72
N=8	1.57	2.12	3.14	3.30	4.25	4.72
N=9	1.57	2.12	3.14	3.30	4.25	4.71
N=10	1.57	2.12	3.14	3.30	4.25	4.71
Ref. [14]	1.57	2.00	3.13	3.28	4.23	4.66
Ref. [15]	1.57	2.11	3.16	3.30		

*Order was not sufficient to obtain reliable computations.

Table 5.10
Percentage of the Matrix That Is Diagonal for the TE Case

	Dimension of system matrix	Dimension of the diagonal block	% Diagonal
N=1	27	4	14.8%
N=2	52	16	30.8%
N=3	85	36	42.3%
N=4	126	64	50.8%
N=5	175	100	57.1%
N=6	232	144	62.1%
N=7	297	196	66%
N=8	370	256	69.2%
N=9	451	324	71.8%
N=10	540	400	74%

Table 5.11
Cut-Off Wavenumbers of the TM Modes of a Vaned Rectangular Waveguide

	Mode 1	Mode 2	Mode 3	Mode 4	Mode 5	Mode 6
N=1	3.74	*	*	*	*	*
N=2	3.72	5.05	6.48	*	*	*
N=3	3.71	5.03	6.48	6.59	7.03	7.87
N=4	3.71	5.01	6.48	6.55	7.03	7.80
N=5	3.71	5.01	6.48	6.53	7.03	7.78
N=6	3.70	5.00	6.48	6.52	7.02	7.76
N=7	3.70	5.00	6.48	6.51	7.02	7.76
N=8	3.70	4.99	6.48	6.50	7.02	7.76
N=9	3.70	4.99	6.48	6.50	7.02	7.75
N=10	3.70	4.99	6.48	6.50	7.02	7.75
Ref. [14]	3.65	4.87	6.31			

*Order was not sufficient to obtain reliable computations.

Table 5.12
Percentage of the Matrix That Is Diagonal for the TM Case

	Dimension of system matrix	Dimension of the diagonal block	% Diagonal
N=1	27	10	37.0%
N=2	52	27	51.9%
N=3	85	52	61.2%
N=4	126	85	67.5%
N=5	175	126	72.0%
N=6	232	175	75.4%
N=7	297	232	78.1%
N=8	370	297	80.3%
N=9	451	370	82.0%
N=10	540	451	83.5%

For the fourth example, consider the solution of the TM modes of a single ridge waveguide as shown in Figure 5.6. The values of the parameters in Figure 5.6 are $a = 1.0$m, $b = 0.5$m, $s = 0.5$m, and $d = 0.25$m. The single ridge waveguide was divided into five rectangular regions as shown by the dotted lines. Table 5.13 provides the cut-off wavenumbers of the TM modes of the single ridge waveguide. The dots indicate that the results are not available. Table 5.14 shows the percentage of the system matrix that is diagonal. Table 5.15 provides the cut-off wavenumbers of the TE modes of the same single ridge waveguide, and Table 5.16 shows the percentage of the system matrix that is diagonal. Comparison of these results with data published in the literature utilizing a finite difference technique [16] and a surface integral equation [17] shows that the new approach is quite accurate in comparison with other techniques described in [16, 17] and provides fast convergence.

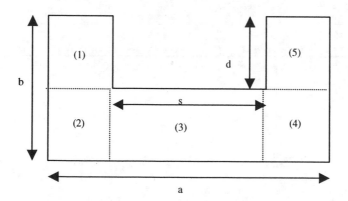

Figure 5.6 A single ridge waveguide.

Table 5.13
Cut-Off Wavenumbers of the TM Modes of a Single Ridge Waveguide

	Mode 1	Mode 2	Mode 3	Mode 4	Mode 5
N=2	12.20	12.5	14.02	15.62	16.67
N=4	12.16	12.45	14.02	15.60	16.66
N=6	12.15	12.44	14.01	15.60	16.66
N=8	12.14	12.43	14.01	15.59	16.65
N=10	12.05	12.43	14.01	15.59	16.65
Ref. [14]	12.05	12.32	13.86	15.34	16.28
Ref. [15]	12.04	12.29	14.00	15.99	*

*Results not available.

Table 5.14
Percentage of the Matrix That Is Diagonal for the TM Modes

	Matrix size	Row size of the diagonal block	% Diagonal
N=2	28	20	71.43%
N=4	96	80	83.33%
N=6	204	180	88.24%
N=8	312	320	90.91%
N=10	540	500	92.59%

Table 5.15
Cut-Off Wavenumbers of the TE Modes of a Single Ridge Waveguide

	Mode 1	Mode 2	Mode 3	Mode 4	Mode 5
N=2	2.26	4.90	6.48	7.53	9.85
N=4	2.25	4.87	6.46	7.52	9.83
N=6	2.25	4.87	6.46	7.52	9.83
N=8	2.25	4.87	6.46	7.72	9.83
N=10	2.25	4.86	6.46	7.52	9.83
Ref. [14]	2.23	4.78	6.40	7.48	9.71
Ref. [15]	2.25	4.94	6.52	7.56	*

*Results not available.

As a final example consider the coaxial rectangular waveguide shown in Figure 5.7, where $a = 1.25$m, $b = 1.0$m, $s = 0.25$m, and $d = 0.25$m. The waveguide has been divided into eight regions. Table 5.17 provides the cut-off wavenumbers of the TM modes of this waveguide. Table 5.18 shows the percentage of the system matrix that is diagonal as the order of the approximation increases. Table 5.19 provides the cut-off wavenumbers for the TE modes, and Table 5.20 shows the percentage of the system matrix that is diagonal as the order of the system increases. Again, good agreement has been obtained with other published results. Here, the results have been compared with those of [16].

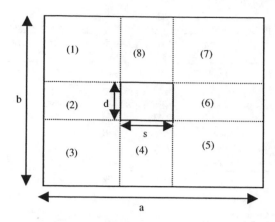

Figure 5.7 A coaxial waveguide.

Table 5.16
Percentage of the Matrix That Is Diagonal for the TE Modes

	Matrix size	Row size of the diagonal block	% Diagonal
N=2	64	20	31.25%
N=4	156	80	51.28%
N=6	288	180	62.50%
N=8	460	320	69.57%
N=10	672	500	74.40%

Table 5.17
Cut-Off Wavenumbers of the TM Modes of a Coaxial Rectangular Waveguide

	Mode 1	Mode 2	Mode 3	Mode 4	Mode 5
N=2	6.95	6.96	8.68	8.72	10.97
N=4	6.94	6.95	8.66	8.70	10.94
N=6	6.94	6.95	8.66	8.70	10.92
N=8	6.94	6.95	8.66	8.70	10.92
N=10	6.94	6.95	8.66	8.70	10.92
Ref. [14]	6.91	6.96	8.5	8.51	10.57

Table 5.18
Percentage of the Matrix That Is Diagonal for the TM Modes

	Matrix size	Row size of the diagonal block	% Diagonal
N=2	48	32	66.67%
N=4	160	128	80.00%
N=6	336	228	85.71%
N=8	576	572	88.89%

Table 5.19
Cut-Off Wavenumbers of the TE Modes of a Coaxial Rectangular Waveguide

	Mode 1	Mode 2	Mode 3	Mode 4	Mode 5
N=2	1.90	2.84	3.91	5.18	5.78
N=4	1.89	2.84	3.91	5.16	5.77
N=6	1.89	2.84	3.91	5.16	5.77
N=8	1.89	2.84	3.91	5.16	5.76
Ref. [14]	1.85	2.81	3.89	5.05	5.68

Table 5.20
Percentage of the Matrix That Is Diagonal for the TE Modes

	Matrix size	Row size of the diagonal block	% Diagonal
N=2	96	32	38.33%
N=4	240	128	53.33%
N=6	448	288	64.29%
N=8	720	512	71.11%

5.5 CONCLUSION

The principle of dilation, used in the wavelet concept, has been introduced into finite element techniques as a way to efficiently choose basis functions. An entire domain basis has thus been introduced for efficient solution of the Helmholtz equation confined to waveguides whose cross-sectional area is composed of rectangular regions. This particular choice of basis in the finite elements transforms the majority of the system matrix into a diagonal one. Therefore, the growth of the condition number can easily be controlled by proper scaling and the computational efficiency can be significantly enhanced over the conventional techniques.

REFERENCES

[1] P. G. Ciarlet and J. L. Lions, Eds., *Handbook of Numerical Analysis Methods: Solution of Equations in R^n*, Vol. 1, North Holland, Amsterdam, 1990.

[2] I. Babuska et al., Eds., *Accuracy Estimates and Adaptive Refinements in Finite Element Computations*, Wiley, New York, 1986.

[3] A. F. Peterson, "Eigenvalue projection theory for linear operator equations of electromagnetics," UILV-ENG-87-2252, Coord. Sci. Lab., University of Illinois, Champaign-Urbana, 1987.

[4] M. A. Krasnosel'skii et al., *Approximate Solution of Operator Equations*, Wolters Noordhoff Publishing, Groningen, The Netherlands, 1972.

[5] S. G. Mikhlin, *The Numerical Performance of Variational Methods*, Wolters-Noordhoff Publishing, Amsterdam, The Netherlands, 1971.

[6] G. Golub and C. F. Van Loan, *Matrix Computations*, Johns Hopkins University Press, Baltimore, MD, 1989.

[7] B. K. Alpert, "Wavelets and other bases for fast numerical linear algebra," in *Wavelets: A Tutorial in Theory and Applications*, C. K. Chui, Ed., Academic Press, San Diego, 1992, pp. 181–216.

[8] S. Jaffard, "Wavelets and analysis of partial differential equations," in *Probabilistic and Stochastic Methods in Analysis with Applications*, J. S. Byrnes et al., Eds., Kluwer Publishers, Dordrecht, 1992, pp. 3–13.

[9] S. Jaffard and P. Laurencot, "Orthonormal wavelets, analysis of operators, and applications to numerical analysis," in *Wavelets: A Tutorial in Theory and Applications*, C. K. Chui, Ed., Academic Press, San Diego, 1992.

[10] T. K. Sarkar, Ed., *Application of Conjugate Gradient Method to Electromagnetics and Signal Analysis*, PIER, Vol. 5, Elsevier, New York, 1991.

[11] T. K. Sarkar, K. R. Siarkiewicz, and R. F. Stratton, "Survey of numerical methods for solution of large systems of linear equations for electromagnetic field problems," *IEEE Trans. Antennas and Propag.*, Vol. AP-29, 1981, pp. 847–856.

[12] H. Chen et al., "Adaptive spectral estimation by the conjugate gradient method," *IEEE Tran. Acoust. Speech, Signal Processing*, Vol. ASSP-34, 1986, pp. 272–284.

[13] L. E. Garcia-Castillo et al., "Efficient solution of the differential form of Maxwell's equations in rectangular regions," *IEEE Trans. Microwave Theory & Tech.*, Vol. MTT-43, No. 3, March 1995, pp. 647–654.

[14] T. K. Sarkar et al., "Utilization of wavelet concepts in finite elements for an efficient solution of Maxwell's equations," *Radio Science*, Vol. 29, No. 4, July – Aug. 1994, pp. 965–977.

[15] R. E. Collin, *Field Theory of Guided Waves*, McGraw-Hill, New York, 1960.

[16] T. K. Sarkar et al., "Computation of the propagation characteristics of TE and TM modes in arbitrarily shaped hollow guides utilizing the conjugate gradient method," *Journal of Electromagnetic Waves and Applications*, Vol. 3, No. 2, 1989, pp. 143–165.

[17] M. Swaminathan et al., "Computation of cutoff wavenumbers of TE and TM modes in waveguides of arbitrary cross sections using a surface integral formulation," *IEEE Trans. Microwave Theory & Tech.*, Vol. MTT-38, No. 2, Feb. 1990, pp. 154–159.

Chapter 6

ADAPTIVE MULTISCALE MOMENT METHOD

In this chapter we illustrate how to use the multiscale methodology to solve operator equations in an electromagnetic scattering problem. Here, using a multiscale basis, we apply the compression on the solution, even though it is an unknown. Compression of the solution leads also to a reduction in the system impedance matrix. The procedure starts by solving the electromagnetic problem on a coarse grid at zero scale using the multiscale technique. Then we extrapolate the solution for a finer grid using a spline methodology. The second derivative of the extrapolated solution at the newly formed nodes obtained through the spline interpolation is then noted. If at a given node the value of the second derivative is below a prespecified threshold, then the unknown solution is approximately linear near that node point and no new multiscale bases are introduced at that point. Otherwise, the scale is increased at that point, and a new solution is obtained. In this way, the multiscale basis is adaptively selected in the method of moment methodology and hence this chapter is so named.

6.1 OVERVIEW

First, we present the concept of multiscale methodology for one-dimensional problems. We then apply this methodology for the solution of electromagnetic scattering problems from materially coated strips. Next, we extend this multiscale concept to deal with two-dimensional rectangular surfaces by using a tensor product of two one-dimensional multiscale bases. This technique is then applied to efficiently solve matrix equations resulting from the use of the piecewise triangular basis functions traditionally utilized in the solution of electromagnetic scattering from arbitrarily shaped three-dimensional conducting structures. Finally, we demonstrate how to extend this multiscale concept to deal with triangular patch discretizations of a structure as they approximate arbitrarily shaped surfaces and develop a multiscale basis for them. Numerical examples are presented to illustrate the efficiency and accuracy of this technique.

6.2 INTRODUCTION OF THE MULTISCALING METHODOLOGY

Consider the problem of approximating a function $f(x)$ using the multiscaling methodology in the interval $0 \le x \le 1$. Suppose the zeroth-order approximation of it is defined through the known basis functions $\psi_i(x)$ weighted by the coefficients $f(x_{j,i})$, where j refers to the scale that has been used, and i refers to the interpolation node. The zeroth-order approximation $f_0(x)$ of $f(x)$ is given by

$$f_0(x) = f(x_{0,0})\psi_0(x) + f(x_{0,1})\psi_1(x) \tag{6.1}$$

where (as observed from Figure 2.9)

$$\psi_0(x) = \psi_{0,1}(x) = \begin{cases} 1-x & x \in [0,1] \\ 0 & \text{otherwise} \end{cases}$$

$$\psi_1(x) = \psi_{0,2}(x-1); \text{ with } \psi_{0,2}(y) = \begin{cases} y+1 & y \in [-1,0] \\ 0 & \text{otherwise} \end{cases}$$

$$x_{0,0} = 0, \quad x_{0,1} = 1$$

Using the multiscaling methodology in the interval $[0,1]$, if we need to increase the scale from zero to one, then a new interpolation node at $x_{1,1} = \frac{x_{0,0} + x_{0,1}}{2} = \frac{1}{2}$ needs to be added. Then the approximation of the function at scale one, defined by $f_1(x)$, can be expressed as follows:

$$f_1(x) = f_0(x) + \tau_{1,1}\psi_{1,1}(x) \tag{6.2}$$

where $\psi_{1,1}(x) = \psi_{0,1}[2(x-x_{1,1})] + \psi_{0,2}[2(x-x_{1,1})]$ where the terms within the brackets of $\psi_{0,1}[\bullet]$ and $\psi_{0,2}[\bullet]$ refer to the argument of the function $\psi_{0,1}$ and $\psi_{0,2}$, respectively. Here $\tau_{1,1}$ is an unknown coefficient. Its significance is shown in Figure 6.1. It represents the amount of deviation of its functional behavior from a linear one. Let us assume $f_1(x_{1,1}) = f(x_{1,1})$; then

$$\tau_{1,1} = f(x_{1,1}) - f_0(x_{1,1}) = f(x_{1,1}) - \tfrac{1}{2}(f(x_{0,0}) + f(x_{0,1})) \tag{6.3}$$

By increasing the scale again to two, we need to introduce new interpolation nodes at $x_{2,1} = \frac{x_{0,0} + x_{1,1}}{2} = \frac{1}{4}$, $x_{2,2} = \frac{x_{1,1} + x_{0,1}}{2} = \frac{3}{4}$. The approximation at scale two represented by $f_2(x)$ is then defined as follows:

$$f_2(x) = f_1(x) + \tau_{2,1}\psi_{2,1}(x) + \tau_{2,2}\psi_{2,2}(x) \tag{6.4}$$

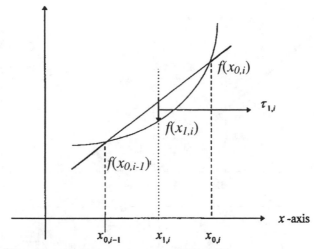

Figure 6.1 Physical significance of the parameter τ.

$$\psi_{2,1}(x) = \psi_{0,1}[2^2(x - x_{2,1})] + \psi_{0,2}[2^2(x - x_{2,1})]$$

$$\psi_{2,2}(x) = \psi_{0,1}[2^2(x - x_{2,2})] + \psi_{0,2}[2^2(x - x_{2,2})]$$

(6.5)

In addition, $\tau_{2,1}, \tau_{2,2}$ are the unknown coefficients. Let us suppose $f_2(x_{2,1}) = f(x_{2,1})$ and $f_2(x_{2,2}) = f(x_{2,2})$; then:

$$\tau_{2,1} = f(x_{2,1}) - f_2(x_{1,1}) = f(x_{2,1}) - \tfrac{1}{2}(f(x_{0,0}) + f(x_{1,1}))$$

$$\tau_{2,2} = f(x_{2,2}) - f_2(x_{2,2}) = f(x_{2,2}) - \tfrac{1}{2}(f(x_{0,1}) + f(x_{1,1}))$$

(6.6)

The plots of some of the multiscale functions $\psi_0(x)$, $\psi_1(x)$, $\psi_{1,1}(x)$, $\psi_{2,1}(x)$, $\psi_{2,2}(x)$, and $\psi_{3,2}(x)$ are shown in Figure 6.2.

In general, by increasing the scale J times in the interval $[0,1]$, the new interpolation nodes will be created at $\{ x_{J,i} = \frac{1}{2^J} + \frac{i-1}{2^{J-1}}, i = 1, 2, ..., 2^{J-1} \}$. This expression is valid for $j \geq 1$, while for $j = 0$ one obtains the end points of the interval. The approximation of the function then at the Jth scale will be $f_J(x)$ and it is expressed by:

$$f_J(x) = f_{J-1}(x) + \sum_{i=1}^{2^{J-1}} \tau_{J,i} \psi_{J,i}(x) = f_0(x) + \sum_{j=1}^{J} \sum_{i=1}^{2^{J-1}} \tau_{j,i} \psi_{j,i}(x)$$

(6.7)

where $\psi_{j,i}(x) = \psi_{0,1}[2^j(x - x_{j,i})] + \psi_{0,2}[2^j(x - x_{j,i})]$ and $\tau_{j,i}$ are the unknown

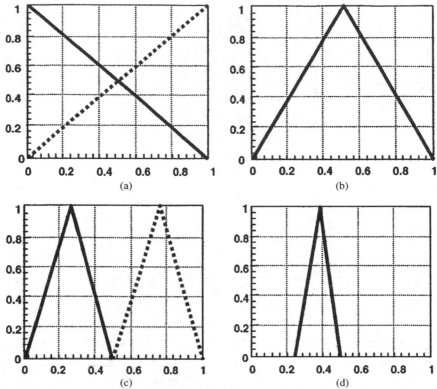

Figure 6.2 Plots of some of the multiscale basis functions: (a) — $\psi(x)$;-- $\psi_{0,1}(x)$; (b) — $\psi_{1,1}(x)$; (c) —

$$\psi_{2,1}(x); --\psi_{2,2}(x); \text{ and (d)} — \psi_{3,2}(x)$$

coefficients. Let $f_J(x_{J,i}) = f(x_{J,i})$, $i = 1, 2, ..., 2^{J-1}$, then the unknown
coefficients are given in terms of the values of the function as:

$$\tau_{J,i} = f(x_{J,i}) - f_{J-1}(x_{J,i}) = f(x_{J,i}) - f_0(x_{J,i}) - \sum_{j=1}^{J} \sum_{n=1}^{2^{J-1}} \tau_{j,n} \psi_{j,n}(x_{J,i})$$

$$= f(x_{J,i}) - \frac{1}{2}\left(f(x_{J,i} - \frac{1}{2^J}) + f(x_{J,i} + \frac{1}{2^J}) \right) \tag{6.8}$$

If $f(x)$ possesses a second order derivative at $x_{J,i}$, defined by $f''(x_{J,i})$, then we
observe that the coefficients are related to the second derivative of the function
through

$$\tau_{J,i} \approx -\frac{1}{2}\frac{1}{2^{J+1}} f''(x_{J,i}) \tag{6.9}$$

The basic philosophy here is that if the function $f(x)$ is linear at the sample node
points then no additional higher order basis functions are necessary at those points.
It is only when the function tends to deviate from the linear variation that more

localized bases at higher scales are necessary for an accurate description of the function.

One can use these multiscale functions { $\psi_0(x)$, $\psi_1(x)$, $\psi_{J,i}(x)$; $J = 1, 2, \dots$; $i = 1, 2, \dots, 2^{J-1}$ } as a basis to represent any function in the interval $[0,1]$. In addition, another basis exists for the same interval where the elements $\phi_i(x)$ are the usual piecewise subdomain triangular functions called the rooftop functions. These functions are shown in Figure 6.3.

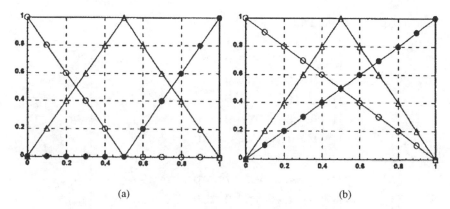

(a)   (b)

Figure 6.3 A representation of (a) the triangular basis functions ϕ_i and (b) the multiscale basis functions ψ_i.

Since both the multiscale and the subdomain triangular bases span the same interval, one can find a matrix transformation that would relate the two systems. Specifically at scale one, the rooftop functions are related to the multiscale basis by

$$\begin{pmatrix} \phi_1(x) \\ \phi_2(x) \\ \phi_3(x) \end{pmatrix} = \begin{pmatrix} 1 & 0 & -\frac{1}{2} \\ 0 & 0 & 1 \\ 0 & 1 & -\frac{1}{2} \end{pmatrix} \begin{pmatrix} \psi_0(x) \\ \psi_1(x) \\ \psi_{1,1}(x) \end{pmatrix}$$

or in a more general form as

$$\phi = T\psi \tag{6.10}$$

where the column matrices $\{\phi\}$ and $\{\psi\}$ represent the subdomain triangular basis and the multiscale basis, respectively. T is the transformation matrix that converts the basis from one form to the other. At scale one, the transformation matrix T_1 between the subdomain triangular basis and the multiscale basis is given by

$$T_1 = \begin{pmatrix} 1 & 0 & -\frac{1}{2} \\ 0 & 0 & 1 \\ 0 & 1 & -\frac{1}{2} \end{pmatrix} \tag{6.11}$$

At scale two, the transformation matrix T_2 between the two bases can be written as

$$T_2 = \begin{pmatrix} 1 & 0 & -\frac{1}{2} & -\frac{1}{2} & 0 \\ 0 & 0 & 0 & 1 & 0 \\ 0 & 0 & 1 & -\frac{1}{2} & -\frac{1}{2} \\ 0 & 0 & 0 & 0 & 1 \\ 0 & 1 & -\frac{1}{2} & 0 & -\frac{1}{2} \end{pmatrix} \tag{6.12}$$

At scale three, the transformation matrix has the following form:

$$T_3 = \begin{pmatrix} 1 & 0 & -\frac{1}{2} & -\frac{1}{2} & 0 & -\frac{1}{2} & 0 & 0 & 0 \\ 0 & 0 & 0 & 0 & 0 & 1 & 0 & 0 & 0 \\ 0 & 0 & 0 & 1 & 0 & -\frac{1}{2} & -\frac{1}{2} & 0 & 0 \\ 0 & 0 & 0 & 0 & 0 & 0 & 1 & 0 & 0 \\ 0 & 0 & 1 & -\frac{1}{2} & -\frac{1}{2} & 0 & -\frac{1}{2} & -\frac{1}{2} & 0 \\ 0 & 0 & 0 & 0 & 0 & 0 & 0 & 1 & 0 \\ 0 & 0 & 0 & 0 & 1 & 0 & 0 & -\frac{1}{2} & -\frac{1}{2} \\ 0 & 0 & 0 & 0 & 0 & 0 & 0 & 0 & 1 \\ 0 & 1 & -\frac{1}{2} & 0 & -\frac{1}{2} & 0 & 0 & 0 & -\frac{1}{2} \end{pmatrix} \tag{6.13}$$

In general, a function $f(x)$ in the interval $[0,1]$ can be approximated by a multiscale basis, which is essentially a dilated and shifted version of the usual subdomain basis. The mutiscale basis at scale zero is the set $\{ \psi_i(x); i = 0, 1, 2, ..., N \}$. For any integer number N, the node points for these functions are located at $\{ x_{0,i} = \frac{i}{N} = ih; i = 0, 1, 2, ..., N; \text{ and } h = \frac{1}{N} \}$. Therefore,

$$\psi_0(x) = \psi_{0,1}(x)$$
$$\psi_N(x) = \psi_{0,2}(x-1) \tag{6.14}$$
$$\psi_i(x) = \psi_{0,1}(x - x_{0,i}) + \psi_{0,2}(x - x_{0,i}) \qquad i = 1, 2, ..., N-1$$

where

$$\psi_{0,1}(x) = \begin{cases} 1 - Nx & [0, h] \\ 0 & otherwise \end{cases}$$

$$\psi_{0,2}(x) = \begin{cases} 1 + Nx & [-h, 0] \\ 0 & otherwise \end{cases}$$

The approximation at the zeroth-scale $f_0(x)$ can then be written as

$$f_0(x) = \sum_{n=0}^{N} f(x_{0,n}) \psi_n(x) \tag{6.15}$$

By multiscaling the interval on [0,1], we obtain the new interpolation nodes for scale one at $x_{1,j} = \frac{x_{0,j-1}+x_{0,j}}{2} = \frac{1}{2N} + \frac{i-1}{N}$. The approximation of the original function at scale one is defined by $f_1(x)$ and it is a sum of the functions at scale zero and a set of a few higher-order functions at scale one. Mathematically, the approximation at scale one is defined through

$$f_1(x) = f_0(x) + \sum_{i=1}^{N} \tau_{1,i} \psi_{1,i}(x) \tag{6.16}$$

where the additional basis functions at scale one are $\psi_{1,i}(x) = \psi_{0,1}[2(x-x_{1,i})] + \psi_{0,2}[2(x-x_{1,i})]$. Here $\tau_{1,i}$ are the unknown coefficients. If we stipulate that $f_1(x_{1,i}) = f(x_{1,i})$, for $i = 1, ..., N$, then

$$\tau_{1,i} = f(x_{1,i}) - f_0(x_{1,i}) = f(x_{1,i}) - \frac{1}{2}(f(x_{0,i-1}) + f(x_{0,i})) \tag{6.17}$$

By multiscaling the interval [0,1] again, we generate new interpolation nodes at scale two. Their positions are given by $x_{2,i} = \frac{1}{4N} + \frac{i-1}{2N}$, ($i = 1, 2, ..., 2N$). The approximation of the function at scale two is $f_2(x)$ and can be written as

$$f_2(x) = f_0(x) + \sum_{i=1}^{N} \tau_{1,i} \psi_{1,i}(x) + \sum_{i=1}^{2N} \tau_{2,i} \psi_{2,i}(x) \tag{6.18}$$

where the additional basis functions introduced at scale two are given by $\psi_{2,i}(x) = \psi_{0,1}[2^2(x-x_{2,i})] + \psi_{0,2}[2^2(x-x_{2,i})]$. The $\tau_{2,i}$ are the unknown coefficients. Equation (6.18) represents the approximation of the function at scale two and it has three separate components: approximation at scale zero, a refined approximation at scale one, and finally a more refined approximation at scale two. The bases for the three scales are shown in Figure 6.4 and will involve only the bottom three layers of that figure defined for the interval [0,4]. If we assume that $f_2(x_{2,i}) = f(x_{2,i})$, $i = 1, 2, ..., 2N$, then

$$\tau_{2,i} = f(x_{2,i}) - (f_0(x_{2,i}) + \sum_{i=1}^{N} \tau_{1,i} \psi_{1,i}(x_{2,i})), \qquad i = 1, 2, ..., 2N \tag{6.19}$$

By J times multiscaling the interval [0,1], new interpolation nodes will be generated at $\{ x_{J,i} = \frac{1}{2^J N} + \frac{i-1}{2^{J-1} N}, i = 1, 2, ..., 2^{J-1} N \}$. The approximation at the Jth scale is defined by $f_J(x)$. Mathematically, it is expressed as follows:

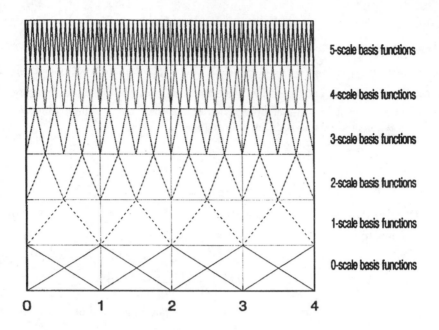

Figure 6.4 Basis functions corresponding to different scales.

$$f_J(x) = f_{J-1}(x) + \sum_{i=1}^{2^{J-1}N} \tau_{J,i}\,\psi_{J,i}(x) = \sum_{n=0}^{N} f(x_{0,n})\,\psi_n(x) + \sum_{j=1}^{J}\sum_{i=1}^{2^{j-1}N} \tau_{j,i}\,\psi_{j,i}(x)$$

(6.20)

where $\psi_{j,i}(x) = \psi_{0,1}[2^j(x-x_{j,i})] + \psi_{0,2}[2^j(x-x_{j,i})$ and $\tau_{J,i}$ are the unknown coefficients. Now if, $f_J(x_{J,i}) = f(x_{J,i}),\ i=1,2,\ldots,2^{J-1}N$, then

$$\tau_{J,i} = f(x_{J,i}) - f_{J-1}(x_{J,i})$$

$$= f(x_{J,i}) - \left(f_0(x_{J,i}) + \sum_{j=1}^{J}\sum_{n=1}^{2^{j-1}N} \tau_{j,n}\psi_{j,n}(x_{J,i}) \right)$$

(6.21)

$$= f(x_{J,i}) - \frac{1}{2}\left(f\left(x_{J,i} - \frac{1}{2^J N}\right) + f\left(x_{J,i} + \frac{1}{2^J N}\right) \right)$$

If $f(x)$ possesses a second-order derivative at $x_{J,i}$, then

$$\tau_{J,i} \approx -\frac{1}{2}\frac{1}{2^{J+1}N} f''(x_{J,i})$$

(6.22)

It can be shown that any function $f(x) \in C[0,1]$ can be uniformly approximated by the set $\{f_J(x),\ J = 1, 2, 3 \ldots\}$, resulting in

$$f(x) = \sum_{i=0}^{N} f(x_{0,i}) \psi_i(x) + \sum_{J=1}^{\infty} \sum_{i=1}^{2^{J-1}N} \tau_{J,i} \psi_{J,i}(x) \quad \text{on } C[0,1] \tag{6.23}$$

where $C[0,1]$ represents the space of continuous functions in the interval $[0,1]$. From (6.22) and (6.23), we can see that if $f(x)$ has a second-order continuous derivative at $x_{J,i}$, then the coefficients $\tau_{J,i}$ will be approximately zero as the scale J increases. So, if $f(x)$ is linear at some interval in $[0,1]$, $\tau_{J,i}$ will be zero. Instead, if $f(x)$ has a jump at x^* in the interval $[0,1]$, the coefficients $\tau_{J,i}$ near x^* will not be small. This property is used to reduce the effective size of the impedance matrix in the Moment Method when using the multiscale basis. The details will be discussed in the next section. In addition, this new basis, which we call a multiscale basis, has local support, but does not have vanishing moment properties like wavelets do.

6.3 USE OF A MULTISCALE BASIS IN SOLVING INTEGRAL EQUATIONS VIA THE MOMENT METHOD (MM)

In this section, we discuss how to use the multiscale basis in the moment method to solve a Fredholm integral equation of the first kind as described by

$$g(x) = \int_0^1 k(x,t) f(t)\, dt \qquad x \in [0,1] \tag{6.24}$$

where x and t represent two different variables. We know from the previous section that any function $f(t) \in C[0,1]$ can be written in the multiscale basis through (6.23). In the MM, we select the multiscale basis functions $\{\Psi\} = \{\psi_i(t), \psi_{J,k}(t)\}$ as both expansion and testing functions resulting in a Galerkin method. Here the set $\{\psi_i(t)\}$ represents the basis at scale zero (i.e., $J = 0$). For numerical computations it is necessary to deal with a finite set of functions, hence (6.23) becomes

$$f(t) \approx \sum_{i=0}^{N} \tau_{0,i} \psi_i(t) + \sum_{j=1}^{J} \sum_{i=1}^{2^{j-1}N} \tau_{j,i} \psi_{j,i}(t) \tag{6.25}$$

By substituting (6.25) into (6.24), we obtain

$$g(x) = \sum_{i=0}^{N} \tau_{0,i} \int_0^1 k(x,t)\, \psi_i(t)\, dt + \sum_{j=1}^{J} \sum_{i=1}^{2^{j-1}N} \tau_{j,i} \int_0^1 k(x,t)\, \psi_{j,i}(t)\, dt \tag{6.26}$$

Since the set $\{\Psi\} = \{\psi_i(x), \psi_{J,k}(x)\}$ are also the weighting functions (i.e., we are using a Galerkin method in the method of moments context), by using the linearity of the inner products, we obtain the following expressions,

$$\int_0^1 g(x)\,\psi_m(x)\,dx =$$

$$\sum_{i=0}^N \tau_{0,i} \int_0^1\int_0^1 k(x,t)\,\psi_i(t)\,\psi_m(x)\,dt\,dx \;+\; \sum_{j=1}^J\sum_{i=1}^{2^{j-1}N} \tau_{j,i} \int_0^1\int_0^1 k(x,t)\,\psi_{j,i}(t)\,\psi_m(x)\,dt\,dx$$

$$(6.27)$$

$$\int_0^1 g(x)\,\psi_{p,n}(x)\,dx =$$

$$\sum_{i=0}^N \tau_{0,i} \int_0^1\int_0^1 k(x,t)\,\psi_i(t)\,\psi_{p,n}(x)\,dt\,dx \;+\; \sum_{j=1}^J\sum_{i=1}^{2^{j-1}N} \tau_{j,i} \int_0^1\int_0^1 k(x,t)\,\psi_{j,i}(t)\,\psi_{p,n}(x)\,dt\,dx$$

$$(6.28)$$

where $m = 0, 1, 2, \ldots, N;\ p = 1, 2, \ldots, J;\ n = 1, 2, \ldots, 2^{p-1}N$. These equations can be written in a compact matrix form as

$$\begin{pmatrix} F_0 \\ F_1 \\ \vdots \\ F_J \end{pmatrix} = \begin{pmatrix} A_{0,0} & A_{0,1} & \cdots & A_{0,J} \\ A_{1,0} & A_{1,1} & \cdots & A_{1,J} \\ \vdots & \vdots & \ddots & \vdots \\ A_{J,0} & A_{J,1} & \cdots & A_{J,J} \end{pmatrix} \begin{pmatrix} X_0 \\ X_1 \\ \vdots \\ X_J \end{pmatrix} \qquad (6.29)$$

where $X_0 = (\tau_{0,0}, \tau_{0,1}, \ldots, \tau_{0,N})^T$, $X_j = (\tau_{j,1}, \tau_{j,2}, \ldots, \tau_{j,2^{j-1}N})^T$, $j = 1, 2, \ldots, J$

$$F_0(i) = \int_0^1 g(x)\psi_i(x)\,dx\,; \qquad i = 0, 1, 2, \ldots, N$$

$$F_j(i) = \int_0^1 g(x)\,\psi_{j,i}(x)\,dx;\ j = 1, 2, \ldots, J;\ i = 0, 1, 2, \ldots, 2^{j-1}N$$

$$A_{0,0}(i,k) = \int_0^1\int_0^1 k(x,t)\psi_i(x)\,\psi_k(t)\,dx\,dt\,;\ i = 0, 1, 2, \ldots, N;\ k = 0, 1, 2, \ldots, N$$

$$A_{0,j}(i,k) = \int_0^1\int_0^1 k(x,t)\psi_i(x)\,\psi_{j,k}(t)\,dx\,dt\,;\ i = 0, 1, 2, \ldots, N;\ k = 1, 2, \ldots, 2^{j-1}N$$

$$A_{j,0}(i,k) = \int_0^1\int_0^1 k(x,t)\psi_{j,k}(x)\,\psi_i(t)\,dx\,dt\,;\ i = 0, 1, 2, \ldots, N;\ k = 1, 2, \ldots, 2^{j-1}N$$

$$A_{j,p}(i,k) = \int_0^1\int_0^1 k(x,t)\,\psi_{j,i}(x)\,\psi_{p,k}(t)\,dx\,dt;\ i = 1, 2, \ldots, 2^{j-1}N;\ k = 1, 2, \ldots, 2^{p-1}N$$

where $j = 1, 2, \ldots, J$ and $p = 1, 2, \ldots, J$. The expressions for $F_0(i)$, $F_j(i)$, $A_{0,0}(i,k)$ and $A_{0,j}(i,k)$, $A_{j,0}(k,i)$, $A_{j,p}(i,k)$ can be written as follows:

$$F_0(i) = \begin{cases} \int\limits_0^{\frac{1}{N}} g(x+\tfrac{i}{N})(1-Nt)\,dx & i=0 \\[2em] \int\limits_0^{\frac{1}{N}} g(x+\tfrac{i}{N})(1-Nx)\,dt + \int\limits_{-\frac{1}{N}}^{0} f(x+\tfrac{i}{N})(1+Nx)\,dx & i=1,2,\dots,N-1 \\[2em] \int\limits_{-\frac{1}{N}}^{0} g(x+\tfrac{i}{N})(1+Nx)\,dx & i=N \end{cases}$$

$$A_{0,0}(m,i) = \int\limits_0^{\frac{1}{N}}(1-Nx)\,dx \int\limits_0^{\frac{1}{N}} k(x+\tfrac{m}{N}, t+\tfrac{i}{N})(1-Nt)\,dt$$

$$+ \int\limits_{-\frac{1}{N}}^{0}(1+Nx)\,dx \int\limits_0^{\frac{1}{N}} k(x+\tfrac{m}{N}, t+\tfrac{i}{N})(1-Nt)\,dt$$

$$+ \int\limits_0^{\frac{1}{N}}(1-Nx)\,dx \int\limits_{-\frac{1}{N}}^{0} k(x+\tfrac{m}{N}, t+\tfrac{i}{N})(1+Nt)\,dt$$

$$+ \int\limits_{-\frac{1}{N}}^{0}(1+Nx)\,dx \int\limits_{-\frac{1}{N}}^{0} k(x+\tfrac{m}{N}, t+\tfrac{i}{N})(1+Nt)\,dt;$$

$$\text{for } i=1,2,\dots,N-1; \quad m=1,2,\dots,N-1$$

$$A_{0,j}(m,i) = \frac{1}{2^j}\int\limits_0^{\frac{1}{N}}(1-Nx)\,dx \int\limits_0^{\frac{1}{N}} k\left(x+\tfrac{m}{N}, \tfrac{1}{2^j}(t+\tfrac{2i-1}{N})\right)(1-Nt)\,dt$$

$$+ \frac{1}{2^j}\int\limits_{-\frac{1}{N}}^{0}(1+Nx)\,dx \int\limits_0^{\frac{1}{N}} k\left(x+\tfrac{m}{N}, \tfrac{1}{2^j}(t+\tfrac{2i-1}{N})\right)(1-Nt)\,dt$$

$$+ \frac{1}{2^j}\int\limits_0^{\frac{1}{N}}(1-Nx)\,dx \int\limits_{-\frac{1}{N}}^{0} k\left(x+\tfrac{m}{N}, \tfrac{1}{2^j}(t+\tfrac{2i-1}{N})\right)(1+Nt)\,dt$$

$$+ \frac{1}{2^j}\int\limits_{-\frac{1}{N}}^{0}(1+Nx)\,dx \int\limits_{-\frac{1}{N}}^{0} k\left(x+\tfrac{m}{N}, \tfrac{1}{2^j}(t+\tfrac{2i-1}{N})\right)(1+Nt)\,dt$$

$$\text{for } i=1,2,\dots,2^{j-1}N; \quad m=1,2,\dots,N-1$$

$$A_{j,0}(m,i) = \frac{1}{2^j} \int\limits_0^{\frac{1}{N}} (1-Nx)\,dx \int\limits_0^{\frac{1}{N}} k\Big(\frac{1}{2^j}(x+\frac{2m-1}{N}),\, t+\frac{i}{N}\Big)(1-Nt)\,dt$$

$$+ \frac{1}{2^j} \int\limits_{-\frac{1}{N}}^0 (1+Nx)\,dx \int\limits_0^{\frac{1}{N}} k\Big(\frac{1}{2^j}(x+\frac{2m-1}{N}),\, t+\frac{i}{N}\Big)(1-Nt)\,dt$$

$$+ \frac{1}{2^j} \int\limits_0^{\frac{1}{N}} (1-Nx)\,dx \int\limits_{-\frac{1}{N}}^0 k\Big(\frac{1}{2^j}(x+\frac{2m-1}{N}),\, t+\frac{i}{N}\Big)(1+Nt)\,dt$$

$$+ \frac{1}{2^j} \int\limits_{-\frac{1}{N}}^0 (1+Nx)\,dx \int\limits_{-\frac{1}{N}}^0 k\Big(\frac{1}{2^j}(x+\frac{2m-1}{N}),\, t+\frac{i}{N}\Big)(1+Nt)\,dt$$

$$\text{for} \quad i=1, 2, ..., N-1; \quad m=1, 2, ..., 2^{j-1}N$$

$$A_{j,p}(m,i) = \frac{1}{2^{j+1}} \int\limits_0^{\frac{1}{N}} (1-Nx)\,dx \int\limits_0^{\frac{1}{N}} k\Big(\frac{1}{2^j}(x+\frac{2m-1}{N}),\, \frac{1}{2^p}(t+\frac{2i-1}{N})\Big)(1-Nt)\,dt$$

$$+ \frac{1}{2^{j+1}} \int\limits_{-\frac{1}{N}}^0 (1+Nx)\,dx \int\limits_0^{\frac{1}{N}} k\Big(\frac{1}{2^j}(x+\frac{2m-1}{N}),\, \frac{1}{2^p}(t+\frac{2i-1}{N})\Big)(1-Nt)\,dt$$

$$+ \frac{1}{2^{j+1}} \int\limits_0^{\frac{1}{N}} (1-Nx)\,dx \int\limits_{-\frac{1}{N}}^0 k\Big(\frac{1}{2^j}(x+\frac{2m-1}{N}),\, \frac{1}{2^p}(t+\frac{2i-1}{N})\Big)(1+Nt)\,dt$$

$$+ \frac{1}{2^{j+1}} \int\limits_{-\frac{1}{N}}^0 (1+Nx)\,dx \int\limits_{-\frac{1}{N}}^0 k\Big(\frac{1}{2^j}(x+\frac{2m-1}{N}),\, \frac{1}{2^p}(t+\frac{2i-1}{N})\Big)(1+Nt)\,dt$$

$$\text{for} \quad i = 1, 2, ..., 2^{p-1}N; \quad m = 1, 2, ..., 2^{j-1}N$$

We next present an efficient iterative method for solving the unknown coefficients X_i in (6.29) by using a multiscale basis. The solution proceeds as follows:

Step 1: By utilizing the conjugate gradient method, we solve the equation at the zeroth-scale $A_{0,0}X_0 = F_0$. The solution is expressed as X_0^0.

Step 2: Using an interpolation technique, we estimate the coefficients $\{\tau_{1,i}\}$ at scale one based on the values of the solution $f_0(x)$ obtained for the uniform subdomain triangular basis at the zeroth scale $(x_{0,i}, X_0^0(x))$ or $(x_{0,i}, f_0(x_{0,i}))$. We use the following rules for interpolating/extrapolating the function between the sampled values:

1. Suppose we know the sampled values of the function at $\{ f(-\frac{3}{2}h),\ f(-\frac{1}{2}h),\ f(\frac{1}{2}h),\ f(\frac{3}{2}h) \}$ and that we need to interpolate the functional value at $f(0)$. The interpolation is carried out by using a cubic polynomial approximation through

$$f(0) = \frac{\left[f(-\frac{1}{2}h) + f(\frac{1}{2}h)\right] - \left[f(-\frac{3}{2}h) + f(\frac{3}{2}h)\right]}{6} \tag{6.30}$$

2. If the point to be extrapolated is located to the left of the region of the given data samples $\{ f(-\frac{1}{2}h), \ f(\frac{1}{2}h), \ f(\frac{3}{2}h) \}$, then we compute $f(-\frac{3}{2}h)$ by using the following quadratic polynomial approximation:

$$f(\frac{-3}{2}h) = \frac{\left[f(-\frac{1}{2}h) + f(\frac{1}{2}h)\right]}{2} - \frac{\left[2f(-\frac{1}{2}h) + f(\frac{3}{2}h)\right]}{3} \tag{6.31}$$

3. If the point to be extrapolated is located to the right of the region of the given data samples $\{ f(-\frac{3}{2}h), \ f(-\frac{1}{2}h), \ f(\frac{1}{2}h) \}$, then the value of the function $f(\frac{3}{2}h)$ can be computed from the following quadratic polynomial approximation:

$$f(\frac{3}{2}h) = \frac{\left[f(-\frac{1}{2}h) + f(\frac{1}{2}h)\right]}{2} - \frac{\left[2f(\frac{1}{2}h) + f(-\frac{3}{2}h)\right]}{3} \tag{6.32}$$

In this way the functional values are interpolated/extrapolated from the given data samples. Next we estimate the second derivatives of the function at the interpolated/extrapolated points through the parameters $\tau_{1,i}$. If $\left|\tau_{1,i}\right| \leq \varepsilon$ for any $i = 1, ..., N$ and ε is a threshold parameter that provides a bound of the value of the second derivative of the function at that point, then we assume that the functional variation of the unknown at that point is linear. Hence, if the second derivative of the function at that point is below ε, the corresponding row i and column i of the matrix $\begin{pmatrix} A_{0,0} & A_{0,1} \\ A_{1,0} & A_{1,1} \end{pmatrix}$ will be eliminated thus producing a reduced system matrix at scale one. The unknowns and the excitations corresponding to row i are also eliminated. Using an initial guess of $\begin{pmatrix} X_0^0 \\ \vec{0} \end{pmatrix}$ for the solution of the conjugate gradient method, we solve the above matrix equation at scale one. The solution, after embedding zero elements in X_1^1 corresponding to row i, is now expressed as $\begin{pmatrix} X_0^1 \\ X_1^1 \end{pmatrix}$, where, as mentioned before, X_1^1 may have some zero elements corresponding to the indices i for which the rows and the columns of the system matrix have been previously eliminated. The physical significance of the zeros in the coefficients of the basis functions is that one does not introduce additional unknowns at the higher scale if the solution to the integral equation has a linear variation.

Step 3: Using the same interpolation technique described earlier, we estimate the coefficients $\{\tau_{2,i}\}$ of the function $f_1(x)$ at the second scale based on the sampled values of the unknown function at $(x_{0,i}, f_1(x_{0,i}))$ and $(x_{1,i}, f_1(x_{1,i}))$. If $\left|\tau_{2,i}\right| \le \varepsilon$ for $i = 1, 2, \dots, 2N$ and $\left|\tau_{1,i}\right| \le \varepsilon$ for $i = 1, \dots, N$, the corresponding row and column of the matrix $\begin{pmatrix} A_{0,0} & A_{0,1} & A_{0,2} \\ A_{1,0} & A_{1,1} & A_{1,2} \\ A_{2,0} & A_{2,1} & A_{2,2} \end{pmatrix}$ are eliminated. The corresponding rows in the excitation and the unknown vectors are also removed. The initial guess for the solution is now $\begin{pmatrix} X_0^1 \\ \vec{X}_1^1 \\ \vec{0} \end{pmatrix}$, where $\vec{X}_1^1$ represents the solution from the previous scale with the zero elements deleted. Also using an estimate for $\{\tau_{2,i}\}$, we know which of the elements of the solution at scale two would be zero. By using the conjugate gradient method, we solve for the solution at scale two as $\begin{pmatrix} X_0^2 \\ X_1^2 \\ X_2^2 \end{pmatrix}$ after reinserting the zero elements of the coefficients. Therefore, X_2^2 may contain some zero elements. This procedure is continued until we reach the highest scale J at which we require the final solution.

A few observations can be made regarding this method:

1. This new method, which utilizes the multiscale basis in the method of moments context, is somewhat different from the conventional MM. It follows from (6.10) that the subdomain triangular basis $\{\phi_i(x)\}$ and the multiscale basis $\{\psi_i(x)\}$ are related by

$$\begin{pmatrix} \phi_1(x) \\ \vdots \\ \phi_N(x) \end{pmatrix} = T \begin{pmatrix} \psi_1(x) \\ \vdots \\ \psi_N(x) \end{pmatrix} \tag{6.33}$$

Therefore, the system matrix for these two bases will have the following relationship:

$$\begin{pmatrix} <\phi_1,\phi_1> & \cdots & <\phi_1,\phi_N> \\ \vdots & \ddots & \vdots \\ <\phi_N,\phi_1> & \cdots & <\phi_N,\phi_N> \end{pmatrix} = T \begin{pmatrix} <\psi_1,\psi_1> & \cdots & <\psi_1,\psi_N> \\ \vdots & \ddots & \vdots \\ <\psi_N,\psi_1> & \cdots & <\psi_N,\psi_N> \end{pmatrix} T^T \quad (6.34)$$

where the superscript T denotes the transpose of a matrix. In addition, the inner products $< \bullet , \bullet >$ are defined by

$$< f(t), g(x) > = \int_0^1 dx \int_0^1 k(x,t) \, f(t) \, g(x) \, dt \, dx \quad (6.35)$$

2. This new method based on the multiscale technique is different from the multigrid method for solving integral equations. This is because, in the multigrid method when one increases the scale all the elements of the coefficient matrix need to be recomputed; whereas, in this multiscale method only the terms introduced by the new scale need to be considered.

3. The initial guess for the iterative solution as we increase the scale can be determined quite easily from the solution obtained at the previous scale.

6.4 DIFFERENCES BETWEEN A MULTISCALE BASIS AND A SUBDOMAIN TRIANGULAR BASIS ON AN INTERVAL [0, L]

Consider a set of piecewise subdomain triangular basis functions $\{\phi\}$ defined on a uniform grid on the interval $[0, L]$ with $N+1$ nodes. The basis can be mathematically written as

$$\begin{aligned} \psi_{0,0}(x) &= \phi(x) \\ \psi_{0,N}(x) &= \phi(x-L) \\ \psi_{0,i}(x) &= \phi(x-x_{0,i}) \qquad i = 1, \, 2,..., \, N-1 \end{aligned} \quad (6.36)$$

where $x_{0i} = \dfrac{iL}{N} = ih$; $i = 0, 1, 2, ..., N$ *with* $h = \dfrac{L}{N}$. In addition,

$$\phi(x) = \begin{cases} 1 - x/h & \text{for } 0 < x < h \\ 1 + x/h & \text{for } -h < x < 0 \end{cases} \quad (6.37)$$

Through the use of the multiscale methodology, we subdivide the interval $[0, L]$, with the nodes located at

$$x_{1,i} = \frac{x_{0,i-1} + x_{0,i}}{2} = (i - \tfrac{1}{2})h \tag{6.38}$$

Then on this interval at the zeroth scale, we define the following basis, which will be a scaled version of the uniform triangular basis, as

$$\psi_{1,i}(x) = \phi[2(x - x_{1,i})] \tag{6.39}$$

By increasing the scale V times in the interval $[0, L]$, we generate new nodes located at

$$x_{V,i} = \left(\frac{1}{2^V} + \frac{i-1}{2^{V-1}} \right)h \; ; \quad \text{with} \quad i = 1, 2, ..., 2^{V-1}N \tag{6.40}$$

Then we have the multiscale basis defined through

$$\psi_{V,i}(x) = \phi[2^V(x - x_{V,i})] \tag{6.41}$$

Here the set $\{\Psi\} = \{\psi_{V,i}(x)\}$ whose members are referred to as the V times multiscale basis functions which have a compact support. Specially, at the zeroth multiscale, on a coarse grid the basis functions are the ordinary triangular basis functions. At the higher scale, the basis consists of dilated and shifted versions of the triangular basis.

The multiscale basis functions are shown in Figure 6.4 for the case of $N = 4$, $L = 4$, and $V = 5$ in the interval $[0,4]$. The term $\psi(x)$ is different from the mother wavelet in the usual wavelet analysis, because it does not have the vanishing moment property in our presentations. The basis functions $\{\Psi\} = \{\psi_{V,i}(x)\}$ on $[0, L]$ can be constructed by shifting and dilating the function $\phi(x)$. The derivative of $\phi(x)$ is similar to the Haar wavelet.

For any piecewise continuous function $f(x)$ on $[0, L]$, the approximation of the function at the Vth scale denoted by $f_V(x)$ is represented by

$$f_V(x) = f_{V-1}(x) + \sum_{i=1}^{2^{V-1}N} \tau_{V,i} \, \psi_{V,i}(x) = \sum_{i=0}^{N} \tau_{0,i} \, \psi_{0,i}(x) + \sum_{j=1}^{V} \sum_{i=1}^{2^{J-1}N} \tau_{j,i} \, \psi_{j,i}(x)$$

where

$$\tau_{V,i} = f(x_{V,i}) - \frac{1}{2}\left(f(x_{V,i} - \frac{1}{2^V}h) + f(x_{V,i} + \frac{1}{2^V}h) \right) \tag{6.42}$$

This means that $\tau_{V,i}$ is the second-order central difference of $f(x)$ at $x_{V,i}$ in the interval $[x_{V,i} - \frac{h}{2^V}; x_{V,i} + \frac{h}{2^V}]$. Now, if $f(x)$ possesses a second-order derivative at $x_{V,i}$, then

$$\tau_{V,i} \approx - \frac{h^2}{2^{V+2}} f''(x_{V,i}) \tag{6.43}$$

From this presentation it is clear that if the unknown function has a linear behavior near the node then no addition of multiscale bases are necessary to approximate that function. Therefore, we stipulate that from a numerical perspective if the second derivative of the function is below a prespecified threshold ε, then no additional basis is necessary. Hence, if $|\tau_{j,i}| \leq \varepsilon$ (ε is the given threshold), $\tau_{j,i}$ can be set to zero.

Suppose $f(x)$ is the function to be approximated by a uniform subdomain triangular basis. Let this approximation in this basis be $f_\phi(x)$. Let the same function be approximated using a multiscale basis and let us term that approximation $f_\psi(x)$. The latter approximation is refined up to the V scale, with the stipulation that if the function is approximately linear in some regions then the corresponding coefficients in the multiscale approximation are set to zero. Then

$$\left\| f(x) - f_\psi(x) \right\| = \int_0^L \left| f(x) - f_\psi(x) \right| dx \leq \left\| f(x) - f_\phi(x) \right\| + \left\| f_\phi(x) - f_\psi(x) \right\|$$

$$\leq \left\| f(x) - f_\phi(x) \right\| + \sum_{v=1}^{V} \int_0^L \left| \varepsilon n_v \right| \Lambda(2^v x) \left| dx \leq \left\| f(x) - f_\phi(x) \right\| + \varepsilon V \tag{6.44}$$

where n_v is the number of terms that have been neglected at the V scale. In a similar fashion, we can show that

$$\left\| f(x) - f_\phi(x) \right\| \leq \left\| f(x) - f_\psi(x) \right\| + \varepsilon V \tag{6.45}$$

Therefore,

$$\left\| \left\| f(x) - f_\phi(x) \right\| - \left\| f(x) - f_\psi(x) \right\| \right\| \leq \varepsilon V \tag{6.46}$$

This means that one can control the accuracy of the approximation using a multiscale basis even when some of the coefficients in the approximation are set to zero in the regions where the function has a linear behavior. So when the function $f(x)$ is almost linear along $[0, L]$, most of the coefficients for the basis functions $\{\tau_{j,i}\}$ will be zero. This will result in a more efficient representation of the function $f(x)$ through using a relatively few number of basis functions as opposed to representing the same function through the use of the piecewise subdomain triangular basis. We now illustrate the efficiency of this methodology in the solution of various 2-D and 3-D electromagnetic field problems.

6.5 ANALYSIS OF ELECTROMAGNETIC SCATTERING FROM MATERIALLY COATED STRIPS

Consider a time-harmonic electromagnetic wave $(\vec{E}^i, \vec{H}^i)$ incident on an infinitely thin, perfectly conducting strip coated with a homogeneous dielectric material as shown in Figure 6.5. The permittivity and permeability of the coating (ε_2, μ_2) can be complex. We assume that the strip width (L) is larger than the thickness (d) of the coating, that is, $d \ll L$.

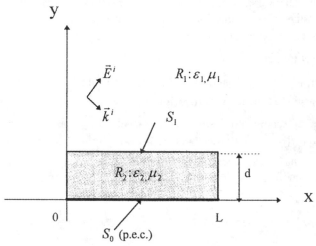

Figure 6.5 Scattering from a materially coated strip.

6.5.1 Integral Equation Relating the Fields to the Excitations

The total electrical and magnetic field in the region R_1, at a field point $\vec{r}$ can be represented by the following coupled system of integrodifferential equations:

$$\begin{cases} \theta(\vec{r})\vec{E}_1(\vec{r}) = \vec{E}^i(\vec{r}) - L_1\vec{J}_1^+(\vec{r}) + K_1\vec{M}_1^+(\vec{r}) - L_1\vec{J}_0^-(\vec{r}) \\ \theta(\vec{r})\vec{H}_1(\vec{r}) = \vec{H}^i(\vec{r}) - K_1\vec{J}_1^+(\vec{r}) - \dfrac{1}{\eta_1^2}L_1\vec{M}_1^+(\vec{r}) - K_1\vec{J}_0^-(\vec{r}) \end{cases} \tag{6.47}$$

where $\eta_1 = \sqrt{\dfrac{\mu_1}{\varepsilon_1}}$ and $\vec{J}_0^+(\vec{r})$, $\vec{J}_0^-(\vec{r})$ are the equivalent surface electric currents defined on the upper and lower surface of S_0. $\vec{J}_1^\pm(\vec{r})$, $\vec{M}_1^\pm(\vec{r})$ are the equivalent electric and magnetic currents defined on the upper and lower surface of S_1, where $\vec{J}_1^+(\vec{r}) = -\vec{J}_1^-(\vec{r})$ and $\vec{M}_1^+(\vec{r}) = -\vec{M}_1^-(\vec{r})$.

The fields in the region R_2 are expressed as

$$\begin{cases} \theta(\vec{r})\vec{E}_2(\vec{r}) = -L_2\vec{J}_1^-(\vec{r}) + K_2\vec{M}_1^-(\vec{r}) - L_2\vec{J}_0^+(\vec{r}) \\ \theta(\vec{r})\vec{H}_2(\vec{r}) = -K_2\vec{J}_1^-(\vec{r}) - \dfrac{1}{\eta_2^2}L_2\vec{M}_1^-(\vec{r}) - K_2\vec{J}_0^+(\vec{r}) \end{cases} \quad (6.48)$$

The integrodifferential operators L_i, K_i in (6.47) and (6.48) are defined as

$$\begin{cases} L_i\vec{X}(\vec{r}) = jk_i\eta_i \displaystyle\int_{\partial R_i}(\vec{X}(\vec{r}') + k_i^{-2}\nabla\nabla'\cdot\vec{X}(\vec{r}'))\, G(k_i|\vec{r}-\vec{r}'|)\, ds' \\ K_i\vec{X}(\vec{r}) = \displaystyle\int_{\partial R_i}\vec{X}(\vec{r}')\times\nabla G(k_i|\vec{r}-\vec{r}'|)\, ds' \end{cases} \quad (6.49)$$

where $G(k_i s) = \frac{1}{4j}H_0^{(2)}(k_i s)$, $k_i = \frac{2\pi}{\lambda_i}$, where λ_i is the wavelength in medium i. Here $H_0^{(2)}$ is the Hankel function of the second kind and $j = \sqrt{-1}$. In addition, $\theta(\vec{r})$ is given by

$$\theta(\vec{r}) = \begin{cases} 1 & \vec{r} \in R_i \\ \frac{1}{2} & \vec{r} \in \partial R_i \\ 0 & \text{otherwise} \end{cases}$$

and ∂R_i is the boundary of the region R_i, for $i = 1, 2$.

By enforcing the boundary conditions (that is, the tangential components of the total fields are continuous on S_1, and the tangential components of the total electric field are zero on the perfectly conducting surface), the basic integrodifferential equations can be obtained as

$$\vec{E}^i(\vec{r})\Big|_{\text{tan}} = \Big\{(L_1+L_2)\vec{J}_1(\vec{r})-(K_1+K_2)\vec{M}_1(\vec{r})+L_1\vec{J}_0^-(\vec{r})-L_2\vec{J}_0^+(\vec{r})\Big\}\Big|_{\text{tan}} \quad (6.50)$$
$$\text{for } \vec{r} \in S_1$$

$$\vec{H}^i(\vec{r})\Big|_{\text{tan}} = \begin{cases} (K_1+K_2)\vec{J}_1(\vec{r})-(\eta_1^{-2}L_1+\eta_2^{-2}L_2)\vec{M}_1(\vec{r}) \\ +K_1\vec{J}_0^-(\vec{r})-K_2\vec{J}_0^+(\vec{r}) \end{cases}\Bigg|_{\text{tan}} \quad \text{for } \vec{r} \in S_1$$

$$\quad (6.51)$$

$$\vec{E}^i(\vec{r})\Big|_{\text{tan}} = \Big\{L_1\vec{J}_1(\vec{r}) - K_1\vec{M}_1(\vec{r}) + L_1\vec{J}_0^-(\vec{r})\Big\}\Big|_{\text{tan}} \quad \text{for } \vec{r} \in S_1^- \quad (6.52)$$

$$0 = \Big\{-L_2\vec{J}_1(\vec{r}) + K_2\vec{M}_1(\vec{r}) + L_2\vec{J}_0^+(\vec{r})\Big\}\Big|_{\text{tan}} \quad \text{for } \vec{r} \in S_1^+ \quad (6.53)$$

where $\vec{J}_1(\vec{r}) = \vec{J}_1^+(\vec{r})$ and $\vec{M}_1(\vec{r}) = \vec{M}_1^+(\vec{r})$ and the subscript "tan" represents the tangential component of the fields. Our goal is to analyze the electromagnetic scattering from the coated flat strip, that is, to solve the above coupled integral equations by the adaptive multiscale moment method and to compute the scattered fields from both the electric and magnetic currents.

6.5.2 Application of the Method of Moments Using a Multiscale Basis

6.5.2.1 Transverse Electric (TE) Case

Let the field incident on the structure be defined by

$$E_x^i = E_0^i \sin \varphi^i \, e^{jk_1(x\cos\varphi^i + y\sin\varphi^i)}$$

$$H_z^i = \frac{E_0^i}{\eta_0} e^{jk_1(x\cos\varphi^i + y\sin\varphi^i)} \tag{6.54}$$

where, $j = \sqrt{(-1)}$ and φ^i denotes angle of arrival of the incident field from the x-axis. Using a multiscale basis, the induced electric current on either side of the surface S_0 can be written as

$$\vec{J}_0^{\pm}(\vec{r}) = \hat{x} J_0^{\pm}(x)$$

$$J_0^{\pm}(x) = \sum_{i=0}^{N} \tau_{0,i}^{\pm} \psi_{0,i}(x) + \sum_{v=1}^{V} \sum_{i=1}^{2^{v-1}N} \tau_{v,i}^{\pm} \psi_{v,i}(x) \tag{6.55}$$

Similarly, the induced electric and magnetic currents on the surface S_1 can be written in the following form using $\hat{x}$ and $\hat{z}$ as the unit vectors:

$$\vec{J}_1(\vec{r}) = \hat{x} J_1(x)$$

$$J_1(x) = \sum_{i=0}^{N} \tau_{0,i}^{1} \psi_{0,i}(x) + \sum_{v=1}^{V} \sum_{i=1}^{2^{v-1}N} \tau_{v,i}^{1} \psi_{v,i}(x) \tag{6.56}$$

$$\vec{M}_1(\vec{r}) = -\hat{z} \eta_0 M_1(x)$$

$$M_1(x) = \sum_{i=0}^{N} \tau_{0,i}^{2} \psi_{0,i}(x) + \sum_{v=1}^{V} \sum_{i=1}^{2^{v-1}N} \tau_{v,i}^{2} \psi_{v,i}(x) \tag{6.57}$$

On S_1, we have the following

$$E_x'(x,h) = \frac{\omega\mu_1}{4}\left\{\int_0^L J_1(x')H_0^{(2)}(k_1|x-x'|)\,dx' + \frac{1}{k_1^2}\frac{\partial}{\partial x}\int_0^L \frac{\partial J_1(x')}{\partial x'}H_0^{(2)}(k_1|x-x'|)\,dx'\right\}$$

$$+ \frac{\omega\mu_2}{4}\left\{\int_0^L J_1(x')H_0^{(2)}(k_2|x-x'|)\,dx' + \frac{1}{k_2^2}\frac{\partial}{\partial x}\int_0^L \frac{\partial J_1(x')}{\partial x'}H_0^{(2)}(k_2|x-x'|)\,dx'\right\}$$

$$+ \frac{\omega\mu_1}{4}\left\{\int_0^L J_0^-(x')H_0^{(2)}(k_1 t)\,dx' + \frac{1}{k_1^2}\frac{\partial}{\partial x}\int_0^L \frac{\partial J_0^-(x')}{\partial x'}H_0^{(2)}(k_1 t)\,dx'\right\}$$

$$- \frac{\omega\mu_2}{4}\left\{\int_0^L J_0^+(x')H_0^{(2)}(k_2 t)\,dx' + \frac{1}{k_2^2}\frac{\partial}{\partial x}\int_0^L \frac{\partial J_0^+(x')}{\partial x'}H_0^{(2)}(k_2 t)\,dx'\right\}$$

$$H_z^i(x,h) = \frac{-\eta_0\,\omega\mu_1}{4\eta_1^2}\left\{\int_0^L M_1(x')H_0^{(2)}(k_1|x-x'|)\,dx'\right\}$$

$$+ \frac{-\eta_0\,\omega\mu_2}{4\eta_2^2}\left\{\int_0^L M_1(x')H_0^{(2)}(k_2|x-x'|)\,dx'\right\}$$

$$+ \frac{d}{4j}\left\{\int_0^L J_0^-(x')\frac{1}{\sqrt{(x-x')^2+d^2}}\frac{dH_0^{(2)}(k_1 t)}{dt}\,dx'\right\}$$

$$- \frac{d}{4j}\left\{\int_0^L J_0^+(x')\frac{1}{\sqrt{(x-x')^2+d^2}}\frac{dH_0^{(2)}(k_2 t)}{dt}\,dx'\right\}$$

with

$$t = \sqrt{(x-x')^2 + d^2} \tag{6.58}$$

On S_0 we have the following formula:

$$E_x^i(x,0) = \frac{\omega\mu_1}{4}\left\{\int_0^L J_1(x')H_0^{(2)}(k_1 t)\,dx' + \frac{1}{k_1^2}\frac{\partial}{\partial x}\int_0^L \frac{\partial J_1(x')}{\partial x'}H_0^{(2)}(k_1 t)\,dx'\right\}$$

$$+ \frac{\eta_0 d}{4j}\left\{\int_0^L M_1(x')\frac{1}{\sqrt{(x-x')^2+d^2}}\frac{\partial H_0^{(2)}(k_1 t)}{\partial t}\,dx'\right\}$$

$$+ \frac{\omega\mu_1}{4}\left\{\begin{array}{l}\int_0^L J_0^-(x')H_0^{(2)}(k_1|x-x'|)\,dx'\\[2mm]+ \frac{1}{k_2^2}\frac{\partial}{\partial x}\int_0^L \frac{\partial J_0^-(x')}{\partial x'}H_0^{(2)}(k_1|x-x'|)\,dx'\end{array}\right\}$$

$$0 = -\frac{\omega\mu_2}{4}\left\{\int_0^L J_1(x')H_0^{(2)}(k_2 t)\,dx' + \frac{1}{k_2^2}\frac{\partial}{\partial x}\int_0^L \frac{\partial J_1(x')}{\partial x'}H_0^{(2)}(k_2 t)\,dx'\right\}$$

$$+ \frac{\eta_0 d}{4j}\left\{\int_0^L M_1(x')\frac{1}{\sqrt{(x-x')^2+d^2}}\frac{\partial H_0^{(2)}(k_2 t)}{\partial t}\,dx'\right\}$$

$$+ \frac{\omega\mu_2}{4}\left\{\begin{array}{l}\int_0^L J_0^+(x')H_0^{(2)}(k_2|x-x'|)\,dx' \\[2mm] + \frac{1}{k_2^2}\frac{\partial}{\partial x}\int_0^L \frac{\partial J_0^+(x')}{\partial x'}H_0^{(2)}(k_2|x-x'|)\,dx'\end{array}\right\}$$

Next, we employ the method of moments using the multiscale basis for the solution of the integral equations. By using the same multiscale expansion and testing functions (that is, a Galerkin method), we obtain

$$\left\langle E_x^i(x,h),\,\phi_{v',i'}(x)\right\rangle = \sum_{v=0}^V \sum_{i=B(v)}^{A(v,N)} \tau_{v,i}^1 \left\langle L_1^1[\phi_{v,i}(x')] + L_2^1[\phi_{v,i}(x')],\,\phi_{v',i'}(x)\right\rangle$$

$$+ \sum_{v=0}^V \sum_{i=B(v)}^{A(v,N)} \tau_{v,i}^{0-} \left\langle L_1^2[\phi_{v,i}(x')],\,\phi_{v',i'}(x)\right\rangle \qquad (6.59)$$

$$- \sum_{v=0}^V \sum_{i=B(v)}^{A(v,N)} \tau_{v,i}^{0+} \left\langle L_2^2[\phi_{v,i}(x')],\,\phi_{v',i'}(x)\right\rangle$$

$$\left\langle H_z^i(x,h),\,\phi_{v',i'}(x)\right\rangle = -\sum_{v=0}^V \sum_{i=B(v)}^{A(v,N)} \tau_{v,i}^2 \left\langle \begin{array}{l}\eta_0\eta_1^{-2}M_1^1[\phi_{v,i}(x')] \\[2mm] + \eta_0\eta_2^{-2}M_2^1[\phi_{v,i}(x')],\,\phi_{v',i'}(x)\end{array}\right\rangle$$

$$+ d\sum_{v=0}^V \sum_{i=B(v)}^{A(v,N)} \tau_{v,i}^{0-} \left\langle K_1^2[\phi_{v,i}(x')],\,\phi_{v',i'}(x)\right\rangle \qquad (6.60)$$

$$- d\sum_{v=0}^V \sum_{i=B(v)}^{A(v,N)} \tau_{v,i}^{0+} \left\langle K_2^2[\phi_{v,i}(x')],\,\phi_{v',i'}(x)\right\rangle$$

$$\left\langle E_x^i(x,0),\,\phi_{v',i'}(x)\right\rangle = \sum_{v=0}^V \sum_{i=B(v)}^{A(v,N)} \tau_{v,i}^1 \left\langle L_1^2[\phi_{v,i}(x')],\,\phi_{v',i'}(x)\right\rangle$$

$$+ d\eta_0 \sum_{v=0}^V \sum_{i=B(v)}^{A(v,N)} \tau_{v,i}^2 \left\langle K_1^2[\phi_{v,i}(x')],\,\phi_{v',i'}(x)\right\rangle \qquad (6.61)$$

$$+ \sum_{v=0}^V \sum_{i=B(v)}^{A(v,N)} \tau_{v,i}^{0-} \left\langle L_1^1[\phi_{v,i}(x')],\,\phi_{v',i'}(x)\right\rangle$$

$$0 \quad = - \sum_{v=0}^{V} \sum_{i=B(v)}^{A(v,N)} \tau_{v,i}^1 \left\langle L_2^2\left[\phi_{v,i}(x')\right], \phi_{v',i'}(x) \right\rangle$$

$$- d\eta_0 \sum_{v=0}^{V} \sum_{i=B(v)}^{A(v,N)} \tau_{v,i}^2 \left\langle K_2^2\left[\phi_{v,i}(x')\right], \phi_{v',i'}(x) \right\rangle \tag{6.62}$$

$$+ \sum_{v=0}^{V} \sum_{i=B(v)}^{A(v,N)} \tau_{v,i}^{0+} \left\langle L_2^1\left[\phi_{v,i}(x')\right], \phi_{v',i'}(x) \right\rangle$$

where

$$v' = 0, 1, 2, \dots, V; \quad i' = B(v'), \dots, A(v',N)$$

$$A(v,N) = \begin{cases} N+1 & v=0 \\ 2^{v-1} N & v \neq 0 \end{cases} \qquad B(v) = \begin{cases} 0 & v=0 \\ 1 & v \neq 0 \end{cases}$$

$$L_i^1[X(x')] = \frac{\omega\mu_i}{4}\left\{\int_0^L X(x')H_0^{(2)}(k_i|x-x'|)dx' + \frac{1}{k_i^2}\frac{\partial}{\partial x}\int_0^L \frac{\partial X(x')}{\partial x'}H_0^{(2)}(k_i|x-x'|)dx'\right\}$$

$$L_i^2[X(x')] = \frac{\omega\mu_i}{4}\left\{\int_0^L X(x')H_0^{(2)}(k_i t)dx' + \frac{1}{k_i^2}\frac{\partial}{\partial x}\int_0^L \frac{\partial X(x')}{\partial x'}H_0^{(2)}(k_i t)dx'\right\}$$

$$M_i^1[X(x')] = \frac{\omega\mu_i}{4}\left\{\int_0^L X(x')H_0^{(2)}(k_i|x-x'|)dx'\right\}$$

$$M_i^2[X(x')] = \frac{\omega\mu_i}{4}\left\{\int_0^L X(x')H_0^{(2)}(k_i\sqrt{(x-x')^2+d^2})dx'\right\}$$

$$K_i^1[X(x')] = \frac{-k_i}{4j}\left\{\int_0^L X(x')\frac{H_1^{(2)}(k_i|x-x'|)}{|x-x'|}dx'\right\}$$

$$K_i^2[X(x')] = \frac{-k_i}{4j}\left\{\int_0^L X(x')\frac{H_1^{(2)}(k_i\sqrt{(x-x')^2+d^2})}{\sqrt{(x-x')^2+d^2}}dx'\right\}$$

The parameter d is defined as the thickness of the dielectric coating as shown in Figure 6.5. Once the above equations are numerically solved for the unknowns J and M, the bistatic radar cross section σ can be computed by

$$\sigma(\varphi, \varphi^i) = \lim_{r \to \infty} [2\pi r \frac{|\vec{H}^s|^2}{|\vec{H}^i|^2}]$$

$$= \frac{k_1 \eta_0^2}{4} \left| \xi_{0,0} \frac{1}{jk_1 \cos\varphi} \left[e^{\frac{1}{2}jk_1 h \cos\varphi} \operatorname{sinc}(0.5\, k_1 h \cos\varphi) - 1 \right] \right.$$

$$+ \xi_{0,N} \frac{e^{jk_1 L \cos\varphi}}{jk_1 \cos\varphi} \left[1 - e^{-\frac{1}{2}jk_1 h \cos\varphi} \operatorname{sinc}(0.5\, k_1 h \cos\varphi) \right]$$

$$+ \sum_{i=1}^{N-1} \xi_{0,i}\, e^{jk_1 x_{0,i} \cos\varphi}\, h \times \operatorname{sinc}^2(0.5\, k_1 h \cos\varphi) \tag{6.63}$$

$$+ \left. \sum_{v=1}^{V} \sum_{i=1}^{2^{v-1}N} \xi_{v,i}\, e^{jk_1 x_{v,i} \cos\varphi}\, \frac{h}{2^v} \times \operatorname{sinc}^2\left(\frac{1}{2^{v+1}} k_1 h \cos\varphi\right) \right|^2$$

where

$$\xi_{v,i} = \sin\varphi(\tau_{v,i}^1 e^{jk_1 d \sin\varphi} + \tau_{v,i}^{0-}) + \frac{\eta_0}{\eta_1} e^{jk_1 d \sin\varphi} \tau_{v,i}^2 \; ; \quad \text{and} \quad \operatorname{sinc}(x) = \frac{\sin x}{x}$$

The scattered magnetic field is given by

$$\vec{H}^s(\vec{r}) = \hat{z} \sqrt{\frac{k_1}{8\pi}}\, e^{\frac{j\pi}{4}} \frac{e^{-jk_1 r}}{\sqrt{r}} \int_0^L \left[\begin{array}{l} \sin\varphi\, \{J_1(x') e^{jk_1 d \sin\varphi} \\ + J_0^-(x')\} + \frac{\eta_0}{\eta_1} e^{jk_1 d \sin\varphi} M_1(x') \end{array} \right] e^{jk_1 x' \cos\varphi}\, dx'$$

$$\tag{6.64}$$

6.5.2.2 Transverse Magnetic (TM) Case

In this case, the incident field is defined by

$$E_z^i = E_0^i\, e^{jk_1(x \cos\varphi^i + y \sin\varphi^i)}$$

$$H_x^i = -\frac{E_0^i}{\eta_0} \sin\varphi^i\, e^{jk_1(x \cos\varphi^i + y \sin\varphi^i)} \tag{6.65}$$

The induced electric current on either side of the surface S_0 can be written in the following form with the Cartesian unit vectors $\hat{x}$ and $\hat{z}$:

$$\vec{J}_0^{\pm}(\vec{r}) = -\hat{z}\, J_0^{\pm}(x)$$

$$J_0^{\pm}(x) = \sum_{i=0}^{N} \tau_{0,i}^{\pm} \psi_{0,i}(x) + \sum_{v=1}^{V} \sum_{i=1}^{2^{v-1}N} \tau_{v,i}^{\pm} \psi_{v,i}(x) \tag{6.66}$$

The induced electric and magnetic currents on the surface S_1 can be written in the following form with the Cartesian unit vectors $\hat{x}$ and $\hat{z}$:

$$\bar{J}_1(\bar{r}) = - \hat{z} J_1(x)$$

$$J_1(x) = \sum_{i=0}^{N} \tau_{0,i}^1 \, \psi_{0,i}(x) + \sum_{v=1}^{V} \sum_{i=1}^{2^{v-1}N} \tau_{v,i}^1 \, \psi_{v,i}(x) \tag{6.67}$$

$$\bar{M}_1(\bar{r}) = \hat{x}\eta_0 \, M_1(x)$$

$$M_1(x) = \sum_{i=0}^{N} \tau_{0,i}^2 \, \psi_{0,i}(x) + \sum_{v=1}^{V} \sum_{i=1}^{2^{v-1}N} \tau_{v,i}^2 \, \psi_{v,i}(x) \tag{6.68}$$

On S_1, we have the following formula:

$$
E'_z(x,h) = \frac{-\omega\mu_1}{4}\left\{\int_0^L J_1(x') \, H_0^{(2)}(k_1|x-x'|) \, dx'\right\}
$$

$$
+ \frac{-\omega\mu_2}{4}\left\{\int_0^L J_1(x') \, H_0^{(2)}(k_2|x-x'|) \, dx'\right\}
$$

$$
- \frac{\omega\mu_1}{4}\left\{\int_0^L J_0^-(x') H_0^{(2)}(k_1 t) \, dx'\right\}
$$

$$
+ \frac{\omega\mu_2}{4}\left\{\int_0^L J_0^+(x') \, H_0^{(2)}(k_2 t) \, dx'\right\}
$$

$$
H_x^i(x,h) = \frac{\eta_0 \, \omega\mu_1}{4\eta_1^2}\left\{\begin{array}{l}\displaystyle\int_0^L M_1(x') H_0^{(2)}(k_1|x-x'|)\,dx'\\[2mm]\displaystyle+\frac{1}{k_1^2}\frac{\partial}{\partial x}\int_0^L \frac{\partial M_1(x')}{\partial x'} H_0^{(2)}(k_1|x-x'|)\,dx'\end{array}\right\}
$$

$$
+ \frac{\eta_0 \, \omega\mu_2}{4\eta_2^2}\left\{\begin{array}{l}\displaystyle\int_0^L M_1(x') H_0^{(2)}(k_2|x-x'|)\,dx'\\[2mm]\displaystyle+\frac{1}{k_2^2}\frac{\partial}{\partial x}\int_0^L \frac{\partial M_1(x')}{\partial x'} H_0^{(2)}(k_2|x-x'|)\,dx'\end{array}\right\}
$$

$$
+ \frac{d}{4j}\left\{\int_0^L J_0^-(x')\frac{1}{\sqrt{(x-x')^2+d^2}}\frac{\partial H_0^{(2)}(k_1 t)}{\partial t}\,dx'\right\}
$$

$$
- \frac{d}{4j}\left\{\int_0^L J_0^+(x')\frac{1}{\sqrt{(x-x')^2+d^2}}\frac{\partial H_0^{(2)}(k_2 t)}{\partial t}\,dx'\right\}
$$

with t defined by (6.58). On S_0, we have the following:

$$E_z^i(x,0) = \frac{\omega\mu_1}{-4} \left\{ \int_0^{L'} J_1(x') H_0^{(2)}(k_1 t)\, dx' \right\}$$

$$+ \frac{\eta_0 d}{4j} \left\{ \int_0^L M_1(x') \frac{1}{\sqrt{(x-x')^2+d^2}} \frac{\partial H_0^{(2)}(k_1 t)}{\partial t}\, dx' \right\}$$

$$+ \frac{\omega\mu_1}{-4} \left\{ \int_0^L J_0^-(x') H_0^{(2)}(k_1|x-x'|)\, dx' \right\}$$

$$0 \quad = \quad \frac{\omega\mu_2}{4} \left\{ \int_0^L J_1(x') H_0^{(2)}(k_2 t)\, dx' \right\}$$

$$+ \frac{-\eta_0 d}{4j} \left\{ \int_0^L M_1(x') \frac{1}{\sqrt{(x-x')^2+d^2}} \frac{\partial H_0^{(2)}(k_2 t)}{\partial t}\, dx' \right\}$$

$$+ \frac{\omega\mu_2}{-4} \left\{ \int_0^L J_0^+(x') H_0^{(2)}(k_2|x-x'|)\, dx' \right\}$$

By using the multiscale basis as an expansion function for the currents and also as testing functions in the method of moment, we obtain the following expressions:

$$\left\langle E_z^i(x,d), \phi_{v',i'}(x) \right\rangle = -\sum_{v=0}^{V} \sum_{i=B(v)}^{A(v,N)} \tau_{v,i}^1 \left\langle M_1^1\big[\phi_{v,i}(x')\big] + M_2^1\big[\phi_{v,i}(x')\big], \phi_{v',i'}(x) \right\rangle$$

$$- \sum_{v=0}^{V} \sum_{i=B(v)}^{A(v,N)} \tau_{v,i}^{0-} \left\langle M_1^2\big[\phi_{v,i}(x')\big], \phi_{v',i'}(x) \right\rangle$$

$$+ \sum_{v=0}^{V} \sum_{i=B(v)}^{A(v,N)} \tau_{v,i}^{0+} \left\langle M_2^2\big[\phi_{v,i}(x')\big], \phi_{v',i'}(x) \right\rangle \tag{6.69}$$

$$\left\langle H_x^i(x,h), \phi_{v',i'}(x) \right\rangle = \sum_{v=0}^{V} \sum_{i=B(v)}^{A(v,N)} \tau_{v,i}^2 \left\langle \begin{array}{l} \eta_0 \eta_1^{-2} L_1^1\big[\phi_{v,i}(x')\big] \\ + \eta_0 \eta_2^{-2} L_2^1\big[\phi_{v,i}(x')\big], \phi_{v',i'}(x) \end{array} \right\rangle$$

$$+ d \sum_{v=0}^{V} \sum_{i=B(v)}^{A(v,N)} \tau_{v,i}^{0-} \left\langle K_1^2\big[\phi_{v,i}(x')\big], \phi_{v',i'}(x) \right\rangle - d \sum_{v=0}^{V} \sum_{i=B(v)}^{A(v,N)} \tau_{v,i}^{0+} \left\langle K_2^2\big[\phi_{v,i}(x')\big], \phi_{v',i'}(x) \right\rangle$$

$$\tag{6.70}$$

$$\left\langle E_z^i(x,0), \phi_{v',i'}(x) \right\rangle = -\sum_{v=0}^{V} \sum_{i=B(v)}^{A(v,N)} \tau_{v,i}^1 \left\langle M_1^2\big[\phi_{v,i}(x')\big], \phi_{v',i'}(x) \right\rangle$$

$$+ d\eta_0 \sum_{v=0}^{V} \sum_{i=B(v)}^{A(v,N)} \tau_{v,i}^2 \left\langle K_1^2\big[\phi_{v,i}(x')\big], \phi_{v',i'}(x) \right\rangle - \sum_{v=0}^{V} \sum_{i=B(v)}^{A(v,N)} \tau_{v,i}^{0-} \left\langle M_1^1\big[\phi_{v,i}(x')\big], \phi_{v',i'}(x) \right\rangle$$

$$\tag{6.71}$$

$$0 = \sum_{v=0}^{V} \sum_{i=B(v)}^{A(v,N)} \tau_{v,i}^{1} \left\langle M_2^2 \left[\phi_{v,i}(x') \right], \phi_{v',i'}(x) \right\rangle$$

$$- d\eta_0 \sum_{v=0}^{V} \sum_{i=B(v)}^{A(v,N)} \tau_{v,i}^{2} \left\langle K_2^2 \left[\phi_{v,i}(x') \right], \phi_{v',i'}(x) \right\rangle \tag{6.72}$$

$$- \sum_{v=0}^{V} \sum_{i=B(v)}^{A(v,N)} \tau_{v,i}^{0+} \left\langle L_2^1 \left[\phi_{v,i}(x') \right], \phi_{v',i'}(x) \right\rangle$$

where

$$v' = 0, 1, 2, ..., V; \quad i' = B'(v), ..., A(v', N)$$

$$A(v, N) = \begin{cases} N+1 & v=0 \\ 2^{v-1} N & v \neq 0 \end{cases} \qquad B(v) = \begin{cases} 0 & v=0 \\ 1 & v \neq 0 \end{cases}$$

$$L_i^1[X(x')] = \frac{\omega\mu_i}{4} \left\{ \begin{aligned} &\int_0^L X(x') H_0^{(2)}(k_i|x-x'|) dx' \\ &+ \frac{1}{k_i^2} \frac{\partial}{\partial x} \int_0^L \frac{\partial X(x')}{\partial x'} H_0^{(2)}(k_i|x-x'|) dx' \end{aligned} \right\}$$

$$L_i^2[X(x')] = \frac{\omega\mu_i}{4} \left\{ \int_0^L X(x') H_0^{(2)}(k_i t) dx' + \frac{1}{k_i^2} \frac{\partial}{\partial x} \int_0^L \frac{\partial X(x')}{\partial x'} H_0^{(2)}(k_i x) dx' \right\}$$

$$M_i^1[X(x')] = \frac{\omega\mu_i}{4} \left\{ \int_0^L X(x') H_0^{(2)}(k_i|x-x'|) dx' \right\}$$

$$M_i^2[X(x')] = \frac{\omega\mu_i}{4} \left\{ \int_0^L X(x') H_0^{(2)}(k_i\sqrt{(x-x')^2+d^2}) dx' \right\}$$

$$K_i^1[X(x')] = \frac{-k_i}{4j} \left\{ \int_0^L X(x') \frac{H_1^{(2)}(k_i|x-x'|)}{|x-x'|} dx' \right\}$$

$$K_i^2[X(x')] = \frac{-k_i}{4j} \left\{ \int_0^L X(x') \frac{H_1^{(2)}(k_i\sqrt{(x-x')^2+d^2})}{\sqrt{(x-x')^2+d^2}} dx' \right\}$$

Once the above integral equations are solved for the unknown currents, the bistatic radar cross section σ can be computed from

$$\sigma(\varphi, \varphi^i) = \lim_{r \to \infty} \left[2\pi r \frac{\left|\vec{E}^s\right|^2}{\left|\vec{E}^i\right|^2} \right]$$

$$= \frac{k_1}{4} \left| \xi_{0,0} \frac{1}{jk_1 \cos\varphi} \left[e^{\frac{1}{2}jk_1 h \cos\varphi} \operatorname{sinc}\left(0.5 k_1 h \cos\varphi\right) - 1 \right] \right.$$

$$+ \xi_{0,N} \frac{e^{jk_1 L \cos\varphi}}{jk_1 \cos\varphi} \left[1 - e^{-\frac{1}{2}jk_1 h \cos\varphi} \operatorname{sinc}\left(0.5 k_1 h \cos\varphi\right) \right] \tag{6.73}$$

$$+ \sum_{i=1}^{N-1} \xi_{0,i} e^{jk_1 x_{0,i} \cos\varphi} h \times \operatorname{sinc}^2\left(0.5 k_1 h \cos\varphi\right)$$

$$+ \left. \sum_{v=1}^{V} \sum_{i=1}^{2^{v-1}N} \xi_{v,i} e^{jk_1 x_{v,i} \cos\varphi} \frac{h}{2^v} \times \operatorname{sinc}^2\left(\frac{1}{2^{v+1}} k_1 h \cos\varphi\right) \right|^2$$

where $\qquad \xi_{v,i} = \eta_1 (\tau_{v,i}^1 e^{jk_1 d \sin\varphi} + \tau_{v,i}^{0-}) - \eta_0 e^{jk_1 d \sin\varphi} \tau_{v,i}^2$

The scattered electric field is given by

$$\vec{E}^s(\vec{r}) = \hat{z} \sqrt{\frac{k_1}{8\pi}} e^{\frac{j\pi}{4}} \frac{e^{-jk_1 r}}{\sqrt{r}} \int_0^L \left[\begin{array}{c} \eta_1 \{ J_1(x') e^{jk_1 d \sin\varphi} + J_0^-(x') \} \\ - \eta_0 e^{jk_1 d \sin\varphi} M_1(x') \end{array} \right] e^{jk_1 x' \cos\varphi} \, dx' \tag{6.74}$$

6.5.3 Solution of the Integral Equations by the Adaptive Multiscale Moment Method (AMMM)

In this section, we discuss how to efficiently solve the linear equations of (6.59)–(6.62) for the TE case and (6.69)–(6.72) for the TM case using the AMMM. For both of these cases, the four equations can be written in the following matrix form:

$$\begin{pmatrix} A_{1,1} & A_{1,2} & A_{1,3} & O \\ A_{2,1} & A_{2,2} & O & A_{2,4} \\ A_{1,3} & O & A_{3,3} & A_{3,4} \\ O & A_{2,4} & A_{3,4} & A_{4,4} \end{pmatrix} \begin{pmatrix} J_0^- \\ J_1 \\ M_1 \\ J_0^+ \end{pmatrix} = \begin{pmatrix} F_1 \\ F_2 \\ F_3 \\ O \end{pmatrix} \tag{6.75}$$

where

$$A_{1,1} = \begin{cases} \left\langle L_1^1 [\phi_{v,i}], \phi_{v',i'} \right\rangle & \text{for TE case} \\ -\left\langle M_1^1 [\phi_{v,i}], \phi_{v',i'} \right\rangle & \text{for TM case} \end{cases}$$

$$A_{1,2} = \begin{cases} \left\langle L_1^2[\phi_{v,i}], \phi_{v',i'} \right\rangle & \text{for TE case} \\ -\left\langle M_1^2[\phi_{v,i}], \phi_{v',i'} \right\rangle & \text{for } TM \text{ case} \end{cases}$$

$$A_{1,3} = d\eta_0 \left\langle K_1^2[\phi_{v,i}], \phi_{v',i'} \right\rangle \qquad \text{for both TE and TM cases}$$

$$A_{2,2} = \begin{cases} \left\langle (L_1^1 + L_2^1)[\phi_{v,i}], \phi_{v',i'} \right\rangle & \text{for TE case} \\ -\left\langle (M_1^1 + M_2^1)[\phi_{v,i}], \phi_{v',i'} \right\rangle & \text{for TM case} \end{cases}$$

$$A_{2,4} = \begin{cases} -\left\langle (\frac{\eta_0^2}{\eta_1^2} M_1^1 + \frac{\eta_0^2}{\eta_2^2} M_2^1)[\phi_{v,i}], \phi_{v',i'} \right\rangle & \text{for TE case} \\ \left\langle (\frac{\eta_0^2}{\eta_1^2} L_1^1 + \frac{\eta_0^2}{\eta_2^2} L_2^1)[\phi_{v,i}], \phi_{v',i'} \right\rangle & \text{for TM case} \end{cases}$$

$$A_{3,4} = -d\eta_0 \left\langle K_2^2[\phi_{v,i}], \phi_{v',i'} \right\rangle \qquad \text{for both TE and TM cases}$$

$$A_{4,4} = \begin{cases} \left\langle L_1^1[\phi_{v,i}], \phi_{v',i'} \right\rangle & \text{for TE case} \\ -\left\langle M_2^1[\phi_{v,i}], \phi_{v',i'} \right\rangle & \text{for TM case} \end{cases}$$

$$F_1 = \begin{cases} \left\langle \sin\varphi^i \exp[jk_1 x \cos\varphi^i], \phi_{v',i'} \right\rangle & \text{for TE case} \\ \left\langle \exp[jk_1 x \cos\varphi^i], \phi_{v',i'} \right\rangle & \text{for TM case} \end{cases}$$

$$F_2 = \begin{cases} \left\langle \sin\varphi^i \exp[jk_1(x\cos\varphi^i + d\sin\varphi^i)], \phi_{v',i'} \right\rangle & \text{for TE case} \\ \left\langle \exp[jk_1(x\cos\varphi^i + d\sin\varphi^i)], \phi_{v',i'} \right\rangle & \text{for TM case} \end{cases}$$

$$F_3 = \begin{cases} \left\langle \exp[jk_1(x\cos\varphi^i + d\sin\varphi^i)], \phi_{v',i'} \right\rangle & \text{for TE case} \\ -\left\langle \sin\varphi^i \exp[jk_1(x\cos\varphi^i + d\sin\varphi^i)], \phi_{v',i'} \right\rangle & \text{for TM case} \end{cases}$$

All elements of the matrix represented by the symbol 0 are zero. It can be demonstrated that all of the matrices $A_{i,j}$, the array of both the excitations F_i,

and the unknowns ($J_0^\pm$, J_1, M_1) can be arranged in a special form using the concepts of multiscales as shown in Figure 6.6. The solution of the problem is carried out following the steps presented earlier. It is summarized here for convenience:

Step 1: At the scale $V = 0$, the system of linear equations generated by the multiscale basis is identical to that obtained by the standard MM using a piecewise triangular basis. The unknown coefficients at this scale can be solved for using LU decomposition or by the iterative conjugate gradient method.

Step 2: At the next step we extrapolate the solution using a quadratic interpolation formula as presented in the previous section of this chapter. This would be the initial estimate of the solution at scale $V + 1$. We now compute the solution at the $V + 1$ scale from the solution obtained at the V scale by the quadratic interpolation methodology described earlier in this chapter. Each of the unknown functions can be represented in the same form that we denote (for convenience) simply by the function $X_{V+1}(x)$. The unknown coefficients related to the unknown currents $J_0^\pm, J_1, M_1$ at the Vth scale are given by $X_v = \left(\tau_{v,1}, \tau_{v,2}, ..., \tau_{v,2^{v-1}N}\right)^T$. Next, we compute the approximate solution at the $V + 1$ scale as it is related to the solution at the Vth scale through

$$X_{V+1}(x) = X_V(x) + \sum_{i=1}^{2^V N} \tau_{V+1,i} \phi_{V+1,i}(x) \qquad (6.76)$$

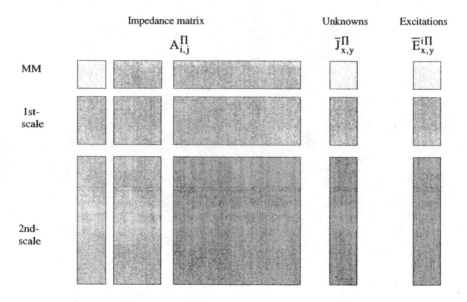

Figure 6.6 Decomposition of the impedance matrix, unknown coefficients, and the excitation in a multiscale basis.

Therefore, the known solution $X_V(x)$ at the V scale can be used to generate the solution for the unknown $X_{V+1}(x)$ at the $V+1$ scale. Now, we need to find the coefficients $\{\tau_{V+1,i}\}$ in (6.76). But before we do that, we make an initial investigation to see if all the unknown coefficients are really necessary.

To address this issue we extrapolate the solution $X_V(x)$ at the Vth scale to the function $X_{V+1}(x)$ using a quadratic interpolation. At the new nodes, we estimate the second derivative of the extrapolated function $X_{V+1}(x)$, which is identical to $\{\tau_{V+1,i}\}$. Now if the extrapolated solution is approximately linear near the new nodes at scale $V+1$, then the second derivative will be close to zero. Therefore, if we prespecify a numerical threshold ε below which the unknown coefficients are set to zero, it will not be necessary to refine the multiscale basis near those nodes. In summary, $\tau_{V+1,i}$ can be estimated from $X_V(x)$ through an interpolation since $X_{V+1}^{(0)} = (\tau_{V+1,1}^{(0)}, \tau_{V+1,2}^{(0)}, ..., \tau_{V+1,2^V N}^{(0)})^T$. Therefore, the array at the $V+1$ scale

$(X_0, X_1, ..., X_{V+1})^T$ can be approximated from the already computed array

$(X_0, X_1, ..., X_V)^T$. Next we eliminate the relatively smaller components of the predicted solution components and omit the corresponding rows and columns from the system matrix obtained using the Galerkin method at scale $V+1$.

If $\left| \tau_{v,i}^{(0)} \right| \le \varepsilon$ $(v = 1, 2, ..., V+1;\ i = 1, 2, ..., 2^V N)$, then we set the

numerical values for $\tau_{v,i}^{(0)} = 0$, and delete the corresponding rows and columns of the system matrix with respect to i at scale $V+1$. This is an important step in reducing the size of the linear equations. In the actual computations, we choose the following criteria $\left| \tau_{v,i}^{(0)} \right| \le \varepsilon\, Q$ where $Q = \max \left| \tau_{v,i}^{(0)} \right|$ for eliminating the elements of the system matrix.

Step 3: At the third step we solve the modified systems of linear equations using the conjugate gradient method and $[X_V\ \overrightarrow{O}]^T$ as the initial guess, where O represents a vector with zero elements.

Step 4: Before transforming the computed solution obtained using the conjugate gradient method to the solution $X_{V+1}(x)$ at the $V+1$ scale, we add back the zero coefficients.

Then we go back to Step 2 and continue the iterative process until we reach the highest scale. Because the size of the computed system of equations is smaller than the original system of linear equations obtained from the discretization of the operator equation by the Galerkin method using a multiscale basis, the new technique will improve the computational efficiency. The flowchart for the adaptive multiscale moment method from the V scale to the $V+1$ scale is given in Figure 6.7.

6.5.4 Numerical Results

Numerical calculations have been performed by using the AMMM procedure with the multiscale basis representing the unknown.

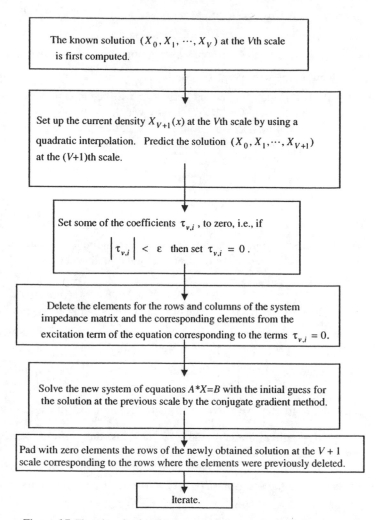

The known solution $(X_0, X_1, \cdots, X_V)$ at the Vth scale is first computed.

Set up the current density $X_{V+1}(x)$ at the Vth scale by using a quadratic interpolation. Predict the solution $(X_0, X_1, \cdots, X_{V+1})$ at the $(V+1)$th scale.

Set some of the coefficients $\tau_{v,i}$, to zero, i.e., if $\left| \tau_{v,i} \right| < \varepsilon$ then set $\tau_{v,i} = 0$.

Delete the elements for the rows and columns of the system impedance matrix and the corresponding elements from the excitation term of the equation corresponding to the terms $\tau_{v,i} = 0$.

Solve the new system of equations $A*X=B$ with the initial guess for the solution at the previous scale by the conjugate gradient method.

Pad with zero elements the rows of the newly obtained solution at the $V + 1$ scale corresponding to the rows where the elements were previously deleted.

Iterate.

Figure 6.7 Flowchart for the adaptive multiscale moment method.

As a first example, we choose a 5.221λ long perfectly conducting strip. An "artificial" dielectric of $\varepsilon_r = 1$ with a thickness of 0.01λ is assumed to coat the strip. The size of the linear equation for AMMM for different threshold values and with normal incidence are listed in Tables 6.1 to 6.3 for the TE case and Tables 6.4 to 6.6 for the TM case, respectively. The number of unknowns for each variable (J_0^+, J_0^-, J_1, M_1) and the total number of unknowns (Σ) are given for each scale. The number of the initial division N is taken as 8. We can see from Tables 6.1 to 6.6 that the size of the moment matrix ($\varepsilon = 0$) can be significantly reduced by utilizing different scales V and setting different threshold values ε for the coefficients $(\tau_{v,i}^{0\pm}, \tau_{v,i}^1 \tau_{v,i}^2)$ in extrapolating the function to the next higher scale.

Table 6.1
Normal Incidence: TE case, $\varepsilon = 0$

$N = 8$	J_0^+	J_0^-	J_1	M_1	Σ
$V = 0$	7	7	7	9	30
$V = 1$	15	15	15	17	62
$V = 2$	31	31	31	33	126
$V = 3$	63	63	63	65	254
$V = 4$	127	127	127	129	510
$V = 5$	255	255	255	257	1,022

Table 6.2
Normal Incidence: TE case, $\varepsilon = 0.01$

$N = 8$	J_0^+	J_0^-	J_1	M_1	Σ
$V = 0$	7	7	7	9	30
$V = 1$	15	15	15	17	62
$V = 2$	31	31	31	31	124
$V = 3$	63	63	63	27	216
$V = 4$	119	127	121	26	393
$V = 5$	170	234	171	28	603

Table 6.3
Normal Incidence: TE case, $\varepsilon = 0.1$

$N = 8$	J_0^+	J_0^-	J_1	M_1	Σ
$V = 0$	7	7	7	9	30
$V = 1$	11	14	11	13	49
$V = 2$	15	23	15	16	69
$V = 3$	20	40	21	16	97
$V = 4$	20	48	20	18	106
$V = 5$	19	44	19	20	102

Table 6.4
Normal Incidence: TM case, $\varepsilon = 0$

$N = 8$	J_0^+	J_0^-	J_1	M_1	Σ
$V = 0$	9	9	9	7	34
$V = 1$	17	17	17	15	66
$V = 2$	33	33	33	31	130
$V = 3$	65	65	65	63	258
$V = 4$	129	129	129	127	514
$V = 5$	257	257	257	255	1,026

Table 6.5
Normal Incidence: TM case, $\varepsilon = 0.01$

$N = 8$	J_0^+	J_0^-	J_1	M_1	Σ
$V = 0$	9	9	9	7	34
$V = 1$	15	16	15	15	61
$V = 2$	19	23	20	30	92
$V = 3$	25	25	25	58	133
$V = 4$	30	24	29	93	176
$V = 5$	33	28	34	184	278

Table 6.6
Normal Incidence: TM case, $\varepsilon = 0.1$

$N = 8$	J_0^+	J_0^-	J_1	M_1	Σ
$V = 0$	7	7	7	9	34
$V = 1$	11	13	11	14	49
$V = 2$	14	17	13	29	73
$V = 3$	13	17	13	45	88
$V = 4$	13	19	13	76	121
$V = 5$	13	21	13	143	190

The magnitudes and the phases of the equivalent currents ($J_0^\pm, J_1, M_1$) for the case of the incident angle $\phi = 22°$ and $\varepsilon = 0.01$ are plotted in Figure 6.8. The curves for the monostatic radar cross section (RCS) for the TE and TM cases are plotted in Figures 6.9 and 6.10, respectively. We show two results in each figure. One of them is due to a perfectly conducting strip; the other one is also for a strip but with a 0.01λ of free-space coating. Therefore, if there is any difference between the two results, then it must be due to the numerical approximations. The monostatic RCS of the structure for both the cases is as shown in Figure 6.11 for the TE and TM cases. The effect of the threshold $\varepsilon = 0.01$ in reducing the number of equations for the unknowns ($J_0^\pm, J_1, M_1$) for the TM case for different incident angles is plotted in Figure 6.12.

As a second example, we consider an 8.23λ long strip coated with a dielectric of $\varepsilon_{2r} = 2$. The thickness of the dielectric is 0.057λ. We consider the TM illumination for this case. The sizes of the systems of linear equations for the AMMM for different values of threshold and normal incidence are listed in Table 6.3. The number of the initial division N is taken as 8. We can see from Tables 6.7 to 6.9 that the size of the moment matrix for this problem can also be significantly reduced by utilizing different scales and by using different values for the threshold ε. The monostatic RCS for the TE case is plotted in Figure 6.13. The computational results at the fifth scale have been presented for all angles of incidence.

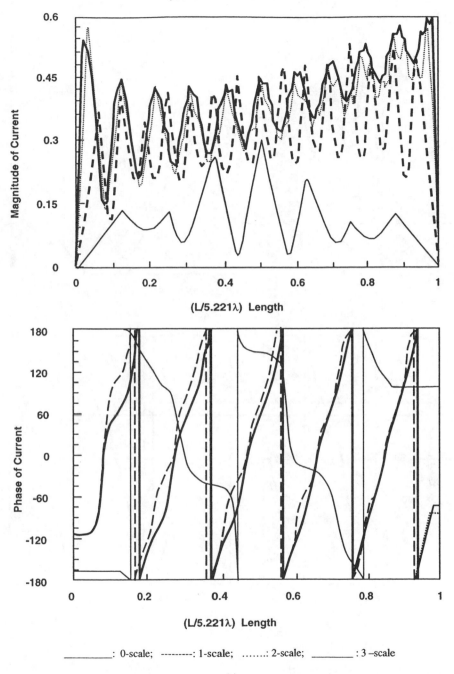

: 0-scale;   ---------: 1-scale;   ....: 2-scale;   _____ : 3 –scale

(a)

Figure 6.8 The magnitude and phase of the electric and magnetic currents at different scales of the approximation: (a) J_0^-, (b) J_0^+, (c) J_1, and (d) M_1.

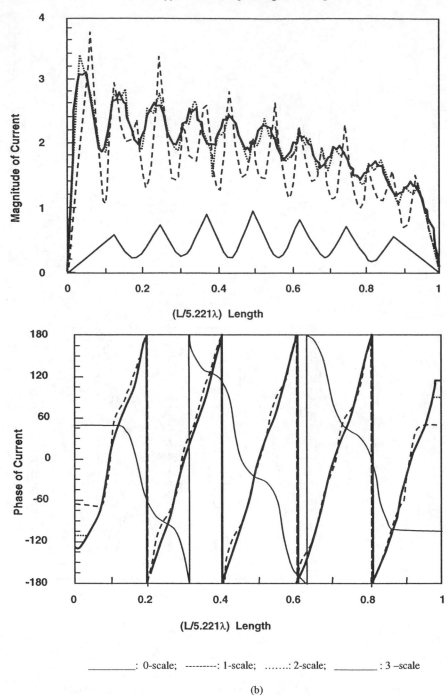

_____ : 0-scale;   ---------: 1-scale;   ....: 2-scale;   _____ : 3 –scale

(b)

Figure 6.8 Continued.

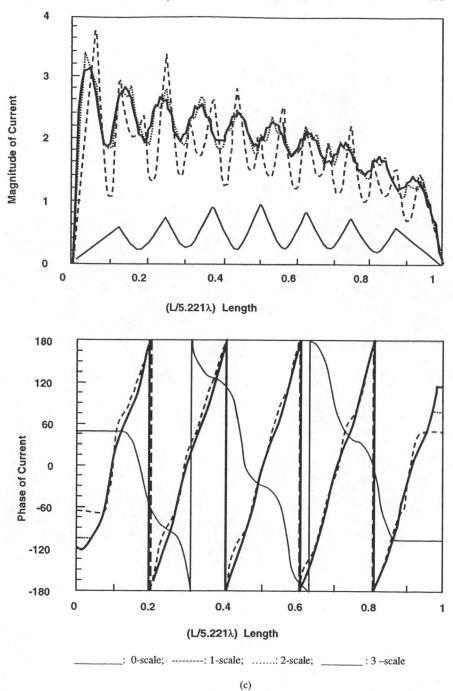

(L/5.221λ) Length

(L/5.221λ) Length

_____ : 0-scale; --------- : 1-scale; ....: 2-scale; _____ : 3 –scale

(c)

Figure 6.8 Continued.

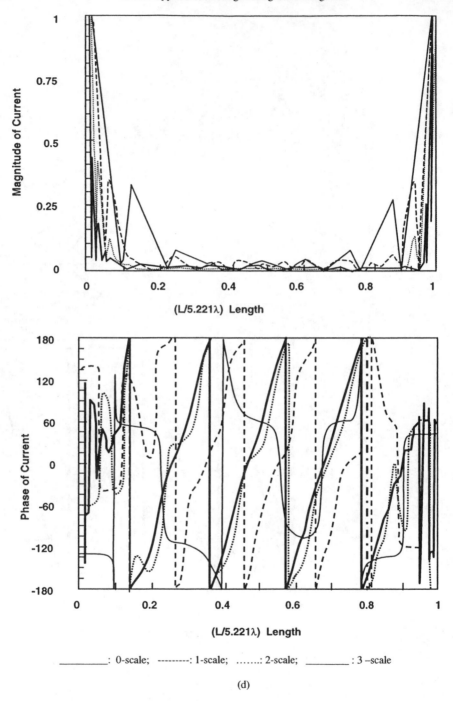

(d)

Figure 6.8 Continued.

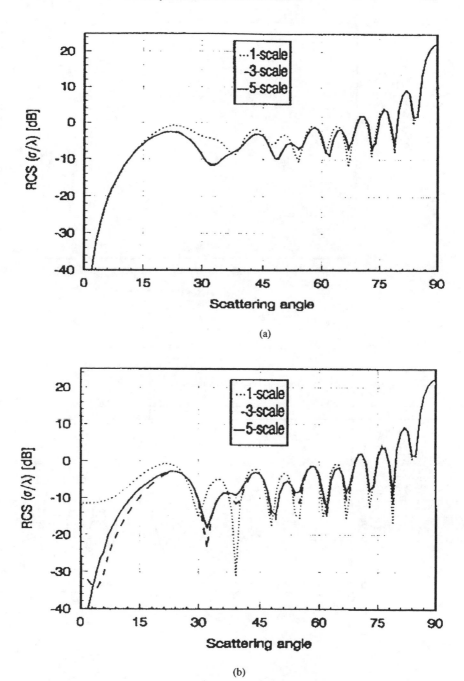

(a)

(b)

Figure 6.9 Monostatic RCS for the TE case: (a) conducting strip and (b) free-space coated strip.

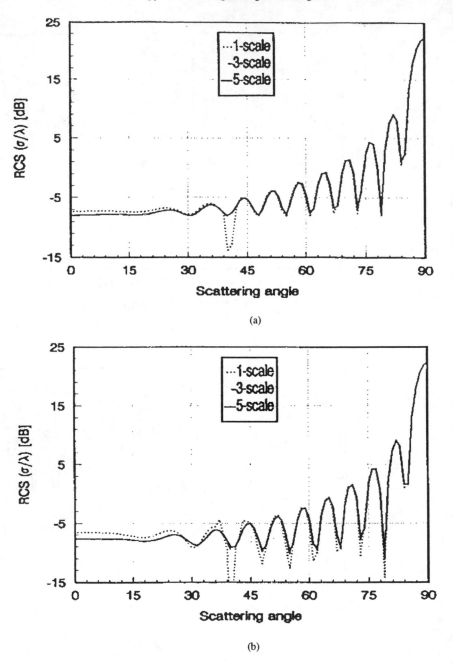

(a)

(b)

Figure 6.10 Monostatic RCS for the TM case: (a) conducting strip and (b) free-space coated strip.

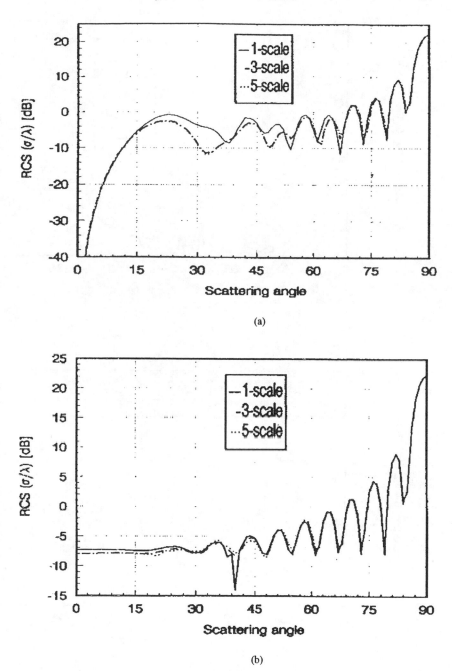

(a)

(b)

Figure 6.11 RCS of the structure for (a) the TE case and (b) the TM case.

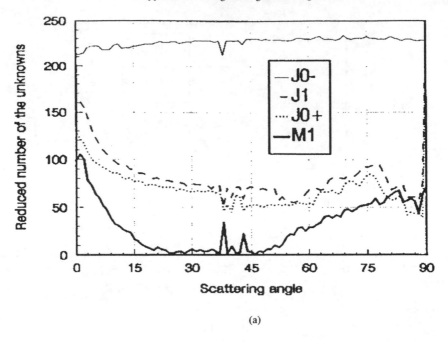

(a)

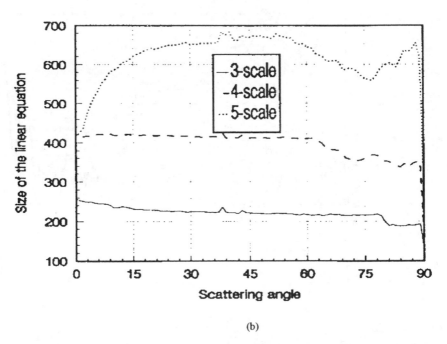

(b)

Figure 6.12 Effect of the threshold $\varepsilon = 0.01$ on the system of equations for the TM case: (a) the reduced number of unknowns at the fifth scale and (b) the size of the system of linear equations.

Table 6.7
Normal Incidence, $\varepsilon = 0$

$N = 8$	J_0^+	J_0^-	J_1	M_1	Σ
$V = 0$	9	9	9	7	34
$V = 1$	17	17	17	15	66
$V = 2$	33	33	33	31	130
$V = 3$	65	65	65	63	258
$V = 4$	129	129	129	127	514
$V = 5$	257	257	257	255	1,026

Table 6.8
Normal Incidence, $\varepsilon = 0.01$

$N = 8$	J_0^+	J_0^-	J_1	M_1	Σ
$V = 0$	9	9	9	7	34
$V = 1$	15	15	15	15	60
$V = 2$	21	26	21	31	99
$V = 3$	28	23	29	53	133
$V = 4$	34	26	32	89	181
$V = 5$	40	31	33	155	259

Table 6.9
Normal Incidence, $\varepsilon = 0.1$

$N = 8$	J_0^+	J_0^-	J_1	M_1	Σ
$V = 0$	7	7	7	9	34
$V = 1$	11	13	11	14	49
$V = 2$	15	17	13	27	72
$V = 3$	15	17	13	43	88
$V = 4$	15	19	13	77	124
$V = 5$	15	21	11	143	190

6.6 EXTENSION OF THE MULTISCALE CONCEPTS TO 2-D PROBLEMS

We have observed that one-dimensional multiscale basis functions ψ defined on the interval $[0,1]$ are related to the usual subsectional triangular basis ϕ at zero scale by

$$\begin{pmatrix} \phi_1(x) \\ \phi_2(x) \\ \phi_3(x) \end{pmatrix} = \begin{pmatrix} 1 & 0 & -\frac{1}{2} \\ 0 & 0 & 1 \\ 0 & 1 & -\frac{1}{2} \end{pmatrix} \begin{pmatrix} \psi_1^{(0)}(x) \\ \psi_2^{(0)}(x) \\ \psi_1^{(1)}(x) \end{pmatrix} \qquad (6.77)$$

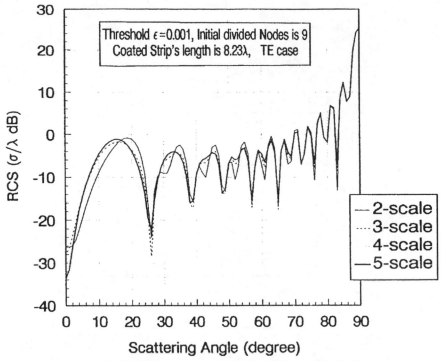

Figure 6.13 Backscatter cross section of the coated strip.

where the superscripts on the functions ψ represent the order of the scale. Values of zero and one represent the function at the zeroth scale and the first scale, respectively. The plots of the various functions $\psi_1^{(0)}(x)$, $\psi_2^{(0)}(x)$, $\psi_1^{(1)}(x)$, $\phi_1(x)$, $\phi_2(x)$, $\phi_3(x)$ are shown in Figure 6.14. From this and 1-D basis, one can obtain a 2-D basis on the region $[0,1] \times [0,1]$ by forming a tensor product of the two 1-D basis functions. Therefore, the 2-D triangular basis functions are obtained as

$$[\phi_1(x)\phi_1(y), \phi_2(x)\phi_1(y), \phi_3(x)\phi_1(y), \phi_1(x)\phi_2(y),$$

$$\phi_2(x)\phi_2(y), \phi_3(x)\phi_2(y), \phi_1(x)\phi_3(y), \phi_2(x)\phi_3(y), \phi_3(x)\phi_3(y)]. \qquad (6.78)$$

These functions are shown in Figure 6.14(a). The 2-D multiscale triangular basis functions are then given by

$$[\psi_1^{(0)}(x)\psi_1^{(0)}(y), \psi_2^{(0)}(x)\psi_1^{(0)}(y), \psi_1^{(0)}(x)\psi_2^{(0)}(y), \psi_2^{(0)}(x)\psi_2^{(0)}(y), \psi_1^{(1)}(x)\psi_1^{(0)}(y),$$

$$\psi_1^{(1)}(x)\psi_2^{(0)}(y), \psi_1^{(0)}(x)\psi_1^{(1)}(y), \psi_2^{(0)}(x)\psi_1^{(1)}(y), \psi_1^{(1)}(x)\psi_1^{(1)}(y)] \qquad (6.79)$$

They are shown in Figure 6.14(b). Let us assume that a function $f(x,y)$ is approximated in the region $[0,1] \times [0,1]$ by the 2-D triangular basis.

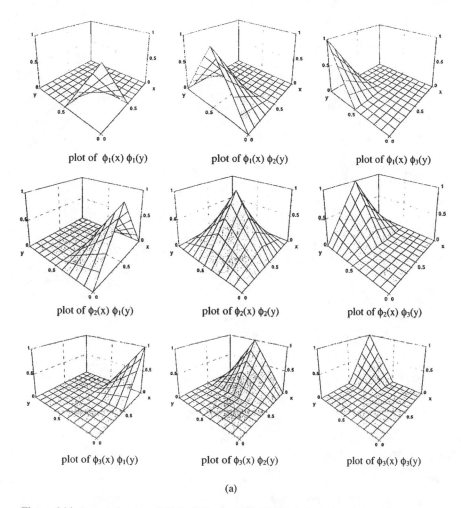

plot of $\phi_1(x)\,\phi_1(y)$ plot of $\phi_1(x)\,\phi_2(y)$ plot of $\phi_1(x)\,\phi_3(y)$

plot of $\phi_2(x)\,\phi_1(y)$ plot of $\phi_2(x)\,\phi_2(y)$ plot of $\phi_2(x)\,\phi_3(y)$

plot of $\phi_3(x)\,\phi_1(y)$ plot of $\phi_3(x)\,\phi_2(y)$ plot of $\phi_3(x)\,\phi_3(y)$

(a)

Figure 6.14 A representation of (a) the 2-D triangular subdomain basis functions and (b) the 2-D multiscale basis functions.

Mathematically, this will be expressed as

$$f_{\text{approx}}(x, y) = \begin{pmatrix} \phi_1(x) & \phi_2(x) & \phi_3(x) \end{pmatrix} \begin{pmatrix} f_{1,1} & f_{1,2} & f_{1,3} \\ f_{2,1} & f_{2,2} & f_{2,3} \\ f_{3,1} & f_{3,2} & f_{3,3} \end{pmatrix} \begin{pmatrix} \phi_1(y) \\ \phi_2(y) \\ \phi_3(y) \end{pmatrix} \quad (6.80)$$

where $f_{i,j}$ represents the corresponding amplitudes of the various basis functions. The same function $f(x, y)$ can be represented in the same region $[0,1] \times [0,1]$ by the 2-D multiscale triangular basis functions as

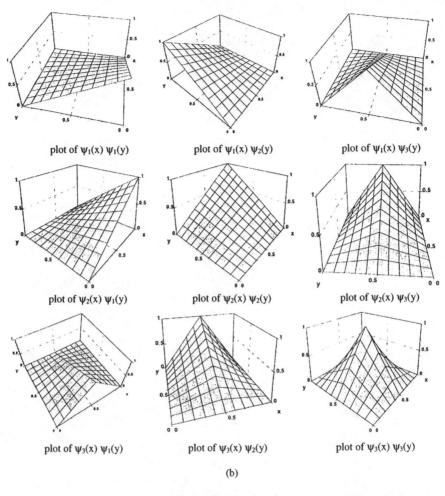

plot of $\psi_1(x)\,\psi_1(y)$ plot of $\psi_1(x)\,\psi_2(y)$ plot of $\psi_1(x)\,\psi_3(y)$

plot of $\psi_2(x)\,\psi_1(y)$ plot of $\psi_2(x)\,\psi_2(y)$ plot of $\psi_2(x)\,\psi_3(y)$

plot of $\psi_3(x)\,\psi_1(y)$ plot of $\psi_3(x)\,\psi_2(y)$ plot of $\psi_3(x)\,\psi_3(y)$

(b)

Figure 6.14 Continued.

$$f_{approx}(x,y) = \left(\psi_1^{(0)}(x) \quad \psi_2^{(0)}(x) \quad \psi_1^{(1)}(x)\right) \begin{pmatrix} \tau_{1,1} & \tau_{1,2} & \tau_{1,3} \\ \tau_{2,1} & \tau_{2,2} & \tau_{2,3} \\ \tau_{3,1} & \tau_{3,2} & \tau_{3,3} \end{pmatrix} \begin{pmatrix} \psi_1^{(0)}(y) \\ \psi_2^{(0)}(y) \\ \psi_1^{(1)}(y) \end{pmatrix}$$

$$\text{(6.81)}$$

$$= f_{(0)}(x,y) + \tau_{3,1}\,\psi_1^{(1)}(x)\psi_1^{(0)}(y) + \tau_{3,2}\,\psi_1^{(1)}(x)\psi_2^{(0)}(y)$$

$$+ \tau_{1,3}\,\psi_1^{(0)}(x)\,\psi_1^{(1)}(y) + \tau_{2,3}\,\psi_2^{(0)}(x)\,\psi_1^{(1)}(y) + \tau_{3,3}\,\psi_1^{(1)}(x)\,\psi_1^{(1)}(y)$$

where

$$f_{(0)}(x, y) = \tau_{1,1}\psi_1^{(0)}(x)\psi_1^{(0)}(y) + \tau_{1,2}\psi_1^{(0)}(x)\psi_2^{(0)}(y) + \tau_{2,1}\psi_2^{(0)}(x)\psi_1^{(0)}(y)$$
$$+ \tau_{2,2}\psi_2^{(0)}(x)\psi_2^{(0)}(y) \tag{6.82}$$

and $\tau_{i,j}$ are the appropriate amplitudes associated with the corresponding basis functions. Through (6.77), we can obtain the following relationship between $f_{i,j}$ and $\tau_{i,j}$:

$$\begin{pmatrix} \tau_{1,1} & \tau_{1,2} & \tau_{1,3} \\ \tau_{2,1} & \tau_{2,2} & \tau_{2,3} \\ \tau_{3,1} & \tau_{3,2} & \tau_{3,3} \end{pmatrix} = \begin{pmatrix} 1 & 0 & 0 \\ 0 & 0 & 1 \\ -\frac{1}{2} & 1 & -\frac{1}{2} \end{pmatrix} \begin{pmatrix} f_{1,1} & f_{1,2} & f_{1,3} \\ f_{2,1} & f_{2,2} & f_{2,3} \\ f_{3,1} & f_{3,2} & f_{3,3} \end{pmatrix} \begin{pmatrix} 1 & 0 & -\frac{1}{2} \\ 0 & 0 & 1 \\ 0 & 1 & -\frac{1}{2} \end{pmatrix} \tag{6.83}$$

Therefore,

$$\tau_{1,1} = f_{1,1} \; ; \; \tau_{1,2} = f_{1,3} \; ; \; \tau_{2,1} = f_{3,1} \; ; \; \tau_{2,2} = f_{3,3} \; ;$$

$$\tau_{1,3} = f_{1,2} - \frac{1}{2}(f_{1,1} + f_{1,3}) \; ; \; \tau_{2,3} = f_{3,2} - \frac{1}{2}(f_{3,1} + f_{3,3})$$

$$\tau_{3,1} = f_{2,1} - \frac{1}{2}(f_{1,1} + f_{3,1}) \; ; \; \tau_{3,2} = f_{2,3} - \frac{1}{2}(f_{1,3} + f_{3,3}) \tag{6.84}$$

$$\tau_{3,3} = f_{2,2} - \frac{1}{2}(f_{1,2} + f_{3,2}) - \frac{1}{2}(f_{2,1} + f_{2,3}) + \frac{1}{4}(f_{1,1} + f_{1,3} + f_{3,1} + f_{3,3})$$

$$= f_{2,2} - \frac{1}{2}(\tau_{1,3} + \tau_{3,1} + \tau_{2,3} + \tau_{3,2}) - \frac{1}{4}(f_{1,1} + f_{1,3} + f_{3,1} + f_{3,3})$$

$$= \tau'_{3,3} - \frac{1}{2}(\tau_{1,3} + \tau_{3,1} + \tau_{2,3} + \tau_{3,2})$$

where $\tau'_{3,3} = f_{2,2} - \frac{1}{4}(f_{1,1} + f_{1,3} + f_{3,1} + f_{3,3})$. Geometrically, $\tau_{i,j}$ represents the difference in amplitude between a linear function in [0,1] and the actual function $f(x, y)$ at the points (x_i, y_j) (see Figure 6.15), which happen to be the middle points of the intervals. If the function $f(x, y)$ has two orders of continuous differentiability in the rectangular region, then

$$\tau_{1,3} \approx -\frac{h^2}{2}\frac{\partial^2 f(0, \frac{1}{2})}{\partial y^2}; \; \tau_{2,3} \approx -\frac{h^2}{2}\frac{\partial^2 f(1, \frac{1}{2})}{\partial y^2}; \; h = \frac{1}{2}; \; \tau_{3,1} \approx -\frac{h^2}{2}\frac{\partial^2 f(\frac{1}{2}, 0)}{\partial x^2};$$

$$\tau_{2,3} \approx -\frac{h^2}{2}\frac{\partial^2 f(\frac{1}{2}, 1)}{\partial x^2}; \quad \tau'_{3,3} \approx -\frac{h^2}{4}\left(\frac{\partial^2 f(\frac{1}{2}, \frac{1}{2})}{\partial x^2} + \frac{\partial^2 f(\frac{1}{2}, \frac{1}{2})}{\partial y^2}\right); \tag{6.85}$$

$$\tau_{3,3} \approx \frac{h^2}{4}\left(\frac{\partial^2 f(\frac{1}{2}, 0)}{\partial x^2} + \frac{\partial^2 f(\frac{1}{2}, 1)}{\partial x^2} + \frac{\partial^2 f(\frac{1}{2}, 0)}{\partial y^2} + \frac{\partial^2 f(\frac{1}{2}, 1)}{\partial y^2} - \frac{\partial^2 f(\frac{1}{2}, \frac{1}{2})}{\partial x^2} - \frac{\partial^2 f(\frac{1}{2}, \frac{1}{2})}{\partial y^2}\right)$$

When $f(x, y)$ is a planar linear function of the form $\{ f(x, y) = ax + by + c \}$ in the region [0,1]×[0,1], then all coefficients $\tau_{1,3}, \tau_{2,3}, \tau_{3,1}, \tau_{3,2} \; \tau_{3,3}$ will be zero.

6.6.1   Functional Approximation Using a Multiscale Basis

The multiscale basis functions developed here are based on the use of a uniform grid and the usual subdomain triangular basis functions. They can be generated using the usual subdomain triangular basis through a matrix transformation. Mathematically, the multiscale basis and the usual subdomain triangular basis will be related by

$$[\Psi_{\Pi}^{N,V}(x)] = [S(N,V)][\Phi_{\Delta}^{1+2^{V}N}(x)] \tag{6.86}$$

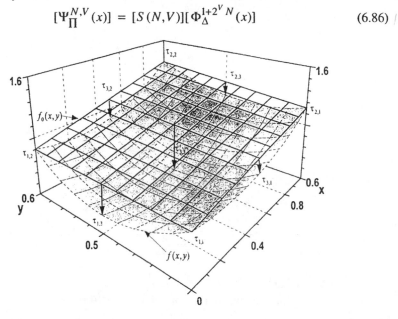

Figure 6.15 Significance of the parameters for the 2-D case.

where $[\Phi_{\Delta}^{1+2^{V}N}(x)] = (\phi_1(x),\cdots,\phi_{1+2^{V}N}(x))^{T}$ represents the set of the uniform subdomain triangular basis functions; $[\Psi_{\Pi}^{N,V}(x)] = (\psi_1(x),\cdots,\psi_{1+2^{V}N}(x))^{T}$ are the multiscale basis functions; N is the number of the initial subdivision of the interval [0,1]; and V is the value of the scale. Therefore, the total number of functions approximating the interval is then given by $1+2^{V}N$. As an example consider the set $\Psi_{\Pi}^{4,5}(x)$, which is depicted in Figure 6.4. Here the initial discretization at the zeroth scale is 4. Hence, there are 4 + 1 basis functions. When the scale $V = 1$, then the total number of multiscale basis functions is $1 + 4 \times 2 = 9$. At scale $V = 5$, the total number of multiscale basis functions used to approximate any function in the interval will be $1 + 4 \times 2^5 = 129$ as illustrated in Figure 6.4. It can be shown that this multiscale basis is related to the uniform 129 triangular basis through a full rank transformation matrix $S(N,V)$. For $N = 2$ and $V = 1$ this transformation matrix S is given by (6.10).

The coefficient vector $[F_\Pi]$ representing the amplitudes corresponding to approximating $f(x)$ using a multiscale basis are then related to the coefficient vectors $[F_\Lambda]$, which are the corresponding amplitudes of approximating the same function $f(x)$ using the uniform triangular basis through the following equation:

$$[F_\Lambda] = [S(N,V)][F_\Pi] \tag{6.87}$$

For the two-dimensional case, since the basis functions can be constructed using a tensor product, an equivalent relationship between the amplitudes using the two different bases can be constructed in a similar fashion as follows. Consider a one-dimensional multiscale basis function and a triangular basis function for the x variation of the solution. Then the x component of the solution can be represented either by the multiscale basis $[\Psi_\Pi^{N,V}(x)] = [\psi_1(x), \cdots, \psi_{1+2^V N}(x)]^T$ or by the subdomain triangular basis $[\Phi_\Lambda^{1+2^V N}(x)] = [\phi_1(x), \cdots, \phi_{1+2^V N}(x)]^T$. The multiscale basis and the triangular basis for the y variation of the solution can similarly be expressed through the following sets of functions: $[\Psi_\Pi^{M,V}(y)] = [\psi_1(y), \cdots, \psi_{1+2^V M}(y)]^T$; $[\Phi_\Lambda^{1+2^V M}(y)] = [\phi_1(y), \cdots, \phi_{1+2^V M}(y)]^T$, respectively, where M is the number of initial subdivisions of the interval $[0,1]$ along the y-direction. Therefore, the basis functions in two dimensions can be expressed through the tensor product (represented by $\otimes$) of the respective one dimensional functions

$$[\Phi_\Lambda^V(x,y)] = [\Phi_\Lambda^{1+2^V N}(x)] \otimes [\Phi_\Lambda^{1+2^V M}(y)] = \begin{pmatrix} \phi_1(x) \\ \vdots \\ \phi_{1+2^V N}(x) \end{pmatrix} \left(\phi_1(y) \quad \cdots \quad \phi_{1+2^V M}(y) \right)$$

$$\tag{6.88}$$

for the uniform triangular functions, and

$$[\Psi_\Pi^V(x,y)] = [\Psi_\Pi^{N,V}(x)] \otimes [\Psi_\Pi^{M,V}(y)] = \begin{pmatrix} \psi_1(x) \\ \vdots \\ \psi_{1+2^V N}(x) \end{pmatrix} \left(\psi_1(y) \quad \cdots \quad \psi_{1+2^V M}(y) \right)$$

$$\tag{6.89}$$

for the multiscale functions. The relationship between the weighting coefficients representing the amplitudes of the solution with the two different bases, namely, the multiscale representation and the triangular basis representation, is given by

$$[\Psi_{\Pi}^{N,V}(x) \otimes \Psi_{\Pi}^{M,V}(y)] = [S(N,V)] \begin{pmatrix} \phi_1(x) \\ \vdots \\ \phi_{1+2^V N}(x) \end{pmatrix} \left(\phi_1(y) \quad \cdots \quad \phi_{1+2^V M}(y) \right) [S(M,V)]^T$$

$$\tag{6.90}$$

$$[\Psi_{\Pi}^{N,V}(x)] \otimes [\Psi_{\Pi}^{M,V}(y)] = [S(N,V)] [\Phi_{\Delta}^{1+2^V N}(x)] \otimes [\Phi_{\Delta}^{1+2^V M}(y)] [S(M,V)]^T$$

$$\tag{6.91}$$

Therefore, the function $f(x, y)$ can be approximated by either the multiscale basis or the uniform triangular basis using either of the following forms:

$$f(x,y) = \sum_{i,j} f_{i,j}^{\Delta} \phi_i(x) \phi_j(y) = \left(\Phi_{\Delta}^{1+2^V N}(x) \right)^T [F_{\Delta}][\Phi_{\Delta}^{1+2^V M}(y)]$$

$$= \sum_{i,j} f_{i,j}^{\Pi} \psi_i(x) \psi_j(y) = \left(\Psi_{\Pi}^{N,V}(x) \right)^T [F_{\Pi}][\Psi_{\Pi}^{M,V}(y)]$$

$$\tag{6.92}$$

Therefore, if

$$[U] = [S]^{-1} \tag{6.93}$$

then the coefficient matrices representing the amplitudes of the various basis functions representing the function f are related to each other by

$$[F_{\Pi}] = [U(N,V)]^T [F_{\Delta}][U(M,V)] \tag{6.94}$$

The function $f(x, y)$ can then be represented in an approximate fashion using the two-dimensional triangular basis functions as

$$f_{approx}(x, y) =$$

$$\left(\phi_1(x), \ ..., \ \phi_{1+2^V N}(x) \right) \begin{pmatrix} f(x_1, y_1) & \cdots & f(x_1, y_{1+2^V M}) \\ \vdots & \ddots & \vdots \\ f(x_{1+2^V N}, y_1) & \cdots & f(x_{1+2^V N}, y_{1+2^V M}) \end{pmatrix} \begin{pmatrix} \phi_1(y) \\ \vdots \\ \phi_{1+2^V M}(y) \end{pmatrix}$$

$$\tag{6.95}$$

The same function $f(x, y)$ can alternately be represented using the multiscale basis as

$$f_{approx}(x,y) = \left(\Psi_{\Pi}^{N,V}(x)\right)^T [U(N,V)]^T$$

$$\times \begin{pmatrix} f(x_1,y_1) & \cdots & f(x_1,y_{1+2^V M}) \\ \vdots & \ddots & \vdots \\ f(x_{1+2^V N},y_1) & \cdots & f(x_{1+2^V N},y_{1+2^V M}) \end{pmatrix} [U(M,V)][\Psi_{\Pi}^{M,V}(y)]$$

(6.96)

The matrix $[F_\Pi]$ representing the coefficients for approximating the function $f(x,y)$ using the two-dimensional triangular basis functions is defined through

$$[F_\Delta] = \begin{pmatrix} f(x_1,y_1) & \cdots & f(x_1,y_{1+2^V M}) \\ \vdots & \ddots & \vdots \\ f(x_{1+2^V N},y_1) & \cdots & f(x_{1+2^V N},y_{1+2^V M}) \end{pmatrix}$$

$$= \begin{pmatrix} f_{1,1} & \cdots & f_{1,1+2^V M} \\ \vdots & \ddots & \vdots \\ f_{1+2^V N,1} & \cdots & f_{1+2^V N,1+2^V M} \end{pmatrix}$$

(6.97)

Similarly, the matrix $[F_\Pi]$ representing the coefficients approximating the function $f(x,y)$ corresponding to the two-dimensional multiscale basis is defined by

$$[F_\Pi] = \begin{pmatrix} f_{1,1}^{\Pi} & \cdots & f_{1,1+2^V M}^{\Pi} \\ \vdots & \ddots & \vdots \\ f_{1+2^V N,1}^{\Pi} & \cdots & f_{1+2^V N,1+2^V M}^{\Pi} \end{pmatrix}$$

(6.98)

$$= [U(N,V)]^T \begin{pmatrix} f(x_1,y_1) & \cdots & f(x_1,y_{1+2^V M}) \\ \vdots & \ddots & \vdots \\ f(x_{1+2^V N},y_1) & \cdots & f(x_{1+2^V N},y_{1+2^V M}) \end{pmatrix} [U(M,V)]$$

In the matrix $[F_\Pi^C]$, if the magnitudes of some of the coefficients fall below a certain prespecified threshold value, then those elements are all set to zero, resulting in a modified coefficient matrix that is defined by

$$[F_\Pi^C] = \begin{pmatrix} f_{1,1}^C & \cdots & f_{1,1+2^V N}^C \\ \vdots & \ddots & \vdots \\ f_{1+2^V N,1}^C & \cdots & f_{1+2^V N,1+2^V N}^C \end{pmatrix}; \text{ where } f_{i,j}^C = \begin{cases} f_{i,j}^{\Pi} & \text{if } \left|f_{i,j}^{\Pi}\right| > \varepsilon \\ 0 & \text{if } \left|f_{i,j}^{\Pi}\right| \le \varepsilon \end{cases}$$

(6.99)

where ε is a small positive constant. Typically, its value is between 0.1 and 0.0001. Therefore, the function $f(x,y)$ can be represented through these modified coefficients as

$$f_{\text{approx}}^C(x,y) = \left(\Psi_\Pi^{N,V}(x)\right)^T \begin{pmatrix} f_{1,1}^C & \cdots & f_{1,1+2^V M}^C \\ \vdots & \ddots & \vdots \\ f_{1+2^V N,1}^C & \cdots & f_{1+2^V N,1+2^V M}^C \end{pmatrix} [\Psi_\Pi^{M,V}(y)]$$

$$(6.100)$$

Because the matrix $[F_\Pi^C]$ is now sparse since some of the elements have been set to zero, we can define a compression ratio as the number of elements $f_{i,j}^C$, which are zero over the total number of elements $(1+2^V N)^2$. In addition, we can define some error quantities associated with the representation of the function in either basis. The average error and the maximum error between $\{ f_{\text{approx}}(x_i, y_j) \}$ and $\{ f_{\text{approx}}^C(x_i, y_j) \}$ are defined as follows:

$$\text{AverErr}(N,V) = \sqrt{\frac{\sum_i \sum_j (f_{\text{approx}}(x_i, y_j) - f_{\text{approx}}^C(x_i, y_j))^2}{\sum_i \sum_j (f_{\text{approx}}(x_i, y_j))^2}} \qquad (6.101)$$

$$\text{MaxErr}(N,V) = \max \left| f_{\text{approx}}(x_i, y_j) - f_{\text{approx}}^C(x_i, y_j) \right| \qquad (6.102)$$

6.6.2   A Multiscale Moment Method for Solving Fredholm Integral Equations of the First Kind in Two Dimensions

Consider the solution of the following Fredholm integral equation of the first kind in two dimensions, where the unknown function $f(x, y)$ will be expanded in terms of the multiscale basis functions rather than by using the conventional subdomain basis functions.

$$\int_0^1 dy' \int_0^1 k(x,y;x',y') f(x',y') dx' = g(x,y) \quad \text{for} \ (x,y) \in [0,1] \times [0,1] \qquad (6.103)$$

We choose the usual uniform triangular subdomain basis functions defined on a uniform grid with the nodes located at $\{ x_m = y_m = mh; \ m = 0, 1, 2, \cdots, 2^V N;$ $h = \dfrac{1}{2^V N} \}$ along the x- and y-axes. Suppose the unknown function $f(x, y)$ is represented by [from (6.81)]

$$f(x,y) = \sum_{i=0}^{1+2^V N} \sum_{j=0}^{1+2^V N} f(x_i, y_j) \phi_i(x) \phi_j(y) = [\Phi_\Delta^V]^T [F_\Delta] \qquad (6.104)$$

Then if we use the usual triangular subdomain basis in two dimensions as expansion functions for the unknown and use point matching as the testing function, (instead of Galerkin's method, which we used earlier in this section), the integral equation can be approximated by the following matrix equation:

$$[A_\Delta][F_\Delta] = [G_\Delta] \qquad (6.105)$$

where

$$[F_\Delta] = (f(x_0, y_0), \cdots, f(x_{1+2^V N}, y_0), \cdots, f(x_0, y_{1+2^V N}), \cdots, f(x_{1+2^V N}, y_{1+2^V N}))^T \qquad (6.106)$$

$$[G_\Delta] = (g(x_0, y_0), \ldots, g(x_{1+2^V N}, y_0), \ldots, g(x_0, y_{1+2^V N}), \ldots, g(x_{1+2^V N}, y_{1+2^V N}))^T \qquad (6.107)$$

with
$$[A_\Delta] = (a_{i,j})_{(1+2^V N)^2 \times (1+2^V N)^2} \qquad (6.108)$$

$$a_{0,0} = \iint K(x_0, y_0, x', y')\phi_0(x')\phi_0(y')dx'dy',$$

$$a_{0,1} = \iint K(x_0, y_0, x', y')\phi_1(x')\phi_0(y')dx'dy'$$

$$a_{0,1+2^V N} = \iint K(x_0, y_0, x', y')\phi_{1+2^V N}(x')\,\phi_0(y')\,dx'dy'$$

$$a_{0,(1+2^V N)\times(1+2^V N)} = \iint K(x_0, y_0, x', y')\,\phi_{1+2^V N}(x')\,\phi_{1+2^V N}(y')\,dx'\,dy'$$

$$a_{1,0} = \iint K(x_1, y_0, x', y')\,\phi_0(x')\,\phi_0(y')\,dx'\,dy';$$

$$a_{1,1} = \iint K(x_1, y_0, x', y')\,\phi_1(x')\,\phi_0(y')\,dx'\,dy'$$

$$a_{1,1+2^V N} = \iint K(x_1, y_0, x', y')\,\phi_{1+2^V N}(x')\,\phi_0(y')\,dx'\,dy'$$

$$a_{1,(1+2^V N)\times(1+2^V N)} = \iint K(x_1, y_0, x', y')\,\phi_{1+2^V N}(x')\,\phi_{1+2^V N}(y')\,dx'\,dy'$$

$$a_{1+2^V N,0} = \iint K(x_{1+2^V N}, y_0, x', y')\,\phi_0(x')\,\phi_0(y')\,dx'\,dy'$$

$$a_{1+2^V N,1} = \iint K(x_{1+2^V N}, y_0, x', y')\,\phi_1(x')\,\phi_0(y')\,dx'\,dy'$$

$$a_{(1+2^V N)\times(1+2^V N),0} = \iint K(x_{1+2^V N}, y_{1+2^V N}, x', y')\,\phi_0(x')\,\phi_0(y')\,dx'dy'$$

$$a_{(1+2^V N)\times(1+2^V N),1} = \iint K(x_{1+2^V N}, y_{1+2^V N}, x', y')\,\phi_1(x')\,\phi_0(y')\,dx'\,dy'$$

$$a_{1+2^V N,1+2^V N} = \iint K(x_{1+2^V N}, y_0, x', y')\,\phi_{1+2^V N}(x')\,\phi_0(y')\,dx'\,dy'$$

$$a_{1+2^V N,(1+2^V N)\times(1+2^V N)} = \iint K(x_{1+2^V N}, y_0, x', y')\,\phi_{1+2^V N}(x')\,\phi_{1+2^V N}(y')\,dx'dy'$$

$$a_{(1+2^V N)\times(1+2^V N),0} = \iint K(x_{1+2^V N}, y_{1+2^V N}, x', y')\,\phi_0(x')\,\phi_0(y')\,dx'dy'$$

$$a_{(1+2^V N)\times(1+2^V N),1} = \iint K(x_{1+2^V N}, y_{1+2^V N}, x', y')\,\phi_1(x')\,\phi_0(y')\,dx'\,dy'$$

$$a_{(1+2^V N)\times(1+2^V N),1+2^V N} = \iint K(x_{1+2^V N}, y_{1+2^V N}, x', y')\,\phi_{1+2^V N}(x')\,\phi_0(y')dx'dy'$$

$$a_{(1+2^V N)\times(1+2^V N),(1+2^V N)\times(1+2^V N)}$$
$$= \iint K(x_{1+2^V N}, y_{1+2^V N}, x', y')\,\phi_{1+2^V N}(x')\,\phi_{1+2^V N}(y')\,dx'\,dy$$

Now, if we consider expanding the two-dimensional unknown function in terms of a multiscale basis instead of the usual subdomain basis, then from the previous discussion we see that a matrix transformation can be used to change the basis from the triangular subdomain to the multiscale representation. If the two-dimensional subdomain triangular basis is of the following form:

$$[\Phi_\Delta^V(x, y)] = \begin{pmatrix} \phi_0(y)\phi_0(x) \\ \vdots \\ \phi_0(y)\phi_{1+2^V N}(x) \\ \vdots \\ \vdots \\ \phi_{1+2^V N}(y)\phi_1(x) \\ \vdots \\ \phi_{1+2^V N}(y)\phi_{1+2^V N}(x) \end{pmatrix} \tag{6.109}$$

then the two-dimensional multiscale basis can be arranged as follows:

$$\begin{aligned}
[\Psi_\Pi^V(x, y)] \equiv [&\psi_0(y) \otimes \psi_0(x), \psi_0(y) \otimes \psi_1(x), \psi_1(y) \otimes \psi_0(x), \psi_1(y) \otimes \psi_1(x), \\
&\psi_0(y) \otimes \psi_2(x), \psi_2(y) \otimes \psi_0(x), \psi_1(y) \otimes \psi_2(x), \psi_2(y) \otimes \psi_1(x), \\
&\psi_2(y) \otimes \psi_2(x), \cdots, \psi_0(y) \otimes \psi_V(x), \psi_V(y) \otimes \psi_0(x), \\
&\psi_1(y) \otimes \psi_V(x), \psi_V(y) \otimes \psi_1(x), \cdots, \psi_V(y) \otimes \psi_V(x), \\
&\qquad\qquad\qquad\qquad\qquad\qquad \psi_V(y) \otimes \psi_V(x)]^T
\end{aligned} \tag{6.110}$$

where $\psi_V(x)$ and $\psi_V(y)$ represent the basis functions at the Vth scale along the x- and the y-axes, respectively. Therefore, the total number of basis functions at each scale from 0 to 5, for example, is given by $(1+N)^2$, $3N^2+2N$, $12N^2+4N$, $48N^2+8N$, $192N^2+16N$, and $768N^2+32N$, respectively.

Using (6.86), we can define the new transformation matrix $[W(N,V)]$ (for the two-dimensional case whereas $[S]$ was defined for the one-dimensional case) at scale V, which transforms the basis functions $[\Phi_\Delta^V(x, y)]$ to the multiscale basis $[\Psi_\Pi^V(x, y)]$ through

$$[\Psi_\Pi^V(x, y)] = [W(N,V)][\Phi_\Delta^V(x, y)] \tag{6.111}$$

Matrix equation (6.104) can be rewritten using the multiscale basis as

$$[A_\Pi][F_\Pi] = [G_\Pi] \tag{6.112}$$

where

$$[F_\Delta] = [W(N,V)][F_\Pi]$$

$$[G_\Pi] = [W(N,V)]^T [G_\Delta]$$

$$[A_\Pi] = [W(N,V)][A_\Delta][W(N,V)]^T$$

The matrix $[A_\Pi]$, and the column vectors $[F_\Pi]$ and $[G_\Pi]$ representing the unknown and the excitations, respectively, can be arranged along the scaled-block form as illustrated by Figure 6.6. This scaled-block form of the matrix representing the discretized version of the two-dimensional integral equation is similar to the scaled-block form for the one-dimensional case. The only difference is that the size of the matrix relating the unknowns at different scales is not the same.

6.6.3 An Adaptive Algorithm Representing a Multiscale Moment Method

Next we present a methodology on how to generate the solution as we increase the scale through a technique which we call the adaptive multiscale moment method where the coefficients for the basis functions representing the unknown solution are generated in an adaptive fashion. The following steps are carried out in this procedure in order to reach the final solution.

Step 1: We first choose a two-dimensional coarse grid of $N \times N$ points, which we call scale zero. On this grid, we obtain the solution $[F_\Pi^0]$ by solving matrix equation (6.92). One can use either the LU decomposition or the iterative conjugate gradient method.

Step 2: We then transform this two-dimensional multiscale solution at the zeroth scale to the two-dimensional solution using the uniform subdomain triangular basis through the transformation $[F_\Pi^0] = [W(N,0)]^{-1}[F_\Delta^0]$.

Step 3: Next we interpolate the solution between the coarse grid points using a two-dimensional spline interpolator. In the numerical evaluation of the interpolated values of the function, we use the tensor product of a spline interpolant. The tensor product spline interpolant function to the sampled values $\{f(x_i, y_j)\}$, with $x_i = h \times \Delta x$ and $y_j = h \times \Delta y$, with $\{1 \le i \le N_x\}$ and $\{1 \le j \le N_y\}$ is obtained from the product of two spline functions of the

form $\sum\limits_{m=1}^{N_y} \sum\limits_{n=1}^{N_x} c_{nm} B_{n,k_x,t_x}(x)\, B_{m,k_y,t_y}(y)$, where $B_{i,k,t}(s)$ is the vth (normalized)

B-spline of order k for the knot sequence $t[1]$. The coefficients $c_{n,m}$ are obtained from solution of the following system of equations:

$$\sum_{m=1}^{N_y} \sum_{n=1}^{N_x} c_{nm} B_{n,k_x,t_x}(x_i) B_{m,k_y,t_y}(y_j) = f(x_i, y_j) \,;\, \{1 \le i \le N_x, 1 \le j \le N_y\}$$

(6.113)

This problem can be resolved quite efficiently by repeatedly solving two univariate interpolation problems as described in Boor [1]. Through the spline interpolator, we obtain an estimate for the solution at the higher scale between the coarse grid points.

Step 4: This provides an estimate for the coefficients for $[\Phi_\Delta^1]$ at scale one.

Step 5: From the estimated solution using the subdomain triangular basis we obtain the coefficients for the multiscale basis $[\Psi_\Pi^1]$ at scale one.

Step 6: Now if some of the coefficients of the multiscale basis are smaller than some threshold then we not only remove them from the unknown columns but also delete the corresponding rows of matrices $[A_\Pi]$ and $[G_\Pi]$. In addition, the corresponding columns from matrix $[A_\Pi]$ are also deleted. This sparse matrix equation is solved through either LU decomposition or the iterative conjugate gradient method for all the existing unknowns at all the scales simultaneously.

Step 7: By using the solution at scale 1 in the multiscale basis, we now transform the solution to the uniform basis through the matrix transformation of (6.92).

Then we go to *Step 3* and continue this iterative process unless we have reached the highest scale of interest. In summary, we are adaptively choosing the required basis functions at each scale, which are derived from the interpolation of the approximate solution, and in this way, we adaptively introduce more basis functions to increase the accuracy and therefore the resolution of the solution.

As an example, let us consider the kernel in (6.103) to be of the form

$$k(x, y, x', y') = -\ln \sqrt{(x - x')^2 + (y - y')^2}$$

(6.114)

In this case, the solution has a singular behavior at the edges and we would like to study if this adaptive technique correctly reproduces it. We consider the region $[0,1] \times [0,1]$. This region is uniformly subdivided into $N = N_x \times N_y = 33 \times 33$ subdivisions. The excitation function is defined as $g(x, y) = 1$. The difference in the solution $f_\Delta(x_i, y_j)$ obtained using a uniform triangular basis for solving the entire problem simultaneously, and the use of the adaptive multiscale basis to get solution $f_\Pi(x_i, y_j)$, is defined through

$$Err(f_\Delta, f_\Pi) = \sqrt{\frac{\sum_i \sum_j (f_\Delta(x_i, y_j) - f_\Pi(x_i, y_j))^2}{\sum_i \sum_j (f_\Delta(x_i, y_j))^2}}$$

(6.115)

For different values of the threshold ε, the reduced size of the matrix equations that one needs to solve, the condition number of the system of linear equations at the largest scale $V = 3$, the error functional $\mathrm{Err}(f_\Delta, f_\Pi)$, and the CPU time to compute the solution on a DELL OptiPlex Gxi 166-MHz personal computer are given in Table 6.10. CPU time is the time required to solve the linear equation for the conventional moment method or the adaptive multiscale moment method. This does not include the time spent to compute the coefficient matrix and the source term. CN is the condition number. MM represents the method of moment solution utilizing the usual triangular sub domain basis in two dimensions. The solutions $f_\Delta(x_i, y_j)$ and $f_\Pi(x_i, y_j)$ for the different threshold at the largest scale $V = 3$ are shown in Figure 6.16. From the computed results, it is seen that this adaptive multiscale moment method can reproduce well the singularity of the function at the four corners of the boundary. In addition, it requires solution of a smaller size matrix than the method of moments and for different threshold values of $\varepsilon = 0.001$, 0.01, and 0.1 achieves the reductions of 38, 72, and 93% in the size of the matrix equations, respectively.

Table 6.10
Summary of the Various Parameters

Threshold	Size	CN	Err	CPU (sec)
0.0	1,089	8,763,167	2.676E-5	56.46
0.001	677	5,242,464	0.1665	40.54
0.01	303	1,973,944	0.4810	27.02
0.1	73	142,653	0.7277	25.38
MM	1,089	6,294	0	56.19

A few points to observe:
1. This new method is different from the one used to solve the 2-D integral equations by the multilevel moment method. There are four differences between the multilevel moment method and the adaptive multiscale moment method (AMMM). First, both the methods differ on how the matrix for the linear equation is generated. For the multilevel moment method, the matrix needs to be computed for all levels. For AMMM, the impedance matrix needs to be computed once at the highest scale. Second, for the multilevel moment method, the initial guess is an approximate value of the solution $f(x, y)$ at the finest level. For AMMM, the initial guess relates to the values of the second derivatives $f_{xx}''(x, y)$, $f_{yy}''(x, y)$ and the scale V. If the solution is almost linear, most of the unknown coefficients will be zero, which can help us reduce the size of the linear equations. Third, when we go from a coarse grid to a fine grid, all of the basis functions need to be constructed again for the multilevel moment method. For AMMM, the basis functions at the newly added nodes need to be evaluated, which are the same basis as in the multilevel moment method. Fourth, the multilevel moment method cannot reduce the size of the linear equations. For different levels, the number of linear equations is fixed no matter how the solution function behaves. For

AMMM, the size of the linear equations can be reduced according to the characteristics of the second derivative of the solution and the value of the threshold. So the new method is an adaptive algorithm. If the threshold is taken to be zero, the size of the linear equation will not be reduced.

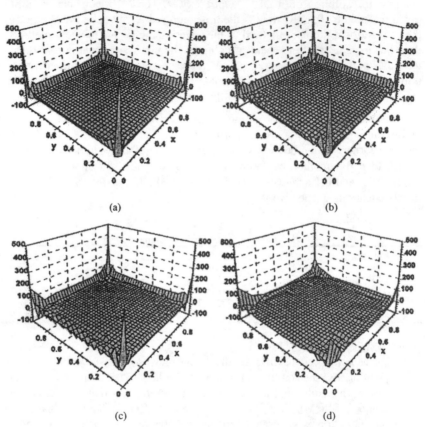

Figure 6.16 The solutions of $f_\Delta (x, y)$ and of $f_\Pi (x, y)$ for different values of the thresholds at the third scale: (a) plot of $f_\Delta(x,y)$, (b) plot of $f_\Pi(x,y)$ for $\varepsilon = 0.001$, (c) plot of $f_\Pi(x,y)$ for $\varepsilon = 0.01$, and (d) plot of $f_\Pi(x,y)$ for $\varepsilon = 0.1$.

2. The AMMM is similar to a wavelet method for solving the Fredholm equation. But there are also some differences between the two methods. For the wavelet moment method, the basis functions are constructed by shifting and dilating the mother wavelet, which has vanishing moment properties. Many of the matrix elements are very small compared to the largest elements and can be dropped without significantly affecting the solution. The matrix is thus rendered sparse. So the method can improve computational efficiency. But it is difficult (almost impossible) to solve the integral equation over an arbitrary domain in two or three dimensions by using the wavelet moment method because it is difficult to define a wavelet on a domain that is

arbitrarily shaped. For the 2-D AMMM, the basis functions are also constructed through a tensor product of the 1-D multiscale triangular basis, which consists of shifted and dilated forms of a function, which has no vanishing moment properties. Therefore, the matrix formed by the moment method at the different scales may not be sparse. However, the 2-D AMMM can improve the computational efficiency in reducing the size of the linear equation if the solution has regions where it has a linear shape for all practical purposes.

3. From (6.92), we see that the 2-D AMMM is very similar to the 1-D AMMM. The only difference is that the transformation matrices are constructed in different ways. Hence, the 2-D AMMM is an extension of the 1-D AMMM.

6.6.4 Application of AMMM for the Solution of Electromagnetic Scattering from Finite-Sized Rectangular Plates

Let S denote the surface of a square, perfectly conducting plate in the xy-plane. Let $\bar{E}^i$ be the electric field, defined by an impressed source in the absence of the scatterer. The electric field is incident on the structure and induces surface currents $\bar{J}$ on S. The induced current $\bar{J}$ can be written as

$$\bar{J}(\bar{r}) = J_x(\bar{r})\,\hat{x} + J_y(\bar{r})\,\hat{y} \tag{6.116}$$

where

$$J_x(\bar{r}) = \sum_{m=1}^{M} J_x^m\,\phi_x^m(\bar{r}); \quad J_y(\bar{r}) = \sum_{m=1}^{M} J_y^m\,\phi_y^m(\bar{r}) \tag{6.117}$$

and $\{\phi_x^m(\bar{r})\}$, $\{\phi_y^m(\bar{r})\}$ are the basis functions. By the use of the Galerkin scheme and through the choice of $\{\phi_x^m(\bar{r})\}$, $\{\phi_y^m(\bar{r})\}$ as the tensor product of two triangular basis functions as the expansion and the weighting functions, the matrix equation will be

$$\begin{pmatrix} A_{11} & A_{12} \\ A_{21} & A_{22} \end{pmatrix} \begin{pmatrix} \bar{J}_x \\ \bar{J}_y \end{pmatrix} = \frac{4\pi}{jk\eta_0} \begin{pmatrix} \bar{E}_x^i \\ \bar{E}_y^i \end{pmatrix} \tag{6.118}$$

$$A_{11}(i,j) = \int_S \phi_i(\bar{r})\,ds \int_S L_{11}\left[G\phi_j(\bar{r}')\right]ds'\,;\ A_{12}(i,j) = \int_S \phi_i(\bar{r})\,ds \int_S L_{12}\left[G\phi_j(\bar{r}')\right]ds'$$

$$A_{21}(i,j) = \int_S \phi_i(\bar{r})\,ds \int_S L_{21}\left[G\phi_j(\bar{r}')\right]ds'\,;\ A_{22}(i,j) = \int_S \phi_i(\bar{r})\,ds \int_S L_{22}\left[G\phi_j(\bar{r}')\right]ds'$$

$$L_{11} = 1 + \frac{1}{k^2}\frac{\partial^2}{\partial x^2};\ L_{21} = L_{12} = \frac{1}{k^2}\frac{\partial^2}{\partial x\,\partial y};\ L_{22} = 1 + \frac{1}{k^2}\frac{\partial^2}{\partial y^2};$$

$$G = \frac{\exp\left[-jk\sqrt{(x-x')^2+(y-y')^2}\right]}{\sqrt{(x-x')^2+(y-y')^2}} \tag{6.119}$$

$$\vec{E}^i = (E_\theta \hat{\theta} + E_\varphi \hat{\varphi}) \exp[jk(x\sin\theta\cos\varphi + y\sin\theta\sin\varphi)],$$

$$E_x^i = \hat{x} \bullet \vec{E}^i; \quad E_y^i = \hat{y} \bullet \vec{E}^i \quad \text{with } (i=1,2,...,M; \ j=1,2,...,M) \tag{6.120}$$

where E_θ and E_φ denote the θ and the φ components of the electric field, respectively. The variables with a "$\wedge$" on the top represent the direction of the unit vectors along that direction. The different parameters are expressed through

$$\bar{J}_x = \left(J_x^1,\cdots,J_x^M\right)^T; \quad \bar{J}_y = \left(J_y^1,\cdots,J_y^M\right)^T; \quad \bar{E}_x^i = \left(E_x^i(1),\cdots,E_x^i(M)\right)^T;$$

$$\bar{E}_y^i = \left(E_y^i(1),\cdots,E_y^i(M)\right)^T; \quad E_x^i(j) = \int_S \phi_j(\vec{r})E_x^i(\vec{r})ds; \quad E_y^i(j) = \int_S \phi_j(\vec{r})E_y^i(\vec{r})ds$$

Here, M denotes the number of subdivisions in both the x- and the y-directions. The RCS can be shown to have the following form:

$$\sigma(\theta,\phi) = \frac{k^2\eta_0^2}{4\pi}\left[\left|\sum_{n=1}^M \left(J_x^n\cos\theta\cos\phi + J_y^n\cos\theta\sin\phi\right)\zeta(\theta,\phi,n)\right|^2\right.$$

$$\left. + \left|\sum_{n=1}^M \left(-J_x^n\sin\phi + J_y^n\cos\phi\right)\zeta(\theta,\phi,n)\right|^2\right] A^2(\theta,\phi) \tag{6.121}$$

where $\zeta(\theta,\phi,n) = \exp[jk(x_n\sin\theta\cos\phi + y_n\sin\theta\sin\phi)]$ and

$$A(\theta,\phi) = h^2 \cdot \text{sinc}^2\left[\frac{h \cdot k \cdot \sin\theta\cos\phi}{2}\right] \cdot \text{sinc}^2\left[\frac{k \cdot h \cdot \sin\theta\sin\phi}{2}\right]$$

Here x_n and y_n represent the center of the coordinate of the nth subdivision. Now consider the solution of (6.117) using the multiscale basis functions where h is the spatial discretization. Suppose the two-dimensional triangular basis $[\vec{\Phi}_\Delta(x,y)]$ and the two-dimensional multiscale basis $[\vec{\Psi}_\Pi^V(x,y)]$ can be rearranged according to (6.109) and (6.110), respectively. Then we know that they will be related by (6.111). Therefore, (6.117) can be transformed using the multiscale basis functions through

$$\begin{pmatrix} A_{11}^\Pi & A_{12}^\Pi \\ A_{21}^\Pi & A_{22}^\Pi \end{pmatrix}\begin{pmatrix} \bar{J}_x^\Pi \\ \bar{J}_y^\Pi \end{pmatrix} = \frac{4\pi}{jk\eta_0}\begin{pmatrix} \bar{E}_x^{i\Pi} \\ \bar{E}_y^{i\Pi} \end{pmatrix} \tag{6.122}$$

where

$$[\overline{E}_{x,y}^{i\Pi}] = [W(N,V)]^T [\overline{E}_{x,y}^i],$$

$$[\overline{J}_{x,y}] = [W(N,V)] [\overline{J}_{x,y}^{\Pi}],$$

$$[A_{i,j}^{\Pi}] = [W(N,V)] [A_{i,j}] [W(N,V)]^T$$

The elements of the impedance matrix $A_{i,j}^{\Pi}$ { $A_{i,j}^{\Pi}(0)$, $\cdots$, $A_{i,j}^{\Pi}(V)$ }, the unknowns $\overline{J}_{x,y}^{\Pi}$ { $\overline{J}_{x,y}^{\Pi}(0)$, $\cdots$, $\overline{J}_{x,y}^{\Pi}(V)$ }, and the excitations $\overline{E}_{x,y}^{i\Pi}$ { $\overline{E}_{x,y}^{i\Pi}(0)$, $\cdots$, $\overline{E}_{x,y}^{i\Pi}(V)$ } are arranged in the scaled-block form as can be seen from Figure 6.6. The scheme of the adaptive multiscale moment method to solve matrix equations (6.122) from the Vth scale to the $(V+1)$th scale is given as follows:

Step 1: Suppose the solutions { $\overline{J}_{x,y}^{\Pi}(0)$, $\cdots$, $\overline{J}_{x,y}^{\Pi}(V)$ } on the Vth scale are given. Then the actual solution $\{\overline{J}_{x,y}(V)\}$ on the coarse grid can be obtained by the use of the following formula

$$[\overline{J}_{x,y}(V)] = [W(N,V)] \begin{pmatrix} \overline{J}_{x,y}^{\Pi}(0) \\ \overline{J}_{x,y}^{\Pi}(1) \\ \vdots \\ \overline{J}_{x,y}^{\Pi}(V) \end{pmatrix} \qquad (6.123)$$

Step 2: Estimate the solution $[\overline{J}_{x,y}(V+1)]$ on the finer grid through the use of a two-dimensional interpolant formula, such as a tensor product spline interpolant [1]. The tensor product spline interpolant function to the data $\{f(x_i, y_j)\}$, where

$\{1 \le i \le N_x\}$ and $\{1 \le j \le N_y\}$, has the form $\sum\limits_{m=1}^{N_y} \sum\limits_{n=1}^{N_x} c_{nm} B_{n,k_x,t_x}(x) B_{m,k_y,t_y}(y)$

where $B_{i,k,t}(s)$ is the Vth (normalized) B-spline of order k for the knot sequence t. The coefficients $c_{n,m}$ can be computed from the solution of the system of equations $\sum\limits_{m=1}^{N_y} \sum\limits_{n=1}^{N_x} c_{nm} B_{n,k_x,t_x}(x_i) B_{m,k_y,t_y}(y_j) = f(x_i, y_j)$ for $\{1 \le i \le N_x, 1 \le j \le N_y\}$.

Step 3: Obtain an initial guess on the finer grid using the following formula:

$$\begin{pmatrix} \overline{J}_{x,y}^{\Pi}(0) \\ \overline{J}_{x,y}^{\Pi}(1) \\ \vdots \\ \overline{J}_{x,y}^{\Pi}(V+1) \end{pmatrix} = [W(N,V+1)]^{-1} [\overline{J}_{x,y}(V+1)] \qquad (6.124)$$

Step 4: If the elements of $\{\overline{J}_{x,y}^{\Pi}(k)\}$ ($k=1, 2, 3, \ldots, V+1$) are less than ε (the given threshold parameter), we set these elements to zero and delete the corresponding rows and columns of the coefficient matrices $A_{i,j}^{\Pi}$ and the corresponding elements of the excitation $\overline{E}_{x,y}^{i\,\Pi}$ on the $(V+1)$th scale. Then after reducing the size of the original linear equation, the modified linear equation and the initial guess are obtained.

Step 5: Solve the modified linear equation by using either the conjugate gradient technique or the LU decomposition. Adding these elements, which are set to be zero in Step 4, the solution $[\overline{J}_{x,y}^{\Pi}(0),\cdots,\overline{J}_{x,y}^{\Pi}(V),\overline{J}_{x,y}^{\Pi}(V+1)]$ on the multiscale triangular basis is obtained. Using Step 1, we can obtain the original solutions on the finer grid.

This procedure continues until the largest scale is reached. Typically, it has been our experience that it is not necessary to go beyond the third scale.

6.6.5 Numerical Implementation of the AMMM Methodology for Solving 3-D Problems Using the Triangular Patch Basis Functions

In this section, we address the problem of efficient solution of large electromagnetic scattering problems from arbitrarily shaped conducting structures using existing MM codes. By large, we mean problems where the matrix solution time far exceeds the matrix fill time. Typically, when there are more than 4,300 unknowns in a conventional MM code using the vector triangular basis functions, the matrix solution time far exceeds the matrix fill time. Hence, the question is one of, can the AMMM methodology be applied to the conventional MM codes that are prevalent, incorporating the triangular patch basis functions? The current distribution on the triangular patch can be categorized as three different classes of currents going across the three edges. So after all the MM matrix elements have been computed, they are arranged in a special form. Application of the usual triangular patch basis functions as expansion and testing functions in a conventional MM to the electric field integral equation reduces the operator equation to a matrix equation, resulting in

$$[Z]_{N\times N}\,[I]_{N\times 1} = [E]_{N\times 1} \qquad (6.125)$$

Here $[Z]$ is the MM impedance matrix of size N. $[I]$ and $[E]$ are $N \times 1$ column matrices containing the unknown amplitudes and the given excitation, respectively.

Next, we apply the AMMM technique to compress the large impedance matrix $[Z]$ of (6.125). It is well known that the vector basis functions for the unknown currents are given by

$$\overline{J}(\overline{r}) = \sum_{n=1}^{N} \alpha_n\,\overline{f}_n(\overline{r}) \qquad (6.126)$$

where the basis functions $\bar{f}_n(\bar{r})$ are the usual linear functions [2, 3]. We now rearrange the vector basis functions into three distinct component forms as outlined next with $[I] = [\alpha_n]$.

As we know, the induced current of $\bar{J}$ on a plate, for example, can be represented by $J_x(\bar{r})\hat{x} + J_y(\bar{r})\hat{y}$, and use of the linear combination of rooftop functions to approximate the currents $J_x(\bar{r})$, $J_y(\bar{r})$ results in a function of two variables. Here x and y represent two local orthogonal Cartesian coordinate axes for the triangle. Therefore, we can classify the current on a plate along two classes of unknowns, namely, one for each component of the current. Because Rao et al. basis functions [2, 3] are constructed across the edges, the unknowns $[I] = [\alpha_n]$ must be classified into three categories $\{J^-, J', J^|\}$ according to the different orientation of the edges (see Figure 6.17). The $\{J^-, J', J^|\}$ terms denote the coefficient sets across the edges, such as: – (horizontal), / (slant), and | (vertical) edges, respectively. This terminology is used in the absence of a more precise form. Therefore, we rewrite matrix equation (6.125) by splitting it into three different forms based on the orientation of the edges. This results in the following representation:

$$\begin{pmatrix} Z_{1,1} & Z_{1,2} & Z_{1,3} \\ Z_{2,1} & Z_{2,2} & Z_{2,3} \\ Z_{3,1} & Z_{3,2} & Z_{3,3} \end{pmatrix} \begin{pmatrix} J^- \\ J' \\ J^| \end{pmatrix} = \begin{pmatrix} E_1 \\ E_2 \\ E_3 \end{pmatrix} \qquad (6.127)$$

where the submatrices $Z_{i,j}$ are from the impedance matrix $[Z]$ of (6.127).

If the unknowns $\{J^-, J', J^|\}$ can be arranged as a $(1+2^v N) \times (1+2^v N)$ matrix, (6.127) can be transformed into the following set by the matrix transformation described in [4]:

$$\begin{pmatrix} Z_{1,1}^{\Pi} & Z_{1,2}^{\Pi} & Z_{1,3}^{\Pi} \\ Z_{2,1}^{\Pi} & Z_{2,3}^{\Pi} & Z_{2,3}^{\Pi} \\ Z_{3,1}^{\Pi} & Z_{3,2}^{\Pi} & Z_{3,3}^{\Pi} \end{pmatrix} \begin{pmatrix} J_-^{\Pi} \\ J_/^{\Pi} \\ J_|^{\Pi} \end{pmatrix} = \begin{pmatrix} E_1^{\Pi} \\ E_2^{\Pi} \\ E_3^{\Pi} \end{pmatrix} \qquad (6.128)$$

where $[E_i^{\Pi}] = [W(N,V)]^T [E_i]$ and $[W]$ is the transformation matrix. The details for this procedure were given earlier in Section 6.6.

The impedance matrix $[Z]$, the unknowns $\{J^-, J', J^|\}$ and the excitation vector $[E]$ are arranged in a scaled-block form. Adaptive multiscale algorithms [4–10] can now be used for solving (6.128). This concept can thus be extended to the MM codes using the conventional triangular patch basis functions used for the solution of electromagnetic scattering from arbitrarily shaped three-dimensional conducting bodies. The method proposed here does not reduce the matrix fill time but can significantly shorten the matrix solution time.

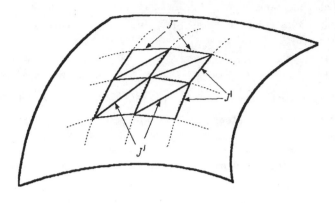

Figure 6.17 Three categories of edge currents.

6.6.6   Numerical Results

In this section, we discuss two numerical examples for analyzing scattering from arbitrarily shaped three-dimensional perfectly conducting objects by AMMM.

First, consider the scattering from a $2\lambda \times 2\lambda$ perfectly conducting plate (as shown in Figure 6.18). The plate is discretized into 22×22 nodes, which are located at $(-1 + ih, 1 + jh)$ with ($h = 2\lambda/21$, for $i, j = 0, 1, 2, ..., 21$). The currents across the horizontal edges (i.e., parallel to the x-axis) can be arranged as a 22×21 matrix. The currents across the vertical edges (i.e., parallel to the y-axis) can be arranged as a 21×22 matrix. The currents across the slanted edges can also be arranged as a 21×21 matrix.

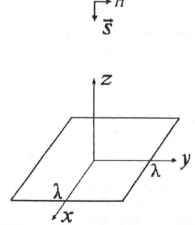

Figure 6.18 A two lambda squared perfectly conducting plate.

When the conducting plate is illuminated by a normally incident plane wave with the magnetic field vector oriented along the y-axis, the currents flowing across the edges can be obtained by solving (6.128).

After deleting the unknowns associated with the horizontal edges on $y = -1$ and the unknowns related to the vertical edges along $x = -1$, the currents across the three different kinds of edges (namely, horizontal, slant, and vertical) are arranged as a 21×22 matrix. Then we can use AMMM to solve the problem.

The largest scale can be taken as 2. So the total number of unknowns for $\{J^-\}, \{J'\}$, and $\{J^|\}$ is $(5 \times 2^0 + 1)^2$, $(5 \times 2^1 + 1)^2$, and $(5 \times 2^2 + 1)^2$ as one moves from the 0 scale to the 2 scale. For the different thresholds, the reduced number of unknowns $\{J^-\}, \{J'\}$, and $\{J^|\}$, the actual size of the linear equations, and the condition number on the 2 scale are given in Table 6.11. The current distribution across the three different types of edges is shown in Figure 6.19. The condition numbers of the impedance matrix before and after the transformations by a multiscale basis are 97.7 and 11,239, respectively. The bistatic RCS is shown in Figure 6.20. The threshold in Table 6.11 defines the value below which the elements are set to zero. The number of unknowns that have been eliminated from the matrix and the size of the reduced matrix are also presented along with the condition number.

Table 6.11
Results at Different Scales

Threshold	0.01	0.02	0.05
Reduction on $J_|$	20	60	183
Reduction on $J_/$	67	127	214
Reduction on J_-	59	117	187
Actual size of matrix	1,177	1,019	739
Condition number	249,771	318,247	80,887

If the incident angle is $\phi = \theta = 30°$, the reduction on the number of unknowns $\{J^-\}, \{J'\}$, and $\{J^|\}$, the actual size of the linear equations, and the condition number on the 2 scale are given in Table 6.12. The bistatic RCS is shown in Figure 6.21.

The threshold in Table 6.12 defines the value below which the elements are set to zero. The number of unknowns that have been eliminated from the matrix and the size of the reduced matrix are also presented along with the condition number. From Tables 6.11 and 6.12, we see that the smaller the threshold, the fewer the number of unknowns that have been eliminated. There is no relationship between the condition number and the actual size of the modified linear equations. When the threshold is taken as 0.01, 0.05, or 0.1, the size of the linear equation is reduced by about 10, 22, or 42%, respectively. The errors in the bistatic RCS obtained by AMMM and the conventional moment method are small.

As a second example, consider the scattering from a 1.5λ length cylinder with two 0.3λ radius half-spheres terminating each side. The length is subdivided uniformly into 16 nodes; the half-circle is subdivided uniformly into 7 nodes. We make 21 divisions along the circumference. The number along the horizontal edges is 21×21. The number along the vertical edges is 20×21 and along the slanted edges is 19×21, as shown in Figure 6.22. The total number of unknowns $\left\{ J^{-}, J^{/}, J^{\mid} \right\}$ is $\{21 \times 21, 19 \times 21, 20 \times 21\}$.

To apply (6.127), we add some additional unknowns for $\left\{ J^{-}, J^{/}, J^{\mid} \right\}$ so that the number of unknowns $\left\{ J^{-}, J^{/}, J^{\mid} \right\}$ becomes $\{21 \times 21, 21 \times 21, 21 \times 21\}$. When the conducting object is illuminated by a normally incident plane wave with the magnetic field vector oriented along the $+ y$-axis, the currents across the edges can be obtained by solving (6.128).

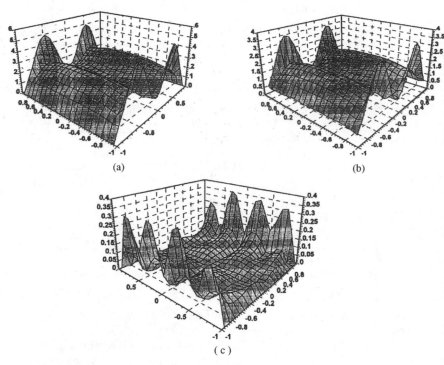

(a)   (b)

(c)

Figure 6.19 Magnitude of the currents across the three different types of edges associated with a triangle for a normally incident plane wave: (a) horizontal edges, (b) slanted edges, and (c) vertical edges.

Table 6.12
Results for Different Parameters

Threshold	0.01	0.02	0.05
Reduction on $J^{\mid}$	24	73	156
Reduction on $J^{/}$	55	103	188
Reduction on J^{-}	59	120	201
Actual size of matrix	1,185	1,027	778
Condition number	275,987	166,769	93,871

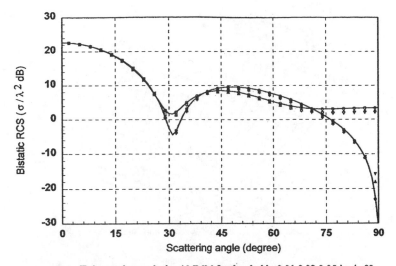

o, Δ, ∇ denote the results by AMMM for threshold =0.01,0.02,0.05 in φ=0°

•, ▲, ▼ denote the results by AMMM for threshold =0.01,0.02,0.05 in φ=90°

Solid lines denote the results by moment method

Figure 6.20 Bistatic RCS of the plate for a normally incident wave in the plane φ = 0° and 90°.

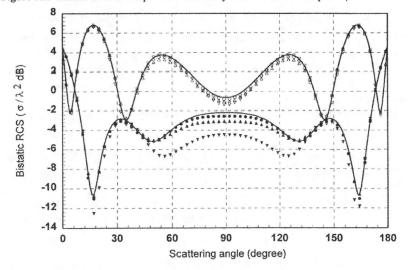

o, Δ, ∇ denote the results by AMMM for threshold = 0.01,0.02,0.05 in φ=0°

•, ▲, ▼ denote the results by AMMM for threshold = 0.01,0.02,0.05 in φ=90°

Solid lines denote the results by moment method

Figure 6.21 Bistatic RCS of the plate with the incident wave arriving from
φ = θ = 30° in the plane φ = 0° and 90°.

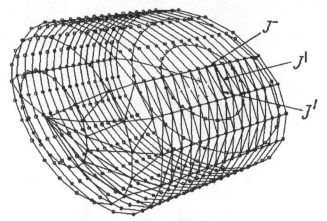

Figure 6.22 Three categories of currents on the surface.

In AMMM, the largest scale is set as 2 and the number of the initial subdivision at scale zero is $N = 5$. So the number of unknowns $\{J^-, J^/, J^|\}$ is (5 $\times 2^0 + 1)^2$, $(5 \times 2^1 + 1)^2$, and $(5 \times 2^2 + 1)^2$ as one goes from the 0 scale to the 2 scale. For the different thresholds, the reduction on the number of unknowns $\{J^-, J^/, J^|\}$, the actual size of the reduced matrix of linear equations, and the condition number on the 2 scale are given in Table 6.13. The current distributions across two of the different types of edges are shown in Figure 6.23. The condition numbers of the original coefficient matrix and the coefficient matrix after the matrix transformation are 397 and 20,911, respectively. The bistatic RCS is shown in Figure 6.24.

If the incident angle is $\phi = \theta = 30°$, the reduction in the number of unknowns $\{J^-, J^/, J^|\}$, the actual size of the reduced matrix of linear equations, and the condition number on the 2 scale are given in the Table 6.14. The bistatic RCS is shown in Figure 6.25. From Tables 6.13 and 6.14, we can see that the size of the linear equation is reduced by about 17, 50, or 72%, respectively, when the threshold is taken as 0.01, 0.05, or 0.1. From Figures 6.24 and 6.25 we can see that the bistatic RCSs obtained by the AMMM and moment method have small errors.

Table 6.13
Results for Different Parameters

Threshold	0.01	0.05	0.10
Reduction on J^-	80	229	318
Reduction on $J^/$	75	268	331
Reduction on $J^|$	73	213	364
Actual size of matrix	1,095	613	778
Condition number	294,893	172,189	46,116

Table 6.14
Results for Different Parameters

Threshold	0.01	0.05	0.10
Reduction on J^-	71	207	310
Reduction on $J^/$	80	202	314
Reduction on $J^|$	86	236	333
Actual size of matrix	1,086	678	366
Condition Number	350,834	94,700	25,310

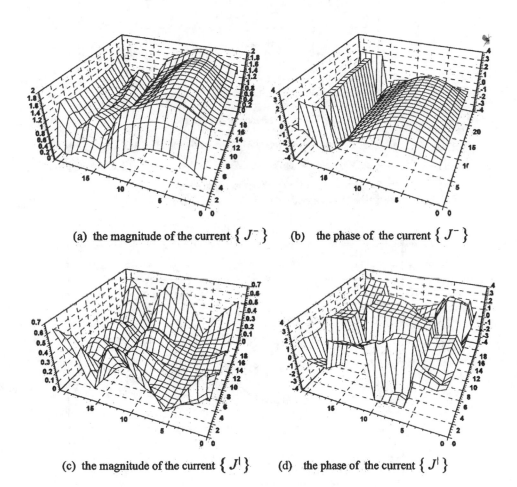

(a) the magnitude of the current $\{ J^- \}$ (b)   the phase of the current $\{ J^- \}$

(c) the magnitude of the current $\{ J^| \}$ (d)   the phase of the current $\{ J^| \}$

Figure 6.23 Currents across two of the different type of edges with a normally incident plane wave: (a) magnitude and (b) phase of the current $\{ J^- \}$; (c) magnitude and (d) phase of the current $\{ J^| \}$.

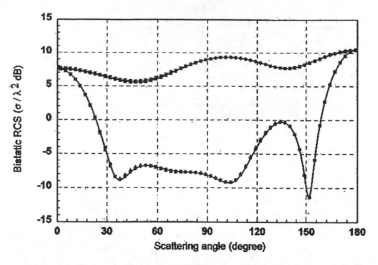

o, Δ, ∇ denote the results by AMMM for threshold =0.01,0.05,0.1 in ϕ=0°

•, ▲, ▼ denote the results by AMMM for threshold =0.01,0.05,0.1 in ϕ=90°

Solid lines denote the results by moment method

Figure 6.24 Bistatic RCS of the cylinder for a normally incident wave in the plane ϕ = 0° and 90°

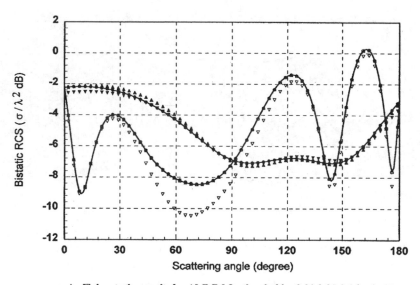

o, Δ, ∇ denote the results by AMMM for threshold = 0.01,0.05,0.1 in ϕ=0°

•, ▲, ▼ denote the results by AMMM for threshold = 0.01,0.05,0.1 in ϕ=90°

Solid lines denote the results by moment method

Figure 6.25 Bistatic RCS of the cylinder with the incident wave arriving from
ϕ = θ = 30° in the plane ϕ = 0° and 90°

6.7 A TWO-DIMENSIONAL MULTISCALE BASIS ON A TRIANGULAR DOMAIN AND THE GEOMETRICAL SIGNIFICANCE OF THE COEFFICIENTS FOR THE MULTISCALE BASIS

In this section, we develop the concept of a multiscale basis to be defined for a triangular domain; that is, each triangular domain is subdivided to a finer higher scale. This is in contrast to the materials of the previous section where the concept of refinement is introduced only at the matrix solution stage so that the actual triangular domains are not physically refined. Only the amplitudes for the unknowns have been appropriately scaled using the multiscale concepts.

6.7.1 Introduction

Consider a triangular domain D constructed by three points $P_1^{(0)}, P_2^{(0)}$, and $P_3^{(0)}$. The shape functions for the three points are denoted by $\phi_1^{\Pi}(P_1^{(0)}, x, y), \phi_2^{\Pi}(P_2^{(0)}, x, y)$, and $\phi_3^{\Pi}(P_3^{(0)}, x, y)$. They are unity at the respective node points, and zero at other nodes. In addition, they vary linearly everywhere on D. This is the situation at scale zero. Increasing the scale by one in this domain, three new nodes are obtained. They are defined by $P_1^{(1)}, P_2^{(1)}$, and $P_3^{(1)}$. The shape functions for these nodes are defined through $\phi_4^{\Pi}(P_1^{(1)}, x, y)$, $\phi_5^{\Pi}(P_2^{(1)}, x, y)$ and $\phi_6^{\Pi}(P_1^{(1)}, x, y)$. They are plotted in Figure 6.26. We call these shape functions the multiscale basis functions or the multiscale shape functions.

For these same nodes constituting a triangular mesh, we can construct the conventional piecewise triangular shape functions in the conventional way through $\phi_1(P_1^{(0)}, x, y)$, $\phi_2(P_2^{(0)}, x, y)$, $\phi_3(P_3^{(0)}(x, y)$, $\phi_4(P_1^{(1)}, x, y)$, $\phi_5(P_2^{(1)}, x, y)$, and $\phi_6(P_1^{(1)}, x, y)$. We refer to these shape functions as conventional shape functions or basis functions. The multiscale shape functions and the conventional shape functions are related through

$$\begin{pmatrix} \phi_1\left(P_1^{(0)}, x, y\right) \\ \phi_2\left(P_2^{(0)}, x, y\right) \\ \phi_3\left(P_3^{(0)}, x, y\right) \\ \phi_4\left(P_1^{(1)}, x, y\right) \\ \phi_5\left(P_2^{(1)}, x, y\right) \\ \phi_6\left(P_3^{(1)}, x, y\right) \end{pmatrix} = \begin{pmatrix} 1 & & & \frac{-1}{2} & \frac{-1}{2} & \\ & 1 & & \frac{-1}{2} & & \frac{-1}{2} \\ & & 1 & & \frac{-1}{2} & \frac{-1}{2} \\ & & & 1 & & \\ & & & & 1 & \\ & & & & & 1 \end{pmatrix} \begin{pmatrix} \psi_1^{\Pi}\left(P_1^{(0)}, x, y\right) \\ \psi_2^{\Pi}\left(P_2^{(0)}, x, y\right) \\ \psi_3^{\Pi}\left(P_3^{(0)}, x, y\right) \\ \psi_4^{\Pi}\left(P_1^{(1)}, x, y\right) \\ \psi_5^{\Pi}\left(P_2^{(1)}, x, y\right) \\ \psi_6^{\Pi}\left(P_3^{(1)}, x, y\right) \end{pmatrix} \qquad (6.129)$$

Let us assume that that the function $f(x, y)$ is approximated in region D with the conventional shape functions. Then it can be described by

$$f_{\text{approx}}(x, y) = f_1 \, \phi_1(P_1^{(0)}, x, y) + f_2 \, \phi_2(P_2^{(0)}, x, y) + f_3 \, \phi_3(P_3^{(0)}, x, y)$$
$$+ f_4 \, \phi_4(P_1^{(1)}, x, y) + f_5 \, \phi_5(P_2^{(1)}, x, y) + f_6 \, \phi_6(P_3^{(1)}, x, y) \qquad (6.130)$$

where $f_i = f(P_i^{(0)})$, $f_{3+i} = f(P_i^{(1)})$, $i = 1, 2, 3$. The same function $f(x, y)$ in region D can be described by the multiscale shape/basis functions as

$$f_{\text{approx}}(x, y) = f_1^{\Pi} \psi_1^{\Pi}(P_1^{(0)}, x, y) + f_2^{\Pi} \psi_2^{\Pi}(P_2^{(0)}, x, y) + f_3^{\Pi} \psi_3^{\Pi}(P_3^{(0)}, x, y)$$
$$+ f_4^{\Pi} \psi_4^{\Pi}(P_1^{(1)}, x, y) + f_5^{\Pi} \psi_5^{\Pi}(P_2^{(1)}, x, y) + f_6^{\Pi} \psi_6^{\Pi}(P_3^{(1)}, x, y)$$

$$(6.131)$$

From (6.129), the coefficients in (6.131) can be determined as

$$
\begin{aligned}
f_i^{\Pi} &= f(P_i^{(0)}) \qquad\qquad i = 1, 2, 3 \\
f_4^{\Pi} &= f(P_1^{(1)}) - \frac{1}{2}\left[f(P_1^{(0)}) + f(P_2^{(0)}) \right] \\
f_5^{\Pi} &= f(P_2^{(1)}) - \frac{1}{2}\left[f(P_1^{(0)}) + f(P_3^{(0)}) \right] \\
f_6^{\Pi} &= f(P_3^{(1)}) - \frac{1}{2}\left[f(P_2^{(0)}) + f(P_3^{(0)}) \right]
\end{aligned}
\qquad (6.132)
$$

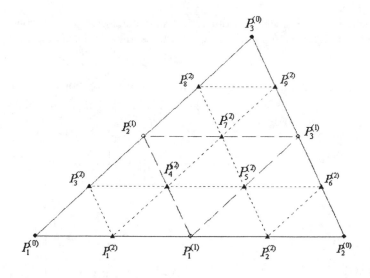

• : initial nodes at scale zero; ○ : nodes corresponding to scale 1; ▲: nodes corresponding to scale 2

Figure 6.26 Nodes corresponding to various scales for a triangular domain.

From a geometrical standpoint, the variables f_4^{Π}, f_5^{Π}, and f_6^{Π} represent the difference between a linear function defined along the lines of $\overline{P_1^{(0)} P_2^{(0)}}$, $\overline{P_1^{(0)} P_3^{(0)}}$, and $\overline{P_2^{(0)} P_3^{(0)}}$ and the original function $f(x,y)$ at the middle points of the triangles denoted by $P_1^{(1)}$, $P_2^{(1)}$, and $P_3^{(1)}$. This is illustrated in Figure 6.27. Therefore, if $f(x, y)$ represents a planar function of the form $\{ f(x,y) = ax + by + c)\}$ defined on region D, the coefficients f_4^{Π}, f_5^{Π}, and f_6^{Π} will be zero.

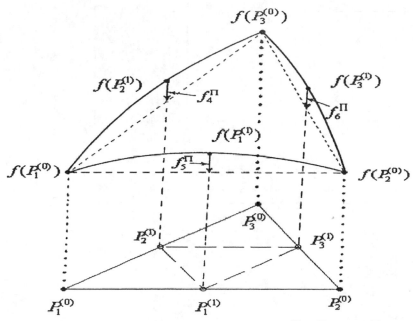

Figure 6.27 Geometrical significance of the coefficients f_4^{Π}, f_5^{Π}, and f_6^{Π}.

After another scaling, nine additional nodes are obtained, that is, $P_i^{(2)}$, $i = 1, 2, ..., 9$, as can be seen in Figure 6.26. The shape functions on these nodes are denoted by $\phi_{6+i}(P_i^{(2)}, x, y)$, $i = 1, 2, ..., 9$. The multiscale shape functions on these nodes are denoted by $\psi_{6+i}^{\Pi}(P_i^{(2)}, x, y)$, $i = 1, 2, ..., 9$. The multiscale shape functions and the conventional shape functions at 2-scale have the following relationship:

$$
\begin{pmatrix}
\phi_1(P_1^{(0)},x,y) \\
\phi_2(P_2^{(0)},x,y) \\
\phi_3(P_3^{(0)},x,y) \\
\phi_4(P_1^{(1)},x,y) \\
\phi_5(P_2^{(1)},x,y) \\
\phi_6(P_3^{(1)},x,y) \\
\phi_7(P_1^{(2)},x,y) \\
\phi_8(P_2^{(2)},x,y) \\
\phi_9(P_3^{(2)},x,y) \\
\phi_{10}(P_4^{(2)},x,y) \\
\phi_{11}(P_5^{(2)},x,y) \\
\phi_{12}(P_6^{(2)},x,y) \\
\phi_{13}(P_7^{(2)},x,y) \\
\phi_{14}(P_8^{(2)},x,y) \\
\phi_{15}(P_9^{(2)},x,y)
\end{pmatrix}
= \aleph \times
\begin{pmatrix}
\psi_1^{\Pi}(P_1^{(0)},x,y) \\
\psi_2^{\Pi}(P_2^{(0)},x,y) \\
\psi_3^{\Pi}(P_3^{(0)},x,y) \\
\psi_4^{\Pi}(P_1^{(1)},x,y) \\
\psi_5^{\Pi}(P_2^{(1)},x,y) \\
\psi_6^{\Pi}(P_3^{(1)},x,y) \\
\psi_7^{\Pi}(P_1^{(2)},x,y) \\
\psi_8^{\Pi}(P_2^{(2)},x,y) \\
\psi_9^{\Pi}(P_3^{(2)},x,y) \\
\psi_{10}^{\Pi}(P_4^{(2)},x,y) \\
\psi_{11}^{\Pi}(P_5^{(2)},x,y) \\
\psi_{12}^{\Pi}(P_6^{(2)},x,y) \\
\psi_{13}^{\Pi}(P_7^{(2)},x,y) \\
\psi_{14}^{\Pi}(P_8^{(2)},x,y) \\
\psi_{15}^{\Pi}(P_9^{(2)},x,y)
\end{pmatrix}
\tag{6.133a}
$$

with

$$
\aleph =
\begin{pmatrix}
1 & & -\tfrac{1}{2} & -\tfrac{1}{2} & & -\tfrac{1}{2} & & -\tfrac{1}{2} & & & & & & & \\
 & 1 & -\tfrac{1}{2} & & -\tfrac{1}{2} & & -\tfrac{1}{2} & & & & & -\tfrac{1}{2} & & & \\
 & & 1 & -\tfrac{1}{2} & -\tfrac{1}{2} & & & & & & & & & -\tfrac{1}{2} & -\tfrac{1}{2} \\
 & & & 1 & & -\tfrac{1}{2} & -\tfrac{1}{2} & & -\tfrac{1}{2} & -\tfrac{1}{2} & & & & & \\
 & & & & 1 & & & -\tfrac{1}{2} & -\tfrac{1}{2} & & & -\tfrac{1}{2} & -\tfrac{1}{2} & & \\
 & & & & & 1 & & & & -\tfrac{1}{2} & -\tfrac{1}{2} & -\tfrac{1}{2} & & & -\tfrac{1}{2} \\
 & & & & & & 1 & & & & & & & & \\
 & & & & & & & 1 & & & & & & & \\
 & & & & & & & & 1 & & & & & & \\
 & & & & & & & & & 1 & & & & & \\
 & & & & & & & & & & 1 & & & & \\
 & & & & & & & & & & & 1 & & & \\
 & & & & & & & & & & & & 1 & & \\
 & & & & & & & & & & & & & 1 & \\
 & & & & & & & & & & & & & & 1
\end{pmatrix}
\tag{6.133b}
$$

So, now at 2-scale, the function $f(x,y)$ can be approximated in region D using the conventional shape functions through

$$
f_{\text{approx}}(x,y) = \sum_{i=1}^{15} f_i\,\phi_i
\tag{6.134}
$$

where $\quad f_i = f(P_i^{(0)}); \; f_{3+i} = f(P_i^{(1)}), \; i = 1, 2, 3; \; f_{6+j} = f(P_j^{(2)}), \; j = 1, 2, ..., 6$.

Let the same function $f(x, y)$ in region D be approximated by the multiscale basis functions as

$$f_{\text{approx}}(x, y) = \sum_{i=1}^{15} f_i^{\Pi} \psi_i^{\Pi} \tag{6.135}$$

According to (6.133) and (6.134), the coefficients in (6.135) can be determined from

$$f_i^{\Pi} = f(P_i^{(0)}) \qquad i = 1, 2, 3$$

$$f_4^{\Pi} = f(P_1^{(1)}) - \frac{1}{2}\left[f(P_1^{(0)}) + f(P_2^{(0)}) \right]; \quad f_5^{\Pi} = f(P_2^{(1)}) - \frac{1}{2}\left[f(P_1^{(0)}) + f(P_3^{(0)}) \right]$$

$$f_6^{\Pi} = f(P_3^{(1)}) - \frac{1}{2}\left[f(P_2^{(0)}) + f(P_3^{(0)}) \right]; \quad f_7^{\Pi} = f(P_1^{(2)}) - \frac{1}{2}\left[f(P_1^{(0)}) + f(P_1^{(1)}) \right]$$

$$f_8^{\Pi} = f(P_2^{(2)}) - \frac{1}{2}\left[f(P_1^{(1)}) + f(P_2^{(0)}) \right]; \quad f_9^{\Pi} = f(P_3^{(2)}) - \frac{1}{2}\left[f(P_1^{(0)}) + f(P_2^{(1)}) \right]$$

$$f_{10}^{\Pi} = f(P_4^{(2)}) - \frac{1}{2}\left[f(P_1^{(1)}) + f(P_2^{(1)}) \right]; \quad f_{11}^{\Pi} = f(P_5^{(2)}) - \frac{1}{2}\left[f(P_1^{(1)}) + f(P_3^{(1)}) \right]$$

$$f_{12}^{\Pi} = f(P_6^{(2)}) - \frac{1}{2}\left[f(P_2^{(0)}) + f(P_3^{(1)}) \right]; \quad f_{13}^{\Pi} = f(P_7^{(2)}) - \frac{1}{2}\left[f(P_2^{(1)}) + f(P_3^{(1)}) \right]$$

$$f_{14}^{\Pi} = f(P_8^{(2)}) - \frac{1}{2}\left[f(P_2^{(1)}) + f(P_3^{(0)}) \right]; \quad f_{15}^{\Pi} = f(P_9^{(2)}) - \frac{1}{2}\left[f(P_3^{(1)}) + f(P_3^{(0)}) \right]$$

If $f(x, y)$ is a planar function of the form $\{ f(x, y) = ax + by + c) \}$ defined on the region D, then all the coefficients f_i^{Π} $(i = 4, 5, ..., 15)$ will be zero. From the above discussions, it is seen that the multiscale basis is related through a matrix transformation to the conventional subdomain triangular basis. For the multiscale basis, we need to construct the shape functions on the newly defined nodes as we increase the scale. The shape functions defined at the previous scale automatically become a part of the multiscale basis. If the function is linear on the region D, the coefficients for the nonzero scales will be zero. This property will prove to be quite useful in compressing representation of functions in an arbitrary domain.

6.7.2 Description of a Multiscale Basis on a Planar Arbitrary Domain

Suppose D is a planar arbitrarily shaped bounded domain as described by Figure 6.28. The nodes at the zero scale are denoted by $\{P_i^{(0)}, \; i = 1, 2, ..., N^{(0)}\}$. At the zero scale, the segments $\{S^{(0)}(N_{i,1}^{(0)}, N_{i,2}^{(0)}), \; i = 1, 2, ..., N_S^{(0)}\}$ connect the nodes. The nodes denoted by $N_{i,1}^{(0)}, N_{i,2}^{(0)}$ connect the ith segment. So at zero scale, the triangular subdomains are defined by $\{T^{(0)}(N_{i,1}^{(0)}, N_{i,2}^{(0)}, N_{i,3}^{(0)}), i = 1, 2, ..., N_T^{(0)}\}$, where $N_{i,1}^{(0)}, N_{i,2}^{(0)}, N_{i,3}^{(0)}$ are the nodes related to the ith triangle. The shape functions then are defined by $\phi_i^{\Pi}(P_i^{(0)}, x, y), i = 1, 2, ..., N^{(0)}$. As the scale in the domain increases, the additional nodes are constructed using the midpoints of all the segments, resulting in

$$P_i^{(1)} = (P_{N_{i,1}}^{(0)} + P_{N_{i,2}}^{(0)})/2 \qquad i = 1, 2, ..., N_S^{(0)} \qquad (6.136)$$

The total number of nodes then becomes $N^{(1)} = N^{(0)} + N_S^{(0)}$. The segments connecting these newly formed nodes are constructed in two stages. At the first stage, a new node is placed at the midpoint of each of the segments. In the second stage, these three new nodes are connected resulting in three new additional segments. The total number of segments then becomes $N_S^{(1)} = 2N_S^{(0)} + 3N_T^{(0)}$. We denote these new segments by $\{S^{(1)}(N_{i,1}^{(1)}, N_{i,2}^{(1)}), i = 1, 2, ..., N_S^{(1)}\}$. Each of the triangles is therefore subdivided into four triangles when one increases the scale by unity. Therefore, the total number of triangles increases by a factor of 4 when the scale is increased by one resulting in $N_T^{(1)} = 4N_T^{(0)}$. We denote the new triangles by $\{T^{(1)}(N_{i,1}^{(1)}, N_{i,2}^{(1)}, N_{i,3}^{(1)}), i = 1, 2, ..., N_T^{(1)}\}$. On this increased scale, the new multiscale functions are denoted by $\psi_{N^{(0)}+i}^{\Pi}(P_i^{(1)}, x, y), i = 1, 2, ..., N_S^{(0)}$. These new functions $\psi_i^{\Pi}(P_i^{(0)}, x, y), i = 1, 2, ..., N^{(0)}$ and $\psi_{N^{(0)}+i}^{\Pi}(P_i^{(1)}, x, y)$, $i = 1, 2, ..., N_S^{(0)}$ can now be used to construct a multiscale basis defined over the new grid points. If the conventional subdomain triangular basis functions over the nodes $\{P_i^{(0)}, P_i^{(1)}\}$ are denoted by $\{\phi_i^{\Delta}(P_i^{(0)}, x, y), \phi_{N^{(0)}+i}^{\Delta}(P_i^{(1)}, x, y)\}$, the following relationship can be established between the multiscale basis and the conventional basis:

$$\begin{pmatrix} \phi_1^{\Delta}(P_1^{(0)}, x, y) \\ \vdots \\ \phi_{N^{(0)}}^{\Delta}(P_{N^{(0)}}^{(0)}, x, y) \\ \phi_{N^{(0)}+1}^{\Delta}(P_1^{(1)}, x, y) \\ \vdots \\ \phi_{N^{(1)}}^{\Delta}(P_{N_S^{(0)}}^{(1)}, x, y) \end{pmatrix} = \begin{pmatrix} 1 & & & \vdots & & \\ & \ddots & & \vdots & A^{(0)} & \\ & & 1 & \vdots & & \\ \cdots\cdots & \cdots & \cdots & \cdots & \cdots\cdots \\ & \bigcirc & & \vdots & 1 & \\ & & & \vdots & & \ddots \\ & & & \vdots & & 1 \end{pmatrix} \begin{pmatrix} \psi_1^{\Pi}(P_1^{(0)}, x, y) \\ \vdots \\ \psi_{N^{(0)}}^{\Pi}(P_{N^{(0)}}^{(0)}, x, y) \\ \psi_{N^{(0)}+1}^{\Pi}(P_1^{(1)}, x, y) \\ \vdots \\ \psi_{N^{(1)}}^{\Pi}(P_{N_S^{(0)}}^{(1)}, x, y) \end{pmatrix}$$

$$(6.137)$$

where $A^{(0)}$ is a $N^{(0)} \times N_S^{(0)}$ sparse matrix.

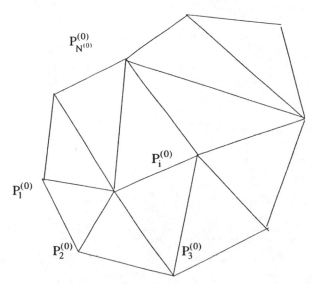

Figure 6.28 An arbitrarily shaped planar bounded domain D.

We can continue with the above subdivisions of the segments by increasing the scale. If we increase the scale by a factor of V, then the domain D, after using a V scaling, will have the following nodes, represented by $\{P_i^{(0)}, i = 1, ..., N^{(0)}\}; ....; \{P_i^{(V)}, i = 1, ..., N_S^{(V-1)}\}$. Now, the matrix relates all elements of the multiscale basis $\psi_V^{\Pi}(x, y)$ with the usual subdomain triangular basis $\Phi_V^{\Delta}(x, y)$ through

$$
\psi_V^{\Pi}(x, y) = \begin{pmatrix} \psi_1^{\Pi}(P_1^{(0)}, x, y) \\ \vdots \\ \psi_{N^{(0)}}^{\Pi}(P_{N^{(0)}}^{(0)}, x, y) \\ \psi_{N^{(0)}+1}^{\Pi}(P_1^{(1)}, x, y) \\ \vdots \\ \psi_{N^{(1)}}^{\Pi}(P_{N_S^{(0)}}^{(1)}, x, y) \\ \vdots \\ \psi_{N^{(V-1)}+1}^{\Pi}(P_1^{(V)}, x, y) \\ \vdots \\ \psi_{N^{(V)}}^{\Pi}(P_{N_S^{(V-1)}}^{(V)}, x, y) \end{pmatrix} \qquad \Phi_V^{\Delta}(x, y) = \begin{pmatrix} \phi_1^{\Delta}(P_1^{(0)}, x, y) \\ \vdots \\ \phi_{N^{(0)}}^{\Delta}(P_{N^{(0)}}^{(0)}, x, y) \\ \phi_{N^{(0)}+1}^{\Delta}(P_1^{(1)}, x, y) \\ \vdots \\ \phi_{N^{(1)}}^{\Delta}(P_{N_S^{(0)}}^{(1)}, x, y) \\ \vdots \\ \phi_{N^{(V-1)}+1}^{\Delta}(P_1^{(V)}, x, y) \\ \vdots \\ \phi_{N^{(V)}}^{\Delta}(P_{N_S^{(V-1)}}^{(V)}, x, y) \end{pmatrix}
$$

with $N^{(i)} = N^{(i-1)} + N_S^{(i-1)}$, $N_S^{(i)} = 2N_S^{(i-1)} + 3N_T^{(i-1)}$, $N_T^{(i)} = 4N_T^{(i-1)}$; $i = 1, 2, ..., V$.

The relation between the multiscale basis and the conventional basis is given through

$$[\Phi_V^{\Delta}(x, y)] = [H(V)][\psi_V^{\Pi}(x, y)] \tag{6.138}$$

where

$$[H(V)] = \begin{pmatrix} I_{N^{(0)} \times N^{(0)}} & A^{(0)}_{N^{(0)} \times N_S^{(0)}} & & & \\ 0 & I_{N_S^{(0)} \times N_S^{(0)}} & A^{(1)}_{N^{(1)} \times N_S^{(1)}} & \cdots & A^{(J-1)}_{N^{(J-1)} \times N_S^{(J-1)}} \\ & 0 & I_{N_S^{(1)} \times N_S^{(1)}} & & \\ & \vdots & 0 & \ddots & I_{N_S^{(J-1)} \times N_S^{(J-1)}} \end{pmatrix}$$

where 0 implies that the elements of the matrix are all zero. Here $[H(V)]$ is a full-rank matrix. The coefficient vector $[F_\Pi]$ of $f(x, y)$ representing the amplitudes corresponding to the multiscale basis and the coefficient vector $[F_\Delta]$ of $f(x, y)$ for the subdomain triangular basis are related by

$$[F_\Pi] = [H(V)][F_\Delta][H]^{-1} \tag{6.139}$$

We can see from the previous discussions that some of the elements of the coefficient vector $[F_\Pi]$ will be small if the function $f(x, y)$ is approximately planar over some region in D. The elements of $[F_\Pi]$ that are below a prespecified threshold are set to zero resulting in the modified vector $[\tilde{F}_\Pi]$:

$$\tilde{F}_\Pi(i) = \begin{cases} F_\Pi(i) \; ; \; \text{if} \;\; i = 1, 2, ..., N^{(0)} \\ F_\Pi(i) \; ; \; \text{if} \; |F_\Pi(i)| > \varepsilon; \; \text{for} \; i = N^{(0)} + 1, ..., N^{(V)} \\ 0 \; ; \; \text{if} \;\; |F_\Pi(i)| \leq \varepsilon \end{cases} \tag{6.140}$$

The new coefficient vector for the conventional subdomain basis will then be

$$[\tilde{F}_\Delta] = [H(V)][\tilde{F}_\Pi] \tag{6.141}$$

The compression ratio is defined as the number of nonzero elements of the vector $[\tilde{F}_\Pi]$ divided by the total number of elements $N^{(V-1)}$. The average error and the maximum error between $[\tilde{F}_\Delta]$ and $[F_\Delta]$ are defined as follows:

$$\text{AverErr}(\tilde{F}_\Delta, F_\Delta) = \sqrt{\frac{\sum\limits_{i=1}^{N^{(V)}} (\tilde{F}_\Delta(i) - F_\Delta(i))^2}{N^{(V)}}} \tag{6.142}$$

$$\text{MaxErr}(\tilde{F}_\Delta, F_\Delta) = \max \left| \tilde{F}_\Delta(i) - F_\Delta(i) \right| \tag{6.143}$$

As an illustrative example, consider the following function

$$f(x,y) = \frac{1}{\sqrt{(x-1)^2 + (y-1)^2 + 0.01}} + \frac{1}{\sqrt{(x+1)^2 + (y-1)^2 + 0.01}}$$
$$+ \frac{1}{\sqrt{(x-1)^2 + (y+1)^2 + 0.01}} + \frac{1}{\sqrt{(x+1)^2 + (y+1)^2 + 0.01}}$$

defined in the domain $D{:}\{(x,y),$ with $1 \leq x \leq 2$ and $1 \leq |y| \leq 2\}$. This function is plotted in Figure 6.29. From the initial segment to the fourth multiscaling, the newly formed nodes are plotted in Figure 6.30. The coefficients for the multiscale basis and the subdomain triangular basis at the highest scale of four are shown in Figure 6.31. The nonzero coefficients for the different thresholds at scale four are given in Table 6.15 along with the average and the maximum value for the errors representing the difference between the reconstructed solution from the threshold coefficients using the multiscale basis and the original function.

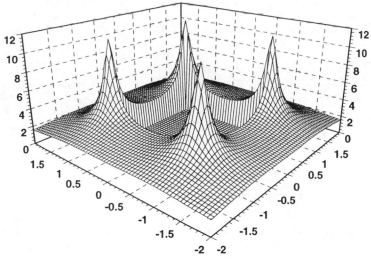

Figure 6.29 Plot of the function $f(x,y)$.

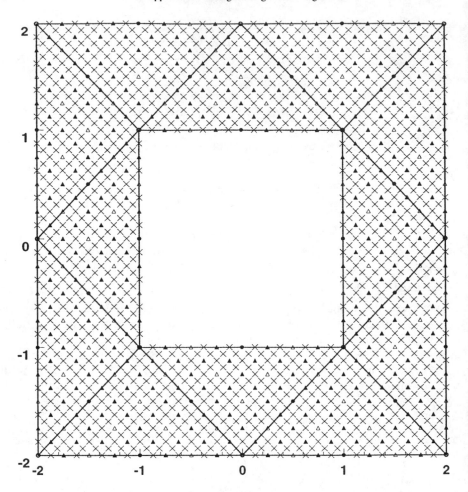

° nodes at scale 0; • nodes at scale 1; Δ nodes at scale 2; ▲ nodes at scale 3; × nodes at scale 4
Total number of nodes including those at the highest scale is 1,632.

Figure 6.30 A multiscale representation of domain *D*.

Table 6.15
Results Obtained After the Fourth Scaling

Threshold ε	Nonzero coefficients	AveErr	MaxErr
0.01	888 (54.4%)	3.19E-03	1.09E-02
0.05	428 (26.2%)	1.79E-02	4.77E-02
0.1	264 (16.2%)	3.43E-02	1.12E-01

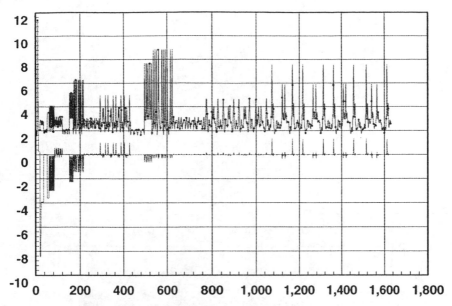

Figure 6.31 The coefficients on a multiscale basis and conventional shaped basis. (Solid line with dots represents the coefficients for the conventional basis; solid line presents the coefficients for the multiscale basis.)

As another example, consider the function

$$f(x, y) = \sum_{i=1}^{8} \frac{1}{\sqrt{(x - x_i)^2 + (y - y_i)^2 + 0.01}}$$

defined on a hexagonal domain D with three regular triangle holes, where the six coordinates of the hexagon are given by

$$(x_i, y_i) = \{(0,0), (1,0), (-1,0), (\tfrac{1}{2}, \tfrac{\sqrt{3}}{2}), (\tfrac{1}{2}, -\tfrac{\sqrt{3}}{2}), (-\tfrac{1}{2}, \tfrac{\sqrt{3}}{2}), (-\tfrac{1}{2}, -\tfrac{\sqrt{3}}{2}), (\tfrac{3}{2}, \tfrac{\sqrt{3}}{2})\}$$

This function is plotted in Figure 6.32, whereas the nodes from the initial segmentation to the fourth scale are shown in Figure 6.33.

The actual values of the coefficients for the multiscale basis and the conventional triangular basis up to 4-scale are plotted in Figure 6.34. For the different thresholds, the number of nonzero coefficients and the errors in the reconstruction of the function using the threshold coefficients at 4-scale are given in Table 6.16.

Table 6.16
Results at Scale 4 Using the Multiscale Basis

Threshold	Nonzero coefficients	AverErr	MaxErr
0.01	1785 (62.5%)	3.41E-03	9.97E-03
0.05	887 (31.1%)	1.82E-02	4.98E-02
0.1	589 (20.6%)	3.88E-02	1.49E-01

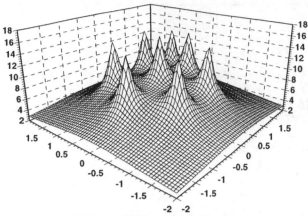

Figure 6.32 Plot of the function $f(x, y)$.

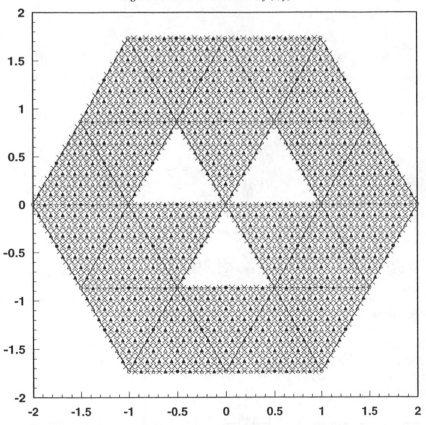

° nodes at scale 0; • nodes at scale 1; △ nodes at scale 2; ▲ nodes at scale 3; × nodes at scale 4
Total number of nodes is 2,854.
Figure 6.33 A multiscale representation for domain D.

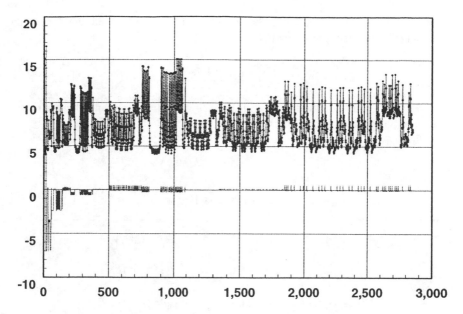

Figure 6.34 The coefficients for the multiscale basis (solid line) and the triangular-shaped basis (solid line with dots).

From these examples, we can see that the reconstruction error between the original function and the function reconstructed from the coefficients that are left after the threshold has been applied to eliminate the small ones in magnitude, can be controlled by the value ε. The ratio between the remaining nonzero elements and the total number of elements varies from about 65% to less than 20% with the threshold values in the range between [0.01, 0.1]. Therefore, it appears that this multiscale basis can be used to approximate any function over an arbitrary planar domain using a prespecified accuracy by specifying the value of the threshold.

6.7.3   A Multiscale Moment Method over an Arbitrary Planar Domain

We consider the following Fredholm integral equation on a planar arbitrary domain

$$\int_D k(x, y; x', y') f(x', y') dx' dy' = g(x, y) \qquad (x, y) \in D \qquad (6.144)$$

Assume that the unknown function f can be represented using the conventional subdomain triangular basis as

$$f(x, y) = [\Phi_V^\Delta(x, y)]^T [F_\Delta] \qquad (6.145)$$

where the superscript T denotes the transpose of a matrix; and that using the multiscale basis it is represented by

$$f(x, y) = [\psi_V^\Pi(x, y)]^T [F_\Pi] \tag{6.146}$$

By using Galerkin's method with both the conventional basis and the multiscale basis, we obtain the following matrix equations:

$$[A_\Delta] [F_\Delta] = [G_\Delta] \tag{6.147}$$

$$[A_\Pi] [F_\Pi] = [G_\Pi] \tag{6.148}$$

where

$$[A_\Delta] = (a_{i,j}^\Delta)_{N^{(V)} \times N^{(V)}} \quad ; \qquad\qquad [A_\Pi] = (a_{i,j}^\Pi)_{N^{(V)} \times N^{(V)}}$$

$$a_{i,j}^\Delta = \int_D \phi_i^\Delta(x, y) \, dx \, dy \int_D K(x, y; x', y') \phi_j^\Delta(x', y') \, dx' \, dy'$$

$$a_{i,j}^\Pi = \int_D \psi_i^\Pi(x, y) \, dx \, dy \int_D K(x, y; x', y') \psi_j^\Pi(x', y') \, dx' \, dy'$$

$$[G_\Delta] = [g_1^\Delta, g_2^\Delta, \cdots, g_{N^{(V)}}^\Delta]^T \quad ; \qquad g_i^\Delta = \int_D g(x, y) \phi_i^\Delta(P_i, x, y) \, dx \, dy$$

$$[G_\Pi] = [g_1^\Pi, g_2^\Pi, \cdots, g_{N^{(V)}}^\Pi]^T \quad ; \qquad g_i^\Pi = \int_D g(x, y) \psi_i^\Pi(P_i, x, y) \, dx \, dy$$

Since the relationship between the conventional basis and the multiscale basis is given by (6.138), the impedance matrix and the source terms for Galerkin's method are related by

$$[G_\Pi] = [H(V)]^T [G_\Delta]$$

$$[A_\Pi] = [H(V)]^{-1} [A_\Delta] [H^{-1}(V)]^T$$

The coefficient matrix $[A_\Pi]$, the unknown $[F_\Pi]$, and the column matrix $[G_\Pi]$ are assumed to be arranged according to the scaled-block form (see Figure 6.6). We utilize the iterative procedure outlined in the previous section to implement the AMMM technique over an arbitrarily shaped planar domain.

6.7.4   Numerical Results

To test the accuracy and applicability of the AMMM for solving two-dimensional Fredholm integral equations of the first kind, we consider two examples. All of the numerical simulations have been performed on a DELL OptiPlex Gxi 166-MHz personal computer in a multitasking environment. The LU decomposition has been used for the solution of the matrix equation at scale zero. The relative error between the solutions $f_\Delta(x_i, y_i)$, obtained by using the conventional subdomain triangular basis, and $f_\Pi(x_i, y_i)$, which represents a multiscale basis, is defined by

$$\text{Err}(f_\Delta, f_\Pi) = \sqrt{\frac{\sum_i (f_\Delta(x_i, y_i) - f_\Pi(x_i, y_i))^2}{\sum_i (f_\Delta(x_i, y_i))^2}} \tag{6.149}$$

As a first example, consider a Fredholm equation of the first kind with the following kernel on a square $[0,1] \times [0,1]$ domain as

$$k(x, y, x', y') = -\ln \sqrt{(x-x')^2 + (y-y')^2} \tag{6.150}$$

The total number of nodes $N = N_x \times N_y = 33 \times 33 = 1,089$ is uniformly distributed over the entire domain. The source function/excitation is given by $g(x, y) = 1$. For the different scales, the actual size of the system matrix and the condition number for the system of linear equations for the threshold parameter $\varepsilon = 0.01$ are given in Table 6.17. In addition, the error $\text{Err}(f_\Delta, f_\Pi)$ is also provided. The data in parentheses are the case for the threshold of $\varepsilon = 0$. The important point to note is that we are trying to make the matrix containing the unknowns sparse by applying the multiscale methodology directly to the solution even though it is unknown, instead of to the MM impedance matrix. The solutions in terms of $f_\Delta(x_i, y_i)$ from the zeroth to the third scale are plotted in Figure 6.35. The solution $f_\Pi(x_i, y_i)$ from zeroth to the third scale for the threshold $\varepsilon = 0.01$ is shown in Figure 6.36.

Table 6.17
Errors at Different Scales

Scale V	Size	CN	Err
0 (MM)	(25) 25	(116) 116	0
1	(81) 74	(1,691) 1,478	0.00796
2	(289) 161	(15,842) 5,488	0.02224
3	(1,089) 345	(496,064) 66,392	0.17455

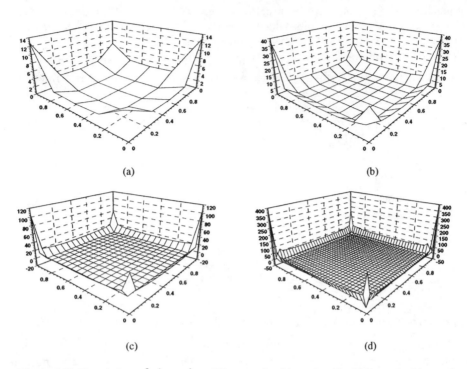

(a)   (b)

(c)   (d)

Figure 6.35 The solutions $f_\Delta(x_i, y_i)$ at different scales: (a) zeroth scale, (b) first scale, (c) second scale, and (d) third scale.

From the computed results, we can see that AMMM can reproduce well the singularity of the function while simultaneously reducing the size of the linear system of equations by $51, 67, 71,$ and 68% for the different scales.

As a second example consider the following kernel:

$$k(x, y, x', y') = 1 / \sqrt{(x - x')^2 + (y - y')^2} \qquad (6.151)$$

in a Fredholm integral equation of the first kind. The domain is a regular hexagon with six regular triangles. At the fourth scale the total number of nodes is 817. This is illustrated in Figure 6.37. The excitation function is $g(x, y) = 1 / \sqrt{x^2 + y^2 + 0.01}$. For the different thresholds, the actual size and the condition number of the linear equations at the largest scale $V = 4$, the error $\mathrm{Err}(f_\Delta, f_\Pi)$ and the CPU time are given in Table 6.18. The solution $f_\Delta(x_i, y_i)$ and the solution $f_\Pi(x_i, y_i)$ for the different thresholds at the largest scale $V = 4$ (the fifth scale) obtained using the AMMM are plotted in Figure 6.38. From the computed results, we can see that the AMMM can reproduce well the peak at the origin (0,0) of the solution although it uses only 45, 26, and 23% of the actual size of the linear system of equations for different thresholds, respectively.

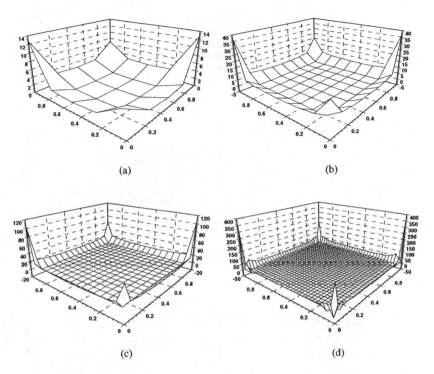

(a)

(b)

(c)

(d)

Figure 6.36 The solutions $f_{\Pi}(x_i, y_i)$ at different scales for the threshold $\varepsilon = 0.01$: (a) zeroth scale, (b) first scale, (c) second scale, and (d) third scale.

Table 6.18
Reduction in the Size of the Matrix as a Function of Threshold and the Error in the Solution for the Fifth Scale.

Threshold	Size	CN	Err	CPU (seconds)
MM	817	233	0	28.84
0.0	817	16,718	0	32.79
0.001	369	5,759	0.141	10.10
0.005	213	3,627	0.234	7.96
0.01	186	2,379	0.309	9.60

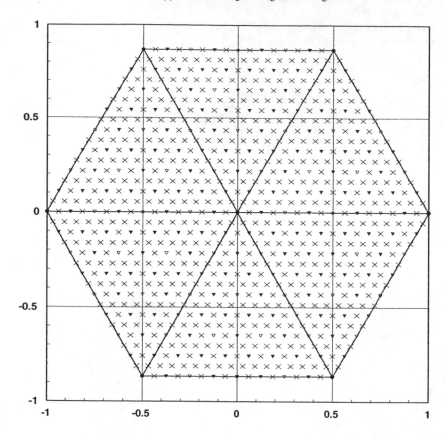

Solid line with •: 0-scale, o : 1-scale, ∇ : 2-scale, ▼ : 3-scale, × 4-scale

Figure 6.37 The domain and the nodes at various scales.

6.8 DISCUSSION

In this chapter, the 2-D adaptive multiscale moment method has been applied for the solution of a Fredholm integral equation of the first kind on planar arbitrarily shaped domains. From the presentation, we can see that:

1. The condition number for the linear system of equations using the multistage basis becomes worse if the transformation is not orthogonal. We can see that the greater the size of the multiscale, the poorer the condition number. Because the condition number becomes worse at higher scales, it is obvious that one cannot go to too large a scale. *Typically, the largest scale should not be greater than five.*

2.   The size of the linear system of equations obtained in the AMMM at the *V*th scale can be reduced after applying a thresholding operation to the interpolated solution obtained from the $(V-1)$th scale.

3.   If the solution of the integral equation is almost a linear function in some regions, the size of the linear equation can be reduced dramatically from the original size of the system of linear equations. This property can significantly reduce the CPU time in the numerical solution of the integral equation.

4.   The AMMM can perform automatically the mesh-refinement procedure locally where the solution is not smooth.

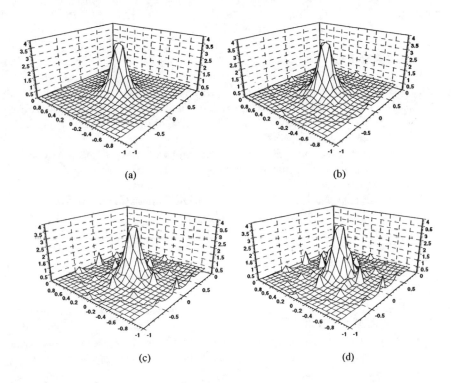

(a)   (b)

(c)   (d)

Figure 6.38 Comparison of the solutions obtained by the moment method and AMMM at the fifth scale: (a) moment method, (b) threshold $\varepsilon = 0.001$, (c) threshold $\varepsilon = 0.005$, and (d) threshold $\varepsilon = 0.01$.

6.9   CONCLUSION

The goal of this chapter is to illustrate how to employ the multiscale basis in the compression of the solution even though it is an unknown. The multiscale basis in higher dimensions is generated from a tensor product of one-dimensional bases. The goal is to compute the MM impedance matrix at the highest desirable scale.

Then we start with the solution at the lowest scale and extrapolate the solution to the next higher scale using a spline interpolation. Based on the value of the second derivative at the nodes of the higher scale, we introduce additional unknowns. Specifically if the value of the double derivative of the interpolant at the higher scales is greater than some prespecified values, additional unknowns are introduced at those nodes. In this way, the solution is iteratively refined.

In addition, we introduce the adaptive methodology using a multiscale basis on a triangular domain and how to refine the triangular patches. We illustrate how to refine the solution as we proceed from a lower scale to a higher scale introducing the multiscale concepts. These methodologies have been incorporated in a traditional moment method resulting in the adaptive multiscale moment method.

REFERENCES

[1] C. de Boor, *A Practical Guide to Splines*, Springer-Verlag, New York, 1978.

[2] S. M. Rao, D. R. Wilton, and A. W. Glisson, "Electromagnetic scattering by surfaces of arbitrary shape," *IEEE Trans. Antennas and Propagat.*, Vol. AP-30, No. 3, May 1982, pp. 409–418.

[3] S. M. Rao, "Electromagnetic scattering and radiation of arbitrarily-shaped surfaces by triangular patch modeling," Ph.D. dissertation, University of Mississippi, 1980.

[4] C. Su and T. K. Sarkar, "Adaptive multiscale moment methods for analyzing EM scattering from electrically large perfectly conducting objects," *J. Electromag. Waves Appl.*, Vol. 12, No. 6, 1998, pp. 753–773.

[5] C. Su and T. K. Sarkar, "A multiscale moment method for solving Fredholm integral equation of the first kind," *Prog. in Electromagnetics Res.*, PIERS 17, 1998, pp. 237–264.

[6] C. Su and T. K. Sarkar, "Scattering from perfectly conducting strips by utilizing an adaptive multiscale moment method," *Prog. in Electromagnetics Res.*, PIERS 19, 1998, pp. 173–197.

[7] C. Su and T. K. Sarkar, "Electromagnetic scattering from coated strips utilizing the adaptive multiscale moment method," *Prog. in Electromagnetics Res.*, PIERS 18, 1998, pp. 173–208.

[8] C. Su and T. K. Sarkar, "Electromagnetic scattering from two-dimensional electrically large perfectly conducting objects with small cavities and humps by use of adaptive multiscale moment method (AMMM)," *J. Electromag. Waves Appl.*, Vol. 12, 1998, pp. 885–906.

[9] C. Su and T. K. Sarkar, "Adaptive multiscale moment method for solving two-dimensional Fredholm integral equation of the first kind," *Prog. in Electromagnetics Res.*, PIERS 21, 1999, pp. 173–201.

[10] C. Su and T. K. Sarkar, "Adaptive multiscale moment method (AMMM) for analysis of scattering from perfectly conducting plates," *IEEE Trans. Antennas and Propagat.*, Vol. 48, No. 6, June 2000, pp. 932–939.

Chapter 7

THE CONTINUOUS WAVELET TRANSFORM AND ITS RELATIONSHIP TO THE FOURIER TRANSFORM

This chapter describes the continuous wavelet transform from the perspective of a Fourier transform. The relationship between the Fourier transform, the Gabor transform, the windowed Fourier transform, and the wavelet transform is described. The differences are also outlined to bring out the characteristics of the wavelet transform. The limitations of the wavelet in localizing the response in various domains is also delineated. The Heissenberg principle of uncertainty limits the ultimate resolution of these techniques. However, since the wavelet transform has redundant features, because it is a function of the two parameters of dilation and shift, it can be advantageous in some cases.

7.1 INTRODUCTION

In nature, one does not encounter pure single-tone signals, but signals whose "frequency content" varies with time. Generally, frequency is defined as a phenomenon where the period of zero crossings of a signal has a fixed duration for all times and frequencies and is related to the period. However, one may define the term "instantaneous frequency" by defining signals whose frequency content changes as a function of time or duration of the signal. A good example of that is an audio signal, where the instantaneous frequency may change from 20 Hz to 20 kHz depending on the system. Hence, it is interesting to develop methodologies that can be introduced to analyze such signals.

In modern times, such concepts have been proven useful in radar system analysis. Conventionally, radar has dealt with pure continuous wave (cw), either frequency modulated or gated (i.e., turned off and on) to generate modulated pulses. However, other types of radar systems deal with wideband pulses. Hence, in understanding how such radar systems work, it is necessary to understand the time-frequency representation of waveforms and how they are characterized and analyzed. In this study, we look at the classical Fourier transforms, Gabor transforms, the windowed Fourier transform, and the wavelet transforms that have been utilized for characterizing waveforms whose frequency content changes with time. One of the main features that distinguishes the various transforms is the choice of a window function. The purpose of a window function is to localize the information about the signal in both frequency and time. Hence, analysis of wave shapes is the main theme of this section.

In this chapter, we summarize the various analytical tools that perform time-frequency analysis.

7.2 CONTINUOUS TRANSFORMS

7.2.1 Fourier Transform

The classical Fourier transform technique is utilized to find the frequency content of a particular wave shape $p(t)$, which has occurred just once and has existed for a time interval $[0,T]$ and is nonexistent for any other times. It is well known that the frequency content of such a signal $p(t)$ is given by the Fourier transform $P_{(\omega)}$ so that

$$P(\omega) = \int_{-\infty}^{\infty} p(t)\ e^{-j\omega t}\ dt \cong \int_{0}^{T} p(t)\ e^{-j\omega t}\ dt \qquad (7.1)$$

Expression (7.1) can be interpreted as modulating the function $p(t)$ by $e^{-j\omega t}$ and then integrating it. The Fourier transform is defined for all values of the variable t (i.e., from $-\infty < t < \infty$). The Fourier series of a function $p(t)$ assumes that such a function is periodic, that is, $p(t)$ repeats itself after every period $T_0 = 2\pi / \omega_0$ (sec), for $\omega_0 = 2\pi f_0$ (rad/sec), and f_0 (Hz). In many cases, T_0 is normalized to 2π (i.e., $\omega_0 = 1$ rad/sec), as we will do in this section. The Fourier series of such a function $p(t)$ is defined as

$$p_F(t) = \sum_{n=-\infty}^{\infty} c_n\ e^{jnt} \qquad (7.2)$$

where

$$c_n = \frac{1}{2\pi} \int_{0}^{2\pi} p(t)\ e^{-jnt}\ dt \qquad (7.3)$$

Therefore, for a Fourier series, the function $p(t)$ is decomposed into a sum of orthogonal functions $c_n e^{jnt}$ Observe that the orthogonal functions into which $p(t)$ is decomposed in (7.2) are generated by integer dilations of a single function e^{jt}. Also, note that the function $p_F(t)$ is periodic with period 2π. In contrast, for the inverse Fourier transform, the spectrum is decomposed into noninteger dilations of the function e^{jt}, as we know:

$$p(t) = \frac{1}{2\pi} \int_{-\infty}^{\infty} P(\omega) \, e^{j\omega t} \, d\omega \tag{7.4}$$

The problem with analyzing signals that are limited in time (i.e., that exist over a finite time window and are zero elsewhere) by Fourier transform techniques is that such functions cannot simultaneously be band limited. Hence, to represent signals $p(t)$ which exist for $0 \le t \le T$, the Fourier transform is generally not the best way to characterize such signals. However, a short time Fourier transform or a windowed Fourier transform has been utilized to analyze such signals in the time-frequency plane.

By its very definition, the Fourier transformation uses the entire signal and permits analysis of only the frequency distribution of the energy of the signal as a whole [1]. To solve this problem, many researchers who perform spectral analysis have taken a piecewise approach. The signal is broken up into contiguous and often overlapping pieces. Each piece is then separately Fourier transformed. The resulting family of Fourier transformations is then treated as if it were the basis for a joint time-frequency energy distribution [1 − 4].

In the short time Fourier transform, the function $p(t)$ is multiplied by a window function $w(t)$ and the Fourier transform is calculated. The window function is then shifted in time and the Fourier transform of the product is computed again. Therefore, for a fixed shift β of the window $w(t)$, the window captures the features of the signal $p(t)$ around β. The window helps to localize the time domain data, before obtaining the frequency domain information. Hence, the short time Fourier transform is given by

$$P_{STFT}(\omega,\beta) = \int_{-\infty}^{\infty} p(t) \, w(t-\beta) \, e^{-j\omega t} \, dt \tag{7.5}$$

where $w(t)$ is defined as the window function, that is, one is observing $p(t)$ through the window $w(t - \beta)$, where β is the shift. When $w(t) = 1$ for all t, one obtains the classical Fourier transform. By observing (7.5), it is clear that the short time Fourier transform is the convolution of the signal $p(t)$ with a filter whose impulse response is of the form $h(t) = w(-t)e^{-j\omega t}$ so that

$$P_{STFT}(\omega,\beta) = e^{-j\omega\beta} \int_{-\infty}^{\infty} p(t) \, w(t-\beta) \, e^{j\omega(\beta-t)} \, dt \tag{7.6}$$

The problem here is that as the window function gets narrower, the localization information in the frequency domain is compromised. In other words, as the frequency domain localization gets narrower, the time window gets wider, so that localization information in the time domain is compromised due to Heissenberg's uncertainty principle [4]. This principle relates uncertainty in time Δt and the uncertainty in angular frequency $\Delta\omega$ through $\Delta\omega \cdot \Delta t \geq 0.5$.

Since the requirements in the time localization and frequency resolution are conflicting, one has to make some judicious choices. The best window $w(t)$ depends on what is the meaning of the term *best*.

The above problem has solutions only under certain conditions. For example, if we require the function $w(t)$ to be symmetric and that its energy be confined to a certain bandwidth B (i.e., $|\omega| \leq B$), then we know the optimum function $w(t)$ is given by the prolate spheroidal functions. Other criteria such as signal-to-noise ratio can also be utilized in designing the function $w(t)$, and there are other "criteria for best" [4–7].

To obtain the original signal back from the short time Fourier transform, we observe that [4]

$$p(t)\ w(t-\beta) = \frac{1}{2\pi} \int_{-\infty}^{\infty} P_{STFT}(\omega,\beta)\ e^{j\omega t}\ d\omega \tag{7.7}$$

If we set $\beta = t$, then

$$p(t)\ w(0) = \frac{1}{2\pi} \int_{-\infty}^{\infty} P_{STFT}(\omega,t)\ e^{j\omega t}\ d\omega \tag{7.8}$$

So that we can recover the original signal $p(t)$ for all t as long as $w(0) \neq 0$. If $w(0) = 0$, then one needs to choose some other value of ß that will guarantee $w(0) \neq 0$. Observe that it is not necessary to know $w(t)$ for all t, in order to recover $p(t)$ from its short time Fourier transforms.

Alternately, one can also recover $p(t)$ from (7.7) by multiplying both sides by $\overline{w}(t-\beta)$ and integrating with respect to ß. The overbar denotes a complex conjugate. Then,

$$p(t) = \frac{1}{2\pi} \cdot \frac{\displaystyle\int_{-\infty}^{\infty}\int_{-\infty}^{\infty} P_{STFT}(\omega,\beta)\ \overline{w}(t-\beta)\ e^{j\omega t}\ d\omega\ d\beta}{\displaystyle\int_{-\infty}^{\infty} \left| w(t-\beta) \right|^2\ d\beta} \tag{7.9}$$

This of course assumes that $\displaystyle\int_{-\infty}^{\infty} \left| w(t-\beta) \right|^2\ d\beta$ is finite. However, if that is not the case, then one can multiply both sides of (7.7) by another sequence $r(t)$, such that

$$\int_{-\infty}^{\infty} \overline{r}(t-\beta)\ \left| w^2(t-\beta) \right|\ d\beta = 1 \tag{7.10}$$

An attempt was made by Gabor [8] in 1946 to develop a methodology where a function can be simultaneously localized in time and frequency. If that is possible, then the "frequency content" of any signal can easily be obtained by observing the response in certain narrow frequency bands. Hence, it is possible to track the instantaneous frequency of a signal.

To simplify the presentation, we introduce the generalized form of Parseval's theorem for two functions $p(t)$ and $q(t)$ and their Fourier transforms $P(\omega)$ and $Q(\omega)$ so that

$$< p\,;q> = \int_{-\infty}^{\infty} p(t)\ \overline{q}(t)\ dt = \frac{1}{2\pi} \int_{-\infty}^{\infty} P(\omega)\ \overline{Q}(\omega)\ d\omega = \frac{1}{2\pi} < P\,;Q> \quad (7.11)$$

where $\langle\, \bullet\, ;\, \bullet\, \rangle$ denotes the inner product between two functions and the overbar denotes the complex conjugate. Note that if [1]

$$q(t) = \delta(t-t_0) \quad \text{and} \quad Q(\omega) = e^{-j\omega t_0} \quad (7.12)$$

then (7.11) along with (7.12) defines the inverse Fourier transform, and if

$$Q(\omega) = 2\pi\,\delta(\omega-\omega_0) \quad \text{and} \quad q(t) = e^{j\omega_0 t} \quad (7.13)$$

then (7.13) when substituted in (7.11) defines the forward Fourier transform.

In the next section we see how Gabor, through the choice of certain window functions, made it possible not only to localize a signal in the time domain, but was also able to localize its frequency content in a narrow frequency band.

7.2.2 Gabor Transform [1, 8, 9]

The objective of the Gabor transform is to expand $p(t)$ into a set of functions that are simultaneously limited in both time and frequency. This is in contrast to the Fourier transform where the expansion is done by functions $e^{-j\omega t}$ which are not time limited, but highly localized in frequency.

Even though from a strict mathematical point of view it is not possible to localize a function simultaneously both in the time and frequency domains, this simultaneous localization can, however, be achieved from a practical standpoint. Let us illustrate this by considering a family of functions $q(t)$ such that a member $q_\alpha(t)$ is defined by

$$q_\alpha(t) = \frac{1}{2\sqrt{\pi\alpha}}\ e^{-\frac{t^2}{4\alpha}} \quad (7.14)$$

Then the product $p(t)\,q_\alpha(t)$ can be localized in time from a practical standpoint if $\alpha > 0$. This is because, beyond a certain value $t = T_1$, the function $q_\alpha(t)$ practically decays down to zero and so will the product $p(t)\,q_\alpha(t)$. Next, we introduce another parameter β, so that the window function $w_{\alpha,\beta}(t)$ (which is real) is defined by

$$w_{\alpha,\beta}(t) = w_\alpha(t-\beta) = \frac{1}{2\sqrt{\pi\alpha}}\, e^{-\frac{(t-\beta)^2}{4\alpha}} = q_{\alpha,\beta}(t) \qquad (7.15)$$

Thus β is a shift parameter and now the functions $w_{\alpha,\beta}(t)$ can span any function $p(t)$ for all possible choices of the parameter α and β. Now if we look at the Fourier transform of the product of the two functions $p(t)\,w_{\alpha,\beta}(t)$, which is called the Gabor transform [$G_{\alpha,\beta}(\omega)$], then

$$G_{\alpha,\beta}(\omega) = \int_{-\infty}^{\infty} p(t)\, w_{\alpha,\beta}(t)\, e^{-j\omega t}\, dt \qquad (7.16)$$

Here $w_{\alpha,\beta}(t)$ has been assumed to be a real function. Also note that

$$\int_{-\infty}^{\infty} w_{\alpha,\beta}(t)\,d\beta = \int_{-\infty}^{\infty} w_\alpha(t-\beta)\,d\beta = 1 \qquad (7.17)$$

Hence integrating (7.16) with respect to β and in addition using (7.17), one obtains

$$\int_{-\infty}^{\infty} G_{\alpha,\beta}(\omega)\,d\beta = \int_{-\infty}^{\infty} p(t)\, e^{-j\omega t}\, dt = P(\omega) \qquad (7.18)$$

So, the Fourier transform of the function $p(t)$, given by $P(\omega)$, results from integrating the Gabor transform with respect to all possible delay parameters β. Since

$$<p\,;q> = \frac{1}{2\pi}<P\,;Q>$$

and if

$$q(t) = w_{\alpha,\beta}(t)e^{j\omega t}$$

then

$$\int_{-\infty}^{\infty} p(t)\,\overline{q}(t)\, dt = \int_{-\infty}^{\infty} p(t)\, w_{\alpha,\beta}(t)\, e^{-j\omega t}\, dt = G_{\alpha,\beta}(\omega) = \frac{1}{2\pi}\int_{-\infty}^{\infty} P(\Omega)\overline{Q}(\Omega)\,d\Omega$$

$$\qquad (7.19)$$

Moreover, since $w_{\alpha,\beta}(t)$ is real, we have

$$Q(\Omega) = \int_{-\infty}^{\infty} e^{j\omega t}\, w_{\alpha,\beta}(t)\, e^{-j\Omega t}\, dt = \int_{-\infty}^{\infty} \frac{1}{2\sqrt{\pi\alpha}}\, e^{-\frac{(t-\beta)^2}{4\alpha}}\, e^{j(\omega-\Omega)t}\, dt \qquad (7.20)$$

$$= e^{-(\omega-\Omega)^2\alpha}\; e^{j\beta(\omega-\Omega)} = e^{-(\Omega-\omega)^2\alpha}\; e^{-j(\Omega-\omega)\beta}$$

Also, as

$$\int_{-\infty}^{\infty} e^{-c^2 x^2 \pm dx} \, dx = e^{c^2/4d^2} \cdot \frac{\sqrt{\pi}}{c} \tag{7.21}$$

we have

$$\begin{aligned}
G_{\alpha,\beta}(\omega) &= \frac{1}{2\pi} \int_{-\infty}^{\infty} P(\Omega) \, e^{+j(\Omega-\omega)\beta} \, e^{-(\Omega-\omega)^2 \alpha} \, d\Omega \\
&= \frac{e^{-j\omega\beta}}{2\pi} \int_{-\infty}^{\infty} P(\Omega) \, e^{-(\Omega-\omega)^2 \alpha} \, e^{j\Omega\beta} \, d\Omega \\
&= \frac{e^{-j\omega\beta}}{2\sqrt{\pi\alpha}} \int_{-\infty}^{\infty} P(\Omega) \, [w_{1/(4\alpha)} (\Omega-\omega)] \, e^{j\Omega\beta} \, d\Omega
\end{aligned} \tag{7.22}$$

It is clear from (7.22) that the Gabor transform also localizes the Fourier transform $P(\omega)$ of $p(t)$, when the window function $w_{\alpha,\beta}(t)$ is highly concentrated, to give its local spectral information. Hence, not only is the function $p(t)$ localized in time by the function $w_{\alpha,\beta}(t)$, but its transform $P(\omega)$ is also localized in frequency by the function $w_{1/(4\alpha),\beta}(\omega)$. The width of the window function in the time domain Δ_t is then obtained as

$$\Delta_t = \frac{1}{\|w_\alpha\|} \left\{ \int_{-\infty}^{\infty} t^2 w_\alpha^2(t) \, dt \right\}^{1/2} \tag{7.23}$$

In addition,

$$\|w_\alpha\|^2 = \int_{-\infty}^{\infty} \frac{1}{4\pi\alpha} e^{-\frac{t^2}{2\alpha}} \, dt = \frac{1}{\sqrt{8\pi\alpha}} \tag{7.24}$$

and

$$\int_{-\infty}^{\infty} \frac{t^2 e^{-\frac{t^2}{2\alpha}}}{4\pi\alpha} \, dt = \sqrt{\frac{\alpha}{8\pi}} \tag{7.25}$$

Hence

$$\Delta_t = \left\{ \frac{1}{\sqrt{8\pi\alpha}} \right\}^{-1/2} \left\{ \sqrt{\frac{\alpha}{8\pi}} \right\}^{1/2} = \sqrt{\alpha} \tag{7.26}$$

and the width of the spectrum of the window function is given by Δ_ω, which is

$$\Delta_\omega = \{\Delta_\omega\}_{w_{1/(4\alpha)}} = \frac{1}{2\sqrt{\alpha}} \tag{7.27}$$

So the product of the functions $p(t)$ $w_{\alpha,\beta}(t)$ is localized in time at $\beta \pm \dfrac{\sqrt{\alpha}}{2}$ and their spectrum; that is, the spectrum of the windowed version of $p(t)$ is localized in frequency at $\Omega \pm \dfrac{1}{4\sqrt{\alpha}}$. In addition, we can observe that

$$\Delta_t \cdot \Delta_\omega = 0.5 \qquad (7.28)$$

Therefore, the width of the time-frequency window is unchanged for observing the spectrum of $p(t)$ at all frequencies. In fact, it is seen that the Gabor transform is essentially a short time Fourier transform with the smallest time-frequency window. It is the smallest time-frequency window because the principle of uncertainty is exactly equal to 0.5 as given by (7.28). So a window with a Gaussian shape used by Gabor satisfies the principle of uncertainty with an exact equality.

7.2.3   Continuous Wavelet Transform

The window functions of the Gabor transform may and do often have "dc" components. The condition of having no "dc" component for the window function is given by

$$\int_{-\infty}^{\infty} w_{\alpha,\beta}(t)\, dt = 0 \qquad (7.29)$$

Namely, because the average value of the window is zero, an extra degree of freedom is given for introducing a dilation (or scale) parameter in the window in order to make the time-frequency window flexible. This is a unique property satisfied by the wavelet transform. With this dilation parameter, the integral wavelet transform provides a flexible time-frequency window, which automatically narrows when observing high-frequency components and widens when studying low-frequency components. Hence, it is in tune with our auditory and visual sense perceptions which at least in the first step process signals in this fashion. This explains the evolution of the musical scale in the West. For example, as illustrated by Vaidyanathan [4], Figure 7.1 shows the location of the notes c and g in the major diatonic scale for several octaves. On a logarithmic scale, they would appear to be nearly equally spaced. Thus, the notes c and g become sparser and sparser as the frequency increases.

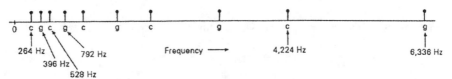

Figure 7.1 Notes as played on a piano keyboard.

The window function in the wavelet transform takes the following form for both positive and negative scale values of α resulting in

$$w_{\alpha,\beta}(t) = \frac{1}{\sqrt{|\alpha|}} \ w\left(\frac{t-\beta}{\alpha}\right) \tag{7.30}$$

The above function is admissible as a window function in the wavelet transform (or succinctly as a wavelet) provided (7.29) is satisfied. The integral wavelet transform of $p(t)$ is defined by $WT[p]$:

$$WT[p] = \int_{-\infty}^{\infty} p(\tau) \ w_{\alpha,\beta}(\tau) \ d\tau = \frac{1}{\sqrt{|\alpha|}} \int_{-\infty}^{\infty} p(\tau) \ w\left(\frac{\tau-\beta}{\alpha}\right) d\tau \tag{7.31}$$

with $\alpha \neq 0$.

The shortcoming of the short time Fourier techniques is taken care of by utilizing the scaled window $w_{\alpha,\beta}(t)$. For a fixed time width of the window $w(t)$ in (7.7), when $p(t)$ is a high-frequency signal, many cycles are captured by the window, whereas if $p(t)$ is a low-frequency signal, only very few cycles are displayed within the same window. Thus, the accuracy of the estimate of the Fourier transform is poor at low frequencies and improves as the frequency increases. One way to correct this problem is to replace the window $w(t)$ with a function of both frequency and time, so that the time domain plot of the window gets wider as frequency decreases. Thus, the window adjusts its width according to frequency. This is accomplished by the wavelet transform as outlined above. For example, if

$$w_{\alpha,\beta}(t) \ = \ a^{-\alpha} \ w\{a^{-\alpha}(t-\beta)\} \quad \text{with a scale factor} \quad a > 1 \tag{7.32}$$

then the Fourier transform of the window function is given by

$$W_{\alpha,\beta}(\omega) \ = \ W(a^{+\alpha}\omega) \ e^{-j\beta\omega} \tag{7.33}$$

Thus, all possible states of the window are obtained by frequency scaling the response of the window function. This is in contrast to the STFT where the various window functions were obtained as $W(\omega) \ e^{-j\beta\omega}$ (i.e., by modulating the spectrum of the window function), where the width of the window is fixed.

Hence if the center of the window function $w(t)$ is given by t^* and the width of the window in the time domain is given by Δ_t^w then the function $w_{\alpha,\beta}(t)$ is a window function with its center at $\beta + \alpha t^*$ and its width as $\alpha \Delta_t^w$. In addition, its spectrum is given by

$$W_{\alpha,\beta}(\omega) = \frac{1}{\sqrt{|\alpha|}} \int_{-\infty}^{\infty} e^{-j\omega t} \, w\!\left(\frac{t-\beta}{\alpha}\right) dt = \frac{\alpha}{\sqrt{|\alpha|}} \, e^{+j\beta\omega} \, W(\alpha\omega) \qquad (7.34)$$

where $W(\omega)$ is the Fourier transform of $w(t)$. From (7.31), one obtains

$$WT[p] = \frac{1}{2\pi} \int_{-\infty}^{\infty} \frac{\alpha}{\left|\sqrt{\alpha}\right|} \, P(\omega) \, e^{-j\beta\omega} \, W(\alpha\omega) \, d\omega \qquad (7.35)$$

The window function in the frequency domain is centered at ω^* and has a width $\frac{2\Delta_\omega^w}{\alpha}$, with the exception of the multiplicative factor $\frac{\alpha}{2\pi\left|\sqrt{\alpha}\right|}$ and the phase factor $e^{\,j\beta\omega}$. The wavelet transform provides local information in the frequency window

$$\left[\frac{\omega^*}{\alpha} - \frac{\Delta_\omega^w}{2\alpha} \; ; \; \frac{\omega^*}{\alpha} + \frac{\Delta_\omega^w}{2\alpha}\right] \qquad (7.36)$$

$$\left[\frac{\text{center frequency}}{\text{bandwidth}} = \frac{\omega^*/\alpha}{2\Delta_\omega^w/\alpha} = \frac{\omega^*}{2\Delta_\omega^w}\right] \qquad (7.37)$$

Note that in the wavelet type of analysis, if ω^* of $W(\alpha\omega)$ is assumed to be positive, then the ratio given by (7.37) is independent of the scaling factor α. The class of bandpass filters represented by (7.36) as a function of α has the property (7.37) and these are called constant-Q filters. This type of processing is done by the human ear at least in the first stage of signal detection [4].

From the wavelet transform given by (7.31) or (7.35), the original function can be recovered by utilizing

$$p(t) = \frac{1}{C_w} \int_{-\infty}^{\infty}\int_{-\infty}^{\infty} WT[p] \, w_{\alpha,\beta}(\alpha) \, \frac{d\alpha}{\alpha^2} \, d\beta \qquad (7.38)$$

where, $\alpha > 0$ and the constant C_w is given by

$$C_w = \int_{-\infty}^{\infty} \frac{|W(\omega)|^2}{|\omega|} \, d\omega \qquad (7.39)$$

Hence (7.39) implies that only certain classes of window functions can be utilized in the wavelet transform, namely, those responses that decay at least as fast as

$\dfrac{1}{\sqrt{|\omega|}}$ as $\omega \to \infty$. The convergence of the integral in (7.38) is defined in a weak

sense [2], that is, taking the inner product of both sides of (7.38) with any function $g(x) \in \mathcal{L}^2$ (where $\mathcal{L}^2$ is the space of square integrable functions). Commuting the inner product with the integral over α and β in the righthand side leads to the true formula. Since any absolutely integrable function $W(\omega)$—the Fourier transform of $w(t)$—is continuous, (7.39) can only be evaluated provided

$$W(\omega = 0) = 0 \qquad (7.40)$$

or equivalently

$$\int_{-\infty}^{\infty} w(x)\, dx = 0 \qquad (7.41)$$

that is, the wavelet $w(x)$ has no dc value as mentioned in (7.29).

Implementation of the wavelet transform is equivalent to filtering the signal $p(t)$ by a bank of bandpass filters, which are of constant Q. The center frequency of the filters is offset from one another by an octave (i.e., multiplied by a factor of 2) and in the limit (7.39) guarantees that the center frequency of the bandpass filters will never be zero.

Finally, it is important to point out that the Bailan-Low theorem proves that for any u_0 and ξ_0, there exists no differentiable window of compact support or which has a slow time decay such that $\{w(t - nu_0)\ \exp(ik\xi_0 t)\}_{n,k}$ is an orthonormal basis of $\mathcal{L}^2(\mathrm{R})$ [10]. This implies that one cannot construct an orthogonal windowed Fourier basis with a differentiable window w of compact support.

In the continuous domain, all three techniques (Fourier, Gabor, and wavelet) have good theoretical properties. The question is what happens in the discrete domain. Do all these properties carry over to the discrete domain or do certain additional constraints need to be imposed?

7.3 DISCRETE TRANSFORMS

The problem of dealing with the discrete transforms has some advantages and disadvantages for each of the three techniques. For example, the Fourier transform for the discrete case changes over to the Fourier series representation. The Gabor transform provides nonphysical representation of the spectrum for a certain class of window functions. Finally, the discrete wavelet transform provides a fast algorithm for computation of the wavelet coefficients from a multirate digital filtering point of view, without even introducing the concept of wavelets.

7.3.1 Discrete Short Time Fourier Transform (DSTFT)

The discrete representation of (7.5) is given by [4]

$$P_{DSTFT}(z,\beta) = \sum_{n=-\infty}^{\infty} p(n)\, w(n-\beta)\, z^{-n} \tag{7.42}$$

where

$$z = e^{j\omega} \tag{7.43}$$

and $p(n)$ and $w(n)$ represent the sampled version of $p(t)$ and $w(t)$, respectively. Here, β is an integer. So, the window function is represented by values that are sampled at $w(n-\beta)$. The inverse DSTFT of P_{DSTFT} is given by

$$p(n)\, w(n-\beta) = \frac{1}{2\pi} \int_0^{2\pi} P_{DSTFT}(z,\beta)\, z^n\, d\omega \tag{7.44}$$

If we set $\beta = n$, then

$$p(n) = \frac{1}{2\pi\, w(0)} \int_0^{2\pi} P_{DSTFT}(z,n)\, z^n\, d\omega \tag{7.45}$$

Hence, one can recover $p(n)$ as long as $w(0) \neq 0$. If $w(0) = 0$ then one chooses a value of β, for which $w(m) = w(n-\beta) \neq 0$ and the procedure continues.

An alternative representation can be made, provided we have

$$\sum_m \left| w(m) \right|^2 = 1 \tag{7.46}$$

For (7.45), $p(n)$ is recovered from

$$p(n) = \frac{1}{2\pi} \int_0^{2\pi} \left(\sum_{m=-\infty}^{\infty} P_{DSTFT}(z,m)\, \overline{w}(n-m) \right) z^n\, d\omega \tag{7.47}$$

where the overbar denotes a complex conjugate. It is interesting to note that the inversion formula is not unique. For example if z_0 is a zero of the Z transform of the conjugate of the window function [i.e., if $W_1(z)$ is in the form of a FIR filter with z_0 as a zero of the polynomial], then

$$W_1(z) = \sum_{k=-\infty}^{\infty} \overline{w}(k)\, z^{-k} \tag{7.48}$$

Now, if we replace $P_{DSTFT}(z, m)$ in (7.47) by $P_{DSTFT}(z, m) + z_0^m$, then (7.47) is still satisfied. This is in contrast to the conventional discrete Fourier transform (or equivalently, the Fourier series), which provides a unique inverse.

7.3.2   Discrete Gabor Transform

For the discrete Gabor transform, the goal is to represent a signal in two dimensions, where time and frequency are the coordinates. Gabor defined the basis functions of approximating a signal $p(t)$ by what may be termed "elementary functions" by

$$\psi_e(t; t_0; f_0) = \frac{1}{2\sqrt{\pi\alpha}} e^{-\frac{(t-t_0)^2}{4\alpha}} e^{-j(2\pi f_0 t + \theta)} \tag{7.49}$$

The Fourier transform Ψ of the elementary signals ψ is given by

$$\Psi_e(f; t_0; f_0) = \int_{\infty}^{\infty} \psi_e(t; t_0; f_0) \, e^{-j2\pi f t} \, dt$$

$$= e^{-\pi^2 4\alpha(f-f_0)^2} e^{-j2\pi t_0(f-f_0)} e^{j\theta} \tag{7.50}$$

These elementary functions then represent complex exponentials, which are a function of frequency f_0, and phase θ whose amplitude is modulated by a Gaussian function centered at t_0. The elementary signal $\psi_e(t; t_0; f_0)$ is called the (0,0) th element and the general (k, ℓ) th element, which we will term $\psi_{k\ell}(t)$, is defined through

$$\psi_{k\ell}(t) = \psi_e\left(t; t_k; f_\ell\right) \tag{7.51}$$

where

$$t_k = k \Delta t \tag{7.52}$$

$$f_\ell = \ell \Delta f \tag{7.53}$$

So the Gabor representation is defined on a two-dimensional grid usually called the "logon" grid by partitioning the time-frequency plane into rectangular grids of equal size, so that the area of each cell of the grid is $\Delta t \times \Delta \omega = 0.5$. The (k, ℓ) th cell, is weighted by the coefficient $a_{k\ell}$ so that the signal $p(t)$ is approximated by a series $\hat{p}(t)$ resulting in

$$p(t) \cong \hat{p}(t) = \sum_{k=-\infty}^{\infty} \sum_{\ell=-\infty}^{\infty} a_{k\ell} \, \psi_{k\ell}(t)$$

$$= \sum_{k=-\infty}^{\infty} \sum_{\ell=-\infty}^{\infty} \frac{a_{k\ell}}{2\sqrt{\pi\alpha}} \, e^{\frac{-(t-t_k)^2}{4\alpha}} \, e^{j(2\pi f_0 t)} \qquad (7.54)$$

In the above expression, the phase term θ has been omitted from the expression (7.54) because this phase coefficient can be taken into account in the coefficient $a_{k\ell}$. However, since the functions $\psi_{k\ell}(t)$ are not orthogonal, it is not a computationally easy task to solve for the coefficient $a_{k\ell}$.

To determine these coefficients $a_{k\ell}$, we rewrite (7.54) as

$$\hat{p}(t) = \sum_{k=-\infty}^{\infty} e^{\frac{-(t-t_k)^2}{4\alpha}} \sum_{\ell=-\infty}^{\infty} \frac{a_{k\ell}}{2\sqrt{\pi\alpha}} \, e^{j(2\pi f_0 t)} \qquad (7.55)$$

As a first approximation, Gabor assumed that there were no contributions from the neighboring signals outside each "logon." Hence, the first approximate $\hat{p}_1(t)$ is given by

$$\hat{p}_1(t) = e^{\frac{-(t-t_N)^2}{4\alpha}} \sum_{\ell=-\infty}^{\infty} \frac{a_{N\ell}}{2\sqrt{\pi\alpha}} \, e^{j(2\pi f_0 t)} \qquad (7.56)$$

for $t_N - \dfrac{\Delta t}{2} < t < t_N + \dfrac{\Delta t}{2}$ and $N = 0, \pm 1, \pm 2, \dots$. Therefore in the sum of (7.55) $k \neq N$ terms are ignored results in the expression for (7.56). Next, Gabor made the following substitution:

$$p(t) e^{\frac{(t-t_N)^2}{4\alpha}} = \sum_{\ell=-\infty}^{\infty} \frac{a_{N\ell}}{2\sqrt{\pi\alpha}} \, e^{j(2\pi f_0 t)} \qquad (7.57)$$

which is equivalent to making a Fourier series representation of the function in the interval centered at t_N. The coefficients $a_{N\ell}$ could then be found by making a Fourier expansion of the left-hand side of (7.57), under the assumption that there is no overlap from neighboring elementary signals.

Gabor presented a heuristic argument for iteratively estimating the coefficients $a_{k\ell}$ so as to obtain an approximation $\hat{p}(t)$. However, no justification of this process was given [8, 9].

Bastiaans [11] used a different set of expansions for $p(t)$ given by

$$g_{k\ell}(t) = e^{\dfrac{-(t-t_k)^2}{4\alpha}} \; e^{j\dfrac{2\pi\ell t}{\Delta T}} \; e^{-j\pi k\ell} \tag{7.58}$$

He demonstrated that when $g_{k\ell}(t)$ replaces $\Psi_{k\ell}(t)$ in (7.54), then indeed the representation of (7.49) and (7.54) forms a complete set and that the approximation error defined by the difference $p(t) - \hat{p}(t)$ goes to zero for a sufficiently large number of basis functions $g_{k\ell}(t)$. He introduced the following notation:

$$t_k = k \times \Delta T \tag{7.59}$$

and

$$f_\ell = \frac{\ell}{\Delta T} \tag{7.60}$$

where ΔT is the sampling interval. Bastiaans' representation demonstrated that a Gabor type of representation is complete over a finite interval as well as over the infinite interval and that the approximation error monotonically decreases as the number of terms in the series increases.

One of the objectives of Gabor's "logon" concept for the representation of a signal $p(t)$ in two dimensions could be to predict where significant amounts of energy are distributed in the time-frequency plane. The coefficients $a_{k\ell}$ are called the Gabor coefficients and $\Psi_{k\ell}(t)$ are called the Gabor logons. It is important to note that this expansion is in terms of basis signals that are discretized in both time and frequency [12 – 15]. A necessary condition for the discrete logons to constitute an arbitrary finite-energy signal is that $\Delta t \times \Delta \omega \leq 0.5$ [12, 13]. If the total bandwidth of the signal is B_ω and its duration is T_D, then we would expect that the total number of nonnegligible Gabor coefficients will be of the order of $U = B_\omega T_D$, the total time-bandwidth product of the signal. It is important to note that if $\Delta t \times \Delta \omega > 0.5$, then the total number of Gabor coefficients will be less than U, which is simply not possible. When $\Delta t \times \Delta \omega < 0.5$, then the number of the Gabor coefficients will be greater than U and this implies that the logons in this case are not linearly independent, and the representation is redundant because it contains more coefficients than are actually needed. When $\Delta t \times \Delta f = 0.5$, then the "logons" are linearly independent and the coefficients contain no redundancy because the number of coefficients in this case will equal U, the time-bandwidth product of the signal [12, 13]. To visualize the time-variant energy over a finite length observation interval $[t_A, t_B]$, we evaluate the total energy within the subinterval centered at $t_N = N \times \Delta T$. The energy is given by

$$\int_{t_N - \frac{\Delta T}{2}}^{t_N + \frac{\Delta T}{2}} |p(t)|^2 \, dt = \int_{-\infty}^{\infty} \int_{t_N - \frac{\Delta T}{2}}^{t_N + \frac{\Delta T}{2}} |p(t) e^{-j2\pi f t} \, dt|^2 \, df \qquad (7.61)$$

where the second part of the expression in (7.61) can be obtained using Parseval's theorem. Next we introduce the following variable:

$$P(t_N, f) = \int_{t_N - \frac{\Delta T}{2}}^{t_N + \frac{\Delta T}{2}} \hat{p}(t) e^{-j2\pi f t} \, dt = \sum_{k=-\infty}^{\infty} \sum_{\ell=-\infty}^{\infty} b_{k\ell} \int_{t_N - \frac{\Delta T}{2}}^{t_N + \frac{\Delta T}{2}} g_{k\ell}(t) e^{-j2\pi f t} \, dt$$

$$= \sum_{k=-\infty}^{\infty} \sum_{\ell=-\infty}^{\infty} b_{k\ell} \, G_{k\ell}(t_N, f) \qquad (7.62)$$

where $b_{k\ell}$ corresponds to the weights associated with the "logon" centered at $t_k = k \times \Delta T$ and $f_\ell = \ell/T$. Therefore, if during the time defined by the subinterval $[t_N - \frac{\Delta T}{2} < t < t_N + \frac{\Delta T}{2}]$ the signal consisted of a pure sinusoid of frequency 100 Hz, then we would hope that the energy density $|P(t_N, f)|^2$ would have a significant amplitude primarily near $f = 100$ Hz. The energy spectral density is given by

$$|\hat{P}(t_N, f)|^2 = \sum_{k, \ell = -\infty}^{\infty} |b_{k\ell}|^2 \, |G_{k\ell}(t_N, f)|^2$$

$$+ \sum_{\substack{k = -\infty \\ \text{with } k \neq m}}^{\infty} \sum_{m = -\infty}^{\infty} \sum_{\substack{\ell = -\infty \\ \text{with } \ell \neq n}}^{\infty} \sum_{n = -\infty}^{\infty} b_{k\ell} \, b_{mn} \, G_{k\ell}(t_N, f) \, G_{mn}(t_N, f) \qquad (7.63)$$

with

$$G_{k\ell}(t_N, f) = e^{-j\pi k\ell} \int_{(t_N - t_k) - \frac{\Delta T}{2}}^{(t_N - t_k) + \frac{\Delta T}{2}} e^{-\frac{(t - t_k)^2}{4\alpha}} e^{j\frac{2\pi \ell t}{\Delta T}} e^{-j2\pi f t} \, dt$$

$$= e^{j\pi k\ell} e^{-j2\pi f t_k} \int_{(t_N - t_k) - \frac{\Delta T}{2}}^{(t_N - t_k) + \frac{\Delta T}{2}} e^{\frac{-y^2}{4\alpha}} e^{-j2\pi(f - f_\ell)y} \, dy$$

$$(7.64)$$

with $y = t - t_k$. If we stipulate further for simplicity that $f_M = \dfrac{M}{\Delta T}$, then with $\ell = M$, we obtain

$$G_{kM}(t_N, f_M) = \int_{\dfrac{(t_N - t_k) - \dfrac{\Delta T}{2}}{\sqrt{4\alpha}}}^{\dfrac{(t_N - t_k) + \dfrac{\Delta T}{2}}{\sqrt{4\alpha}}} e^{-z^2} dz \qquad (7.65)$$

with $z = \dfrac{y}{\sqrt{4\alpha}}$. Since

$$e^{j\pi k M} = (-1)^{k M} \qquad (7.66)$$

and

$$\frac{2}{\sqrt{\pi}} \int_0^t e^{-z^2} dz = \text{erf}(t) \qquad (7.67)$$

then (7.65) becomes

$$G_{kM}(t_N, f_M) = \sqrt{\alpha\pi}\,(-1)^{kM} \left[\text{erf}\left\{ \frac{(t_N - t_k) - \dfrac{\Delta T}{2}}{\sqrt{4\alpha}} \right\} - \text{erf}\left\{ \frac{(t_N - t_k) - \dfrac{\Delta T}{2}}{\sqrt{4\alpha}} \right\} \right]$$

$$(7.68)$$

If we set $\Delta T = 2\sqrt{\pi\alpha}$, as was done by Bastiaans, then

$$G_{kM}(t_N, f_M) = \frac{\Delta T}{2}(-1)^{kM} \left[\text{erf}\left\{ \sqrt{\pi}\,(N - k + 0.5) \right\} - \text{erf}\left\{ \sqrt{\pi}\,(N - k - 0.5) \right\} \right]$$

$$(7.69)$$

It is therefore clear from (7.63) and (7.69) that all the "logons" along the line $t = t_N$ for which $\ell \neq M$ will influence the calculation of energy spectral density at $f_M = M / \Delta T$. In addition, the "logons" away from the line $t = t_N$, for which $k \neq N$, will also contribute to the spectral density. The latter contribution is quite troublesome because it indicates that "logons" along the lines, $t = (t_N \pm \Delta T)$, $(t_N \pm 2 \times \Delta T)$, and so on will "leak" their associated weights $b_{N\pm1,\ell}$; $b_{N\pm2,\ell}$ to the (N, M)th cell/logon in the time-frequency plane. The exact amount of leakage is given by (7.69). Table 7.1 represents $10 \log_{10}[G_{kM}/G_{NM}]$. They are defined by (7.69) with $\Delta T = 2$.

The values in Table 7.1 mean that there are three significant $G_{kM}(t_N, f_M)$ functions. These correspond to $K = N$, and $N \pm 1$ and their associated amplitude coefficients $b_{k\ell}'$. This creates a serious problem in some cases because it is against our expectation that the only "logons" along the line $t = t_N = N \times \Delta T$ should represent the energy corresponding to the time interval centered at t_N. Therefore, the basic philosophy of the Gabor "logon" concept for analyzing and isolating energy densities from individual subintervals is not conceptually correct. Hence, the assumption that there are distinct continuous blocks in the time-frequency plane where energy is concentrated is not true because neighboring cells may smear the results if their corresponding amplitude weightings $b_{k\ell}$ are large. This leakage may also be because the basis $g_{k\ell}(t)$ is not strictly time limited.

Table 7.1
Leakage Factors Along the Line $f = f_M = M/(\Delta T)$

K	Actual magnitude ratio	Ratio (dB)
N	1.5798	0.0
$N \pm 1$	0.2099	-8.76
$N \pm 2$	1.67×10^{-4}	-39.76
$N \pm 3$	3.69×10^{-10}	-96.32

This is why Lerner [12] extended the representation in (7.58) to have the basis functions represent an arbitrary signal. The basis functions are now of the following general form:

$$v_{k\ell}(t) = v(t - t_k) \, e^{j2\pi f_\ell t} \tag{7.70}$$

Here, $v(t)$ is considered to be a "convenient" finite energy function whose energy is concentrated near $t = 0$ and whose energy spectrum $|V(\omega)|^2$ is concentrated near $\omega = 0$. This was later modified by Roach [9] who demonstrated that unless the function $v(t)$ is of finite support in the time domain, representation of the form (7.70) produces an energy spectrum $|P(t_N, f)|^2$, which has no clear relationship to the energy of the signal concentrated in the corresponding interval in time. Note that this precludes the Gaussian function or the prolate spheroidal functions as possible window functions since they are of infinite duration because they do not produce localization in time.

It has been shown that for a proper time-frequency representation, the expansion must be of the form [9]

$$\overline{p}(t) = \sum_{k=k_s}^{k_f} \sum_{\ell=-m}^{m} c_{k\ell} \, \psi_{k\ell}(t) \tag{7.71}$$

where

$$\psi_{k\ell}(t) = \begin{cases} \sqrt{w(t-t_k)} \; e^{j\omega_\ell t} \; ; & |t-t_k| \leq T/2 \\ 0 & ; \quad |t-t_k| \geq T/2 \end{cases} \tag{7.72}$$

Here, T is the duration of the window. The window function $w(t)$ in (7.72) is chosen in such a way that

$$W_T(\omega) = \int_{-T/2}^{T/2} w(t)\; e^{-j\omega t}\; dt \qquad (7.73)$$

which fulfills

$$W_T(\omega = 0) = 1 \qquad (7.74)$$

with

$$W_T\left[n\left(\frac{2\pi Q}{T}\right)\right] = 0 \quad \text{for} \quad n = \pm 1;\; \pm 2;\; \pm 3;\; ... \qquad (7.75)$$

The class of window functions that possess this property is formed by convolving a rectangular pulse of duration $(T - \beta)$, with a symmetric positive pulse $a(t)$ having duration β and an area under the curve of $\dfrac{1}{T-\beta}$. Here

$$\beta = T\left[\frac{Q-1}{Q}\right] \qquad (7.76)$$

where Q is the integer frequency spacing factor. The coefficients $c_{k\ell}$ in (7.71) can be given by

$$c_{k\ell} = \int_{-T/2}^{T/2} \hat{p}(t)\sqrt{w(t-t_k)}\; e^{\frac{-j2\pi Q \ell t}{T}}\; dt \qquad (7.77)$$

provided

$$\int_{-\infty}^{\infty} \psi_{k\ell}(t)\, \overline{\psi}_{mn}(t)\; dt = \delta_{km}\,\delta_{\ell n} \qquad (7.78)$$

where the overbar denotes the conjugate and δ is the Dirac delta function.

Equation (7.78) guarantees that the coefficient $c_{k\ell}$ for a given segment centered at $t = t_k$ is completely independent of the signal outside the kth segment. To confirm this, we observe that

$$\int_{-\infty}^{\infty} \psi_{k\ell}(t)\, \overline{\psi}_{mn}(t)\; dt = \int_{t_k-T/2}^{t_k+T/2} \sqrt{w(t-t_k)\, w(t-t_m)}\; e^{\frac{-j2\pi Q}{T}(\ell-n)t}\; dt \qquad (7.79)$$

Now note that $k = m$ always; otherwise, the window functions $w(t - t_k)$ and $w(t - t_m)$ are disjoint. In addition, we have

$$
\int_{-\infty}^{\infty} \psi_{k\ell}(t) \, \overline{\psi}_{kn}(t) \, dt = \int_{t_k - \frac{T}{2}}^{t_k + \frac{T}{2}} w(t - t_k) \; e^{-j\frac{2\pi Q}{T}(\ell - n)t} \, dt
$$

$$
= e^{-j\frac{2\pi Q}{T}(\ell - n)t_k} \int_{-\frac{T}{2}}^{\frac{T}{2}} w(u) \; e^{-j\frac{2\pi Q}{T}(\ell - n)u} \, du
$$

(7.80)

and finally

$$
= e^{-j\frac{2\pi Q}{T}(\ell - n)t_k} \; w_T\left[\frac{2\pi Q}{T}(\ell - n)\right]
$$

(7.81)

Utilizing properties (7.74) and (7.75) of the window function we can see that (7.81) becomes unity for $\ell = n$ and zero otherwise, satisfying (7.78) and hence (7.77) holds. The interesting point in (7.77) is that the coefficient for a given segment in time is completely independent of the signal outside the kth cell since the coefficients in different time segments are orthonormal, (i.e., the total energy in the signal is then the sum of all the individual coefficients squared).

It is interesting to observe that (7.72) and (7.77) closely resemble the windowed Fourier transform presented in the earlier sections, when the window function becomes a rectangular window. However, any window function that results from the convolution of any rectangular pulse with a symmetric window function will provide the mathematical requirements for $w(t)$ as described by (7.72) to (7.75).

7.3.3 Discrete Wavelet Transform

If in the continuous wavelet transform, one uses integer values for the parameters k, n in (7.32) and further assume that $\beta = 2^k nT$ (one can assume $T = 1$, without loss of generality for the discrete case) and $\alpha = 2^k$, then the discrete wavelet transform of $p(t)$ is given by [4]

$$
WT[p] = \int_{-\infty}^{\infty} p(t) \, 2^{-k/2} \, w(2^{-k}t - nT) \; dt
$$

(7.82)

The inverse transform is given by

$$
p(t) = \sum_{k=-\infty}^{\infty} \sum_{n=-\infty}^{\infty} WT[p] \, 2^{-k/2} \, \overline{w}(2^{-k}t - nT)
$$

(7.83)

under the condition that the window functions given by

$$w_{\alpha,\beta}(t) = w_{k,n}(t) = 2^{-k/2}\, w(2^{-k}t - nT) \tag{7.84}$$

are orthonormal, that is,

$$\int_{-\infty}^{\infty} w_{p,q}(t)\, \overline{w}_{k,n}(t)\, dt = \delta_{pk}\, \delta_{nq} \tag{7.85}$$

The shift integers n are chosen in such a way that $w(2^{-k}t - nT)$ covers the whole line for all values of t. The wavelet transform thus separates the object into different components in its transform domain and studies each component with a resolution matched to its scale. Another interesting property to observe is that the original signal from (7.83) is recovered from the discrete wavelet transform coefficients by filtering with the window $w_{\alpha,\beta}(t)$ of (7.84). Equivalently, the discrete wavelet transform is obtained by filtering the signal p(t) by $w_{\alpha,\beta}(-t)$ or $2^{-k/2}\, w_{\alpha,\beta}(-2^{-k}t)$. The wavelet series amounts to expanding the function $p(t)$ in terms of wavelets $w_{k,n}(t)$ or equivalently by the shifted and dilated versions of the same window function, so that

$$p(t) = \sum_{k,n=-\infty}^{\infty} C_{k,n}\, w_{k,n}(t) \tag{7.86}$$

where $C_{k,n}$ are constants independent of t. If we further assume that the wavelets $w_{k,n}(x)$ are orthogonal [i.e., (7.85) holds], then

$$C_{k,n} = <p\,;\, w_{k,n}> \tag{7.87}$$

By comparing equations (7.83) and (7.86), it is apparent that the (k, n)th wavelet coefficient of p is given by the integral wavelet transform of p, if the same orthogonal wavelets are used in both the integral wavelet transform and in the wavelet series. Therefore, the wavelet series provides an approximation for p that is not necessarily the least-squares ($\mathcal{L}^2$) orthogonal projection that a Fourier series provides. Also, the wavelets provide an unconditional basis for $\mathcal{L}^i$ (i.e., space of functions integrable to the ith power) for $1 < i < \infty$. Since $\mathcal{L}^1$ has no unconditional bases, wavelets cannot do the impossible. However, they can still do a better job than the Fourier expansion, by displaying no Gibb's phenomenon for approximating functions that are discontinuous, by utilizing discontinuous wavelets as a basis. So, if we approximate a discontinuous function by the Haar wavelets, for example, we will have no Gibb's phenomenon. However, if the wavelets used in the approximation are not discontinuous, like the Haar wavelets (or Walsh functions), but utilize continuous functions as a basis instead to approximate the function, then the Gibb's phenomenon will be visible in the wavelet approximation. The problem now at hand is to determine if there are any numerically stable algorithms to compute the wavelet coefficients $C_{k,n}$ in (7.87). Specifically, in real life, the function p is not an analytic function, but is sampled at a few points. Computing the integrals of $< p\,; w_{k,n} >$ then requires a quadrature formula. For the smallest value of k often referred to by the scale parameter (i.e., most negative k values), (7.87) will not involve many samples of p, and one can do the computation quickly. For large scales, however, one faces large integrals, which might considerably slow down the computation of the wavelet transform of any given function. Especially for on-line implementations, one should avoid having to compute these long integrals. One way out is to use the technique used in multirate/multiresolution analysis—that of introducing an auxiliary function, $\phi(x)$, so that [2]

$$w(x) = \sum_{m=-\infty}^{\infty} d_m\ \phi(2x-m) \qquad (7.88)$$

The $\phi(x)$ are called the scaling functions and are defined by the dilation equation

$$\phi(x) = \sum_{m=-\infty}^{\infty} c_m\ \phi(2x-m) \qquad (7.89)$$

where in each case only a finite number of coefficients c_m and d_m are different from zero. Here ϕ has a dc component (i.e., its integral is not zero) but $w(x)$ does not. Moreover ϕ is normalized such that

$$\int_{-\infty}^{\infty} \phi(x)\ dx = 1 \qquad (7.90)$$

We define $\phi_{k,n}$ even though ϕ is not a wavelet:

$$\phi_{k,n} = 2^{-k/2}\,\phi(2^{-k}x - n) \qquad (7.91)$$

The inner products of (7.87) can be evaluated using

$$< p\,; w_{k,n} > = \frac{1}{\sqrt{2}} \sum_{m=-\infty}^{\infty} d_m < p\,; \phi_{k-1,2n+m} > \qquad (7.92)$$

So the problem of finding the wavelet coefficients is reduced to that of computing $<p ; \phi_{j,k}>$. We can see that

$$< p ; \phi_{k,n} > = \frac{1}{\sqrt{2}} \sum_{m=-\infty}^{\infty} c_m < p ; \phi_{k-1, 2n+m} > \qquad (7.93)$$

so that $<p ; \phi_{k,n}>$ can be computed recursively starting from the smallest scale (most negative, i.e., $k = -1$) to the largest. The advantage of this procedure is that it is numerically robust, namely, even though the wavelet coefficients C_{kn} in (7.87) are computed with low precision—say, with a couple of bits—one can still reproduce $p(t)$ with comparatively much higher precision [2].

In summary, what we have done is as follows: Consider $p(t)$ to be a function of time. We have taken the spectrum of $p(t)$ and separated it into octaves of widths $\Delta\omega_k$ such that its frequency band ω has been divided into $[2^k\pi$ to $2^{k+1}\pi]$ for all values of k. We now define wavelets in each frequency interval $\Delta\omega_k$ and approximate $p(t)$ by them. If we choose

$$\phi(t) = \frac{\sin \pi t}{\pi t} \qquad (7.94)$$

then

$$\psi(t) = 2\phi(2t) - \phi(t) \qquad (7.95)$$

and the wavelet expansion of $p(t)$ is given by

$$p(t) = \sum p_k(t) = \sum_{k,n} C_{k,n} w_{k,n}(t)$$

with

$$w_{k,n}(t) = 2^{k/2} \psi(2^k t - n) \quad (\text{with } T = 1) \qquad (7.96)$$

The functions $w_{kn}(t)$ are orthonormal because their bandwidths are nonoverlapping; namely, a fixed k, $P_k(\omega)$, has the bandwidth $\Delta\omega_k$, which is $[2^k\pi$ to $2^{k+1}\pi]$. Therefore, the wavelet expansion of a function is complete in the sense that it makes an approximation by orthogonal functions, which have nonoverlapping bandwidths.

As concluded by Vaidyanathan [4], even though the continuous wavelet transform has a wider scope with deeper mathematical issues, the discrete wavelet transform is quite simple and can be explained in terms of basic filter theory. Even before the development of wavelets, nonuniform filter banks were used in speech processing [15–17]. The motivation was that the nonuniform bandwidths could be used to exploit the nonuniform frequency resolution of the human ear [18, 19].

Therefore, if the wavelet application is already in the digital domain, it really is not necessary to understand the deeper meaning of the concepts of scaling functions $\phi(t)$ and wavelets $w(t)$. In this case, all we need to focus on is the dilation equation and the shift principle to generate a complete basis.

First Word of Caution

Often overzealous researchers, in order to push the advantages of a newly developed tool (which, of course, is quite useful for selective problems) overlook two important factors.

The first one is termed by Strang and Nguyen as a "wavelet crime." The problem has to do with the selection of sample values of the function as the wavelet coefficients. As Strang and Nguyen point out: "Start with a function $x(t)$. Its samples $x(n)$ are often the input to the filter bank (to generate the wavelet decomposition). Is this legal? NO. IT IS A WAVELET CRIME. Some can't imagine doing it; others can't imagine not doing it. Is this crime convenient? YES. We may not know the whole function $x(t)$, it may not be a combination of $\phi(t - k)$, and computing the true coefficients in $\Sigma \; a(k) \; \phi(t - k)$ may take too long. But the crime cannot go unnoticed—we have to discuss it" [20, p. 232].

The problem is in (7.87) and (7.92), where the samples of $x(n)$ are used at the zeroth scale to initiate the pyramid algorithm for doing a discrete wavelet decomposition. As Strang and Nguyen point out "the pyramid algorithm acts on the numbers $x(n)$ as if they were expansion coefficients of its underlying function $x_s(t)$, i.e.,

$$x_s(t) = \Sigma \; x(n) \; \phi(t - n) \qquad (7.97)$$

Does $x_s(t)$ have the correct sample values for $x(n)$? This seems a minimum requirement." They further point out the correct way to solve the problem. There is a perennial problem of relating the sampled values of the function to the continuous one because the mapping from the sampled z domain to the continuous s domain is not unique [21].

Second Word of Caution

It is possible to generate a local orthogonal basis by using the Malvar wavelets, for example [22]. Malvar created an orthogonal basis with smooth windows modulated by the orthogonal family

$$\left\{ \frac{\sqrt{2}}{\sqrt{N}} \; \cos\left[\frac{\pi}{N} \; (k + \frac{1}{2}) \; (n + \frac{1}{2}) \right] \right\} \; 0 \leq k < N \qquad (7.98)$$

Here a signal of N samples is extended into a signal $\hat{p}$ of period $4N$, which is symmetric with respect to -0.5 and antisymmetric with respect to $N-0.5$ and $(-N-0.5)$. The expansion of $\hat{p}$ over a period $4N$ has no sine terms and no cosine terms of each frequency. If we employ $\hat{p}(n) = p[n]$, for $0 = n < N$, we find that p can be written as a linear expansion of the cosine terms. The Malvar wavelets are basically windowed cosine and possibly sine functions. Even though they generate orthogonal bases with some localization, there is a problem, as with the Gabor basis. Unless the windows have the specific properties as outlined by (7.72) to (7.75), the amplitudes of the coefficients corresponding to windowed localized bases of (7.77), which are similar to the Malvar wavelets, do not always provide the correct localization picture in the transformed domain.

7.4 CONCLUSION

This chapter has provided a survey of the short-time Fourier techniques, the Gabor transform, and the wavelet transform. The strengths and the weaknesses of each method have been identified. The unity in diversity between the three transforms is the choice of the appropriate window function. It is seen that a particular transform cannot provide simultaneous localization strictly in time and frequency. The resolution in all the transforms is dictated by Heissenberg's principle of uncertainty. Hence, resolution in time can be achieved at the expense of frequency and vice versa. The windows that have the best localization satisfy the Heissenberg principle of uncertainty $\Delta_t \Delta_\omega \leq \pi$ with equality. However, it is seen that for the Gabor transform, the Gaussian window provides nonphysical power spectral density for a region of the function localized in time. However, one has to be careful when choosing the window function of the proper interpretation for a time-frequency localization to be considered correct in the representation of a signal.

REFERENCES

[1] C. K. Chui, *An Introduction to Wavelets*, Academic Press, New York, 1992.

[2] I. Daubechies, *Ten Lectures on Wavelets*, CBMS, Vol. 61, Philadelphia, SIAM, 1992.

[3] R. Gopinath and C. S. Burrus, "Wavelet transforms and filter banks," in *Wavelets—A Tutorial in Theory and Applications*, C. K. Chui, Ed., Academic Press, New York, 1992, pp. 603–654.

[4] P. P. Vaidyanathan, *Multirate Systems and Filter Banks*, Prentice Hall, Upper Saddle River, NJ, 1993.

[5] F. I. Tseng, T. K. Sarkar, and D. D. Weiner, "A novel window for harmonic analysis," *IEEE Trans. Acoustics, Speech, and Signal Processing*, April 1981,

pp. 177–188.

[6] W. F. Walker et al., "Carrier frequency estimation based on the location of the spectral peak of a windowed sample of carrier plus noise," *IEEE Trans. Inst. and Meas.*, Vol. IM-31, Dec. 1982, pp. 239–249.

[7] W. F. Walker et al., "Optimum windows for carrier frequency estimation," *IEEE Trans. Geoscience Electronics*, Vol. 21, Oct. 1983, pp. 446–454.

[8] D. Gabor, "Theory of communications," *Journal of the Institute for Electrical Engineers*, Nov. 1946, pp. 429–457.

[9] J. E. Roach, "A vector space approach to time-variant energy spectral analysis," Ph.D. Thesis, Syracuse University, Syracuse, NY, 1982.

[10] R. Bailan, "Un principle d' incertitude en theorie du signal ou en mechanique quantique," *C. R. Acad. Sci, Paris*, Vol. 292, Series 2, 1981.

[11] M. Bastiaans, "Gabor's expansion of a signal into gaussian elementary signals," *Proc. of the IEEE*, Vol. 68, April 1980, pp. 538–539.

[12] R. Lerner, "Representation of signals," in *Lectures on Communication System Theory*, E. Baghdady, Ed., McGraw-Hill, New York, 1961, Chapter 10.

[13] G. A. Nelson, L. L. Pfeiffer, and R. C. Wood, "High speed octave band digital filtering," *IEEE Trans. on Audio and Electroacoust.*, Vol. AU-20, March 1972, pp. 8–65.

[14] R. W. Schafer, L. R. Rabiner, and O. Herrmann, "FIR digital filter banks for speech analysis," *Bell System Technical J.*, Vol. 54, March 1975, pp. 531–544.

[15] J. L. Flanagan, *Speech Analysis, Synthesis and Perception*, Springer-Verlag, New York, 1972.

[16] M. Vetterli and J. Kovacevic, *Wavelets and Subband Coding*, Prentice Hall, Upper Saddle River, NJ, 1995.

[17] H. Ozaktas, Z. Zalevsky, and M. A. Kutay, *The Fractional Fourier Transform*, John Wiley and Sons, New York, 2001.

[18] F. Hlawatsch and G. F. Boudreaux-Bartels, "Linear and quadratic time-frequency signal representations," *IEEE Signal Processing Magazine*, April 1992, pp. 21–67.

[19] I. Daubechies, "The wavelet transform, time-frequency localization and signal analysis," *IEEE Trans. Information Theory*, Vol. 36, 1990, pp. 961–1006.

[20] G. Strang and T. Nguyen, *Wavelets and Filter Banks*, Wellesley-Cambridge Press, Wellesley, MA, 1996.

[21] T. K. Sarkar, N. Radhakrishna, and H. Chen, "Survey of various z-domain to s-domain transformations," *IEEE Trans. Instrumentation and Measurements,* Vol. IM-35, Dec. 1986, pp. 508–520.

[22] H. S. Malvar, *Signal Processing with Lapped Transforms*, Artech House, Norwood, MA, 1992.

Chapter 8

T-PULSE: WINDOWS THAT ARE STRICTLY TIME LIMITED AND PRACTICALLY BAND LIMITED

8.1 INTRODUCTION

One of the problems with the wavelet representation of a signal is that its mathematical characterization is not shift-invariant. This implies that if we shift the original waveform in time, we have to recompute the wavelet coefficients representing the waveform. Since the Fourier techniques do not suffer from such problems, they are ideally suited for dealing with shift-invariant linear systems, which arise in most problems in electrical engineering. Furthermore, the Fourier transform of a function is equivalent to modulating the original function with an integer dilated version of an exponential function. This procedure is useful in characterizing narrowband systems and makes computations involving convolutions quite fast. This chapter addresses this strength of the Fourier techniques.

In Chapter 7 the differences among the windowed Fourier transform, Gabor transform, and the wavelet transform were explained in terms of the window that shapes the finite data length.

The objective of this chapter is to address another property of the window function, namely, its effective bandwidth, which may have many practical applications. A flexible methodology is presented to design windows that are strictly limited in time and practically limited in frequency. It is not possible to design windows that are simultaneously limited in time and frequency due to Heissenberg's principle of uncertainty. Using an optimization procedure one can generate time-limited pulses that have extremely narrow spectral content to excite a very limited region of the target spectrum. Hence, it may be useful for target identification because it only excites a very narrow band of the target response, which may contain its characteristic signature. Therefore, we design finite duration discrete signals with a very narrow spectrum. In addition, one can minimize the intersymbol interference by making the waveform orthogonal to its delayed versions. Waveforms that have both zero-mean and nonzero-mean are designed by minimizing a cost function and employing several different optimization techniques. Examples are presented to illustrate these points.

8.2. A DISCUSSION ON VARIOUS CHOICES OF THE WINDOW FUNCTION

The claim that a particular transform provides simultaneous localization strictly in time and frequency is really not feasible under any circumstances because it violates the Heissenberg principle of uncertainty. What we have seen in the previous chapter is that if a function is limited in time, it cannot be simultaneously limited in frequency (i.e., band limited). Hence, resolution in time is accomplished at the expense of the frequency, such that for the best window, the Heissenberg principle of uncertainty $\Delta_t \Delta_\omega \geq \frac{1}{2}$ is satisfied with an exact equality. Equality is achieved only for a Gaussian window function. However, as we have seen for the Gabor transform, the Gaussian window provides nonphysical power spectral density for a region of the function localized in time. The wavelet transform cannot do any better. For all other choices of the window function, $\Delta_t \Delta_\omega \geq \frac{1}{2}$. Other possible choices for the window may be the prolate spheroidal functions. Unfortunately, the prolate spheroidal functions are not limited in time but they are strictly band limited. An alternate choice is to deal with the truncated prolate spheroidal functions, which are strictly time limited, but one does not have any control over the spectrum of such functions [1–3]. That is why a flexible methodology has been developed where one can design a window shape that is finite in time and in addition can focus the energy into a prespecified band [4, 5].

However, what one may try to do is to have a finite time window for localizing a process in time. But its shape can be manipulated in such a way that it is approximately band limited for most practical purposes; that is, 99.9% of its energy is concentrated within a very narrowly prescribed band. Such a construct is termed a T-pulse. A T-pulse is a pulse limited in time of duration T. The data signaling rate, or simply the data rate, is defined as the rate measured in bits per second at which the data are transmitted [6]. In contrast, the modulation rate [6] is defined as the rate at which signal level is changed, depending on the nature of the format to represent digital data. The modulation rate is measured in bauds or symbols per second. A baud is the same as the bit per second if each symbol represents 1 bit (i.e., there are two possible levels of the signals). Therefore, one-baud time is a duration of one symbol. We would like the pulse to be of a few symbols of duration (typically n is between 2 and 4). Moreover, we would like the major spectral content of the pulse, that is, 99.9% of its energy, to be focused within approximately one baud.

In digital transmission systems, it is important to design signals of finite duration whose energy is concentrated in a given bandwidth. If the signal is constrained to have only a single baud, the optimum signal is the well-known "prolate spheroidal wave function" [1]. However, for signals of more than one symbol duration, it is necessary to remove intersymbol interference. This is achieved when a shifted version of the waveform by one symbol duration is orthogonal to itself. Two common constraints are used to remove intersymbol interference. The first constraint is that the desired signal have zero values at all sampling instances but one. This is often referred to as the zero crossing property. Such signals are called "Nyquist signals." The other constraint is that the desired time-limited signal be orthogonal to its shifted versions; that is, its autocorrelation function is a Nyquist signal. Such signals are called orthogonal Nyquist signals.

Several methods have been used to generate Nyquist signals [2–5, 7]. In [2], the computation is straightforward and involves the determination of eigenvectors of a symmetric matrix. The prolate spheroidal wave functions are used to approximate a function. This is an orthogonal expansion. An integral equation describing Nyquist signals with orthogonal constraints has been defined in [3]. Techniques for numerically solving this equation have also been developed. Recently, a new method, which is based on the periodicity of nonuniform sampling theory making use of the linear splines, has been described in [4]. Although methods mentioned above are related to continuous Nyquist signals, discrete sequences have been used in all numerical examples. In [5, 7], a cost function that includes orthogonal constraints has been defined. By optimizing the cost function through numerical optimization, the discrete sequence of orthogonal Nyquist signals is generated using various optimization techniques.

In this section, several methods, which can find the global minimum, are used to optimize the desired cost function. The numerical results are compared with results available in the literature. Both zero- and nonzero-mean cases of the T-pulse have been considered.

8.3 DEVELOPMENT OF THE T-PULSE

The design of the T-pulse is carried out in the discrete domain. Let us assume a discrete signal sequence $f(n)$, which is defined for $n = 1, 2, \cdots, N_n$, and is identically zero outside these N_n values. Let us assume there are N_s samples in one-symbol duration. Then the total number of symbols N_c generated is

$$N_c = N_n / N_s \qquad (8.1)$$

The discrete Fourier transform (DFT) of the signal $f(n)$ is defined by

$$F(p) = \frac{1}{\sqrt{N_p}} \sum_{n=1}^{N_n} f(n) \exp\left[\frac{-j2\pi(n-1)(p-1)}{N_p}\right]; \text{ with } p = 1, 2, \cdots, N_p \qquad (8.2)$$

and its inverse transform is given by

$$f(n) = \frac{1}{\sqrt{N_p}} \sum_{p=1}^{N_p} F(p) \exp\left[\frac{+j2\pi(p-1)(n-1)}{N_p}\right]; \text{ with } n = 1, 2, \cdots, N_n \qquad (8.3)$$

In the frequency domain, N_p is the total number of samples of the DFT sequence $F(p)$. If we assume there are N_r samples per baud (the inverse of baud time), then

$$N_p = N_r N_s \tag{8.4}$$

In addition, we have $N_r \geq N_c$, and increasing N_r increases the resolution in the frequency domain. In the T-pulse construction, the objective is to maximize the in-band energy within the set $\Phi = \{ -N_b \leq p \leq N_b \}$, which corresponds to a frequency band of N_b / N_r baud. Or equivalently, we want to minimize the energy outside the N_b samples. The out-of-band energy is given by

$$g_1 = \sum_{p \notin \Phi} F(p)^2 = \sum_{n=1}^{N_n} f(n)^2 - \sum_{p=-N_b}^{N_b} F(p)^2$$

$$= \sum_{n=1}^{N_n} f(n)^2 - \frac{1}{N_p} \sum_{n=1}^{N_n} \sum_{m=1}^{N_n} f(n) f(m) \frac{\sin\left[\frac{2\pi(n-m)}{N_p}(N_b + \frac{1}{2})\right]}{\sin\left[\frac{\pi(n-m)}{N_p}\right]} \tag{8.5}$$

Secondly, the desired signal must be orthogonal to its shifted version in order to minimize the intersymbol interference. This implies that

$$\sum_{n=1}^{N_n-(p-1)N_s} f(n) f[n+(p-1)N_s] - \delta(p) = 0 ; \quad p = 1, 2, \cdots, N_c \tag{8.6}$$

where $\delta(p)$ is the delta function. Thus, (8.6) guarantees that if the waveform is shifted by a baud time or its multiples, then the wave shape is orthogonal to itself. This is termed zero intersymbol interference. Note that when $p = 1$, it produces the square of the function itself, and no constraint needs to be put on that. To include the zero intersymbol interference constraint in the cost function, the second term can be defined as

$$g_2 = \sum_{p=1}^{N_c} w_p \left[\sum_{n=1}^{N_n-(p-1)N_s} f(n) f[n+(p-1)N_s] - \delta(p) \right]^2 \tag{8.7}$$

where w_p for $p = 1, 2, \cdots, N_c$ are the weights of the various error terms.

Thirdly, we need to apply a dc constraint, that is, the signal should have zero mean. This implies

$$\frac{1}{N_n} \sum_{n=1}^{N_n} f(n) = 0 \tag{8.8}$$

Therefore, the third term of the cost function can be given by

$$g_3 = w_n \left[\frac{1}{N_n} \sum_{n=1}^{N_n} f(n) \right]^2 \tag{8.9}$$

where w_n is the weight associated with this objective function.

Hence, the cost function consists of the sum of the following three terms:

$$J = g_1 + g_2 + g_3 \tag{8.10}$$

resulting in

$$J = w_e \left[\sum_{n=1}^{N_n} f^2(n) - \frac{1}{N_p} \sum_{m=1}^{N_n} \sum_{n=1}^{N_n} f(m) f(n) \frac{\sin\left\{\frac{2\pi(n-m)}{N_p}\left(N_b + \frac{1}{2}\right)\right\}}{\sin\left\{\pi\frac{(n-m)}{N_p}\right\}} \right] +$$

$$\sum_{p=1}^{N_c} w_p \left[\sum_{m=1}^{N_n - (p-1)N_s} f(m) \, f\{m + (p-1)N_s\} - \delta(p) \right]^2 + \frac{w_n}{N_n^2} \left[\sum_{m=1}^{N_n} f(m) \right]^2 \tag{8.11}$$

When the wave shape is chosen to have a nonzero mean, then the third term g_3 is not necessary. Some other constraints can be added to the cost function to introduce additional constraints. For example, one can force the signal to be zero at the beginning due to causality and at the end when it must decay to zero. Minimizing the cost function defined by (8.11) will yield the optimum signal under the specified constraints.

In summary, the following observations are of importance:

1. The primary objective is to minimize the cost function J, such that g_1 is minimum with $g_2 = 0$ and $g_3 = 0$.
2. The weights w_e, w_n, and w_p for $p = 1, \ldots, N_c$ should be adjusted in a search procedure to achieve the above goal.

The minimization process can be outlined as follows:

Step 1: Choose an initial guess for $f(m)$ and initial guess for the weights w_e, w_n, and w_p for $p = 0, 1, \ldots, N_c - 1$.

Step 2: Compute the gradient of the functional J, with respect to $f(m)$. This is given by

$$\frac{\partial J}{\partial f(m)} = 2 w_e \left[f(m) - \frac{1}{N_p} \sum_{n=1}^{N_n} f(n) \frac{\sin\left\{ \frac{2\pi(n-m)}{N_p} \left(N_b + \frac{1}{2} \right) \right\}}{\sin\left\{ \frac{\pi(n-m)}{N_p} \right\}} \right]$$

$$+ 2 \sum_{p=1}^{N_c} w_p \left(\begin{array}{c} \left[\sum_{m=1}^{N_n - (p-1)N_s} f(m)\ f\{m + (p-1)N_s\} - \delta(p) \right] \\ \times [f\{m - (p-1)N_s\} + f\{m + (p-1)N_s\}] \end{array} \right) \quad (8.12)$$

$$+ \frac{2 w_n}{N_n^2} \sum_{m=1}^{N_m} f(m)$$

Here, $f(m)$ is assumed to be real.

Next, an optimum step length to update the signal sequence $f(m)$ is chosen through one-dimensional searches.

Step 3: If the norm of the previous gradient vector is not small enough, go to Step 2. Otherwise, see whether the orthogonality constraints expressed through the error terms $|e_p|$ for $p = 1, 2, ..., N_c$ are small enough. If the errors are small enough, the iterative process is terminated. If the errors are still considered to be large, increase each w_p for $p = 1, 2, ..., N_c$ by multiplying it with a number and then go back to Step 2.

Note that throughout the process, the w_e terms have a fixed nonzero value, since the absolute value of the total energy is not important. However, w_e can be increased during the process if the in-band energy to generate the T-pulse is unexpectedly low. This may happen because the cost function may have more than one local minimum and increasing w_e can force the mathematical optimization procedure to come out of an undesired local minimum.

8.4　SUMMARY OF THE OPTIMIZATION TECHNIQUES

Because the cost function may have more than one local minimum, weights are changed in (8.11) to reach the global minima. Here we present a survey of the various methods that can reach the global minima directly. In an ideal condition, the cost function would be zero, so that the weights can be chosen to emphasize the constraints. The constraints are not changed during optimization. The various optimization methods used are discussed next.

Steepest Descent (SD) Method: In this method, the gradient of the objective function $f(x)$ at any point x is a vector along a local direction of the greatest increase in the value of the function. Clearly, then, one might proceed in the direction opposite that of the gradient of $f(x)$, that is, in the direction of steepest descent. This method is simple to apply but sometimes it may converge to the position of a local minimum. An explicit expression for the gradient is necessary in this methodology.

Hook and Jeeve's (HJ) Method: This is another kind of a direct search method. This algorithm is a derivative-free multidimensional search technique that employs two types of directions, namely, exploratory directions and pattern-search directions, in order to generate updated iterates. It is widely believed that pattern searches are less efficient than a derivative-based method because the search gets trapped in local minimums. Pattern searches can be customized to cast a wide net when searching for better points, so that the search may be able to escape a shallow basin that contains a local minimum and find a deeper basin that hopefully contains the global minimum.

Simulated Annealing (SA) Method: This method is so named because the optimization process strongly resembles the annealing process for creating a durable metal by systematically cooling and reheating it. Simulated annealing techniques were originally employed in thermodynamics. More recently they have been applied in other optimization problems. Simulated annealing can avoid getting stuck in a local minimum. This optimization strategy is based on a stochastic search and is quite suitable when one is searching a functional, which has one deep minimum. However, when the functional has a number of deep minima, then the next algorithm is more suitable.

Genetic Algorithm (GA): This method draws its analogy from nature. It revolves around the genetic reproduction processes and survival of the fittest strategies. The process of optimization in a genetic algorithm can be outlined as follows:

Step 1. Creation of an Initial Population: The initial population is created on a computer using a random number generator.

Step 2. Evaluation: The function values (also called fitness values) are evaluated. The average fitness is calculated.

Step 3. Creation of a Mating Pool: The weaker members are replaced by stronger ones based on the fitness values.

Step 4. Crossover Operation: Offspring are produced using the mating parent pool.

Step 5. Mutation: Mutation is a random operation performed to change the expected fitness.

Step 6. Evaluation: The population is evaluated using the ideas developed in Step 2. A generation is now completed. If the preset number of generations is complete, the process is stopped. If not, go to Step 3.

GA is a powerful tool for optimization of any functions. It provides robust solutions to continuous as well as discrete problems. However, it is quite slow.

8.5 NUMERICAL RESULTS

As a first example, consider designing a T-pulse with the following parameters: $N_c = 5$, $N_s = 20$, and $N_r = 100$. It is necessary to generate a T-pulse with a zero mean. Table 8.1 shows the results for the in-band energy as a function of the percentage of the baud and compared with those obtained in [3]. It can be seen that the HJ method gets the best results and genetic algorithm gets the worst for large numbers of symbol duration. Table 8.2 shows the numerical results when using different methods to synthesize a T-pulse. We deal with both the zero-mean case and the nonzero-mean case. The quality of intersymbol interference is defined through the function

$$orth(p) = \sum_{n=1}^{N_n - (p-1)N_s} f(n) \, f[n + (p-1)N_s]; \qquad p = 1, 2, \cdots, N_c \qquad (8.13)$$

Table 8.1
Comparison of In-band Energy Obtained Using Different Optimization Methods

Band (% in terms of baud)		0.6	0.8	1.0	1.4	1.8
In-Band energy (%)	[3]	—	92.4	99.8	99.94	99.949
	SD	96.595	99.83	99.96	99.999	>99.999
	HJ	99.66834	99.99845	99.999689	99.999933	>99.9999
	SA	99.73582	99.97265	99.999110	99.999600	99.999860
	GA	99.47518	99.69031	99.959785	99.907646	99.367719

Figures 8.1 and 8.2 show the waveforms that are generated by using different optimization methods in generating T-pulses both with zero-mean and nonzero-mean cases, respectively. We can see that the genetic algorithms do not provide a solution, which can be recognized as a signal of four symbols duration for the non-zero-mean case. It can be concluded from these results that the HJ method and the simulated annealing method provide better quality solutions than the GA technique. This is because the GA method uses a discrete code for values of the variables and also chooses the new generation by using a random search procedure. The steepest descent method is the most efficient but it needs the derivative of the cost function. Figures 8.3 and 8.4 show the spectrum of the signals obtained for Figures 8.1 and 8.2, respectively. It is seen quite clearly that the spectrum is highly band limited and is quite sharp.

Table 8.3 presents the results for a two-baud time signal having both zero-mean and nonzero-mean values. Figures 8.5 and 8.6 show the synthesized waveforms. Since the genetic algorithm did not work well, we did not use it in this case. An important point to note is that the plots in Figures 8.5 and 8.6 are strictly time limited and are of finite support.

Table 8.2
Properties of a Signal Stretching Four-Symbol Duration

	Method	In-band energy (%)	Signal	Cost function or its derivation evaluation	p	Orthogonal values orth (p)
$N_s = 25$ $N_c = 4$ $N_r = 100$ $N_b = 130$ Zero-mean case	Simulated annealing (SA)	99.9993	Fig. 8.1	590001	1	1.000008
					2	-0.00000
					3	-0.00000
					4	-0.000565
	Hook and Jeeve's (HJ)	99.9847	Fig. 8.1	121141	1	0.999999
					2	-0.000001
					3	-0.000008
					4	-0.000036
	Steepest descent (SD)	99.9896	Fig. 8.1	870	1	0.999952
					2	-0.000140
					3	-0.000568
					4	-0.000523
	Genetic algorithm (GA)	99.9801	Fig. 8.1	101427	1	0.999364
					2	-0.000065
					3	-0.004984
					4	0.003870
$N_s = 25$ $N_c = 4$ $N_r = 100$ $N_b = 130$ Nonzero-mean case	Simulated annealing (SA)	99.9989	Fig. 8.2	655002	1	1.000004
					2	0.000014
					3	-0.000160
					4	0.001175
	Hook and Jeeve's (HJ)	99.9893	Fig. 8.2	125361	1	1.000000
					2	0.000000
					3	0.000003
					4	0.000162
	Steepest descent (SD)	99.9999	Fig. 8.2	870	1	0.9999872
					2	-0.000000
					3	0.0000074
					4	0.0000031
	Genetic algorithm (GA)	99.9125	Fig. 8.2	101690	1	0.997866
					2	-0.003514
					3	0.010229
					4	0.009844

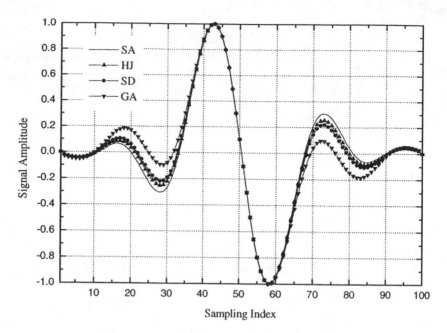

Figure 8.1 Signal of four-symbol duration for a zero-mean case.

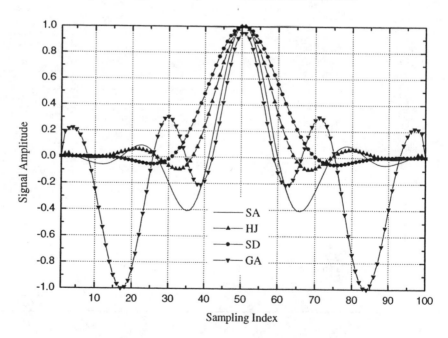

Figure 8.2 Signal of four-symbol duration for a nonzero-mean case.

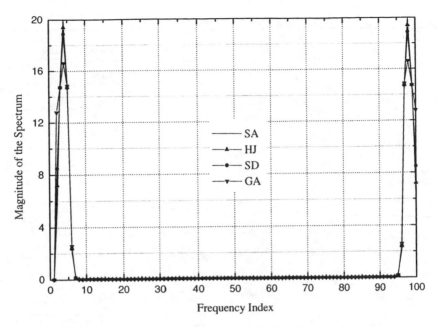

Figure 8.3 Spectrum of the signal of four-symbol duration for a zero-mean case.

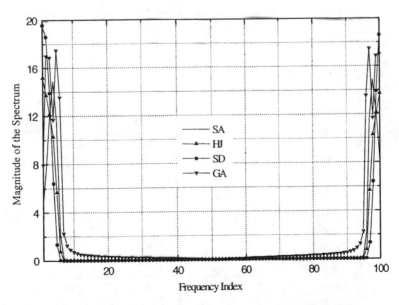

Figure 8.4 Spectrum of the signal of four-symbol duration for a nonzero-mean case.

Table 8.3
Properties of a Two-Baud Times Signal

	Method	In-band energy (%)	Signal	Cost function or its derivation evaluation	Orthogonal values	
					p	orth (p)
$N_s = 50$ $N_c = 2$ $N_r = 100$ $N_b = 160$ Zero-mean case	Simulated annealing (SA)	100.000	Fig. 8.5	610001	1	1.000000
					2	-0.000014
	Hook and Jeeve's (HJ)	99.9700	Fig. 8.5	24979	1	0.999870
					2	-0.000739
	Steepest descent (SD)	99.9700	Fig. 8.5	870	1	0.997292
					2	-0.000781
$N_s = 50$ $N_c = 2$ $N_r = 100$ $N_b = 160$ Nonzero-mean case	Simulated annealing (SA)	100.000	Fig. 8.6	715001	1	1.000025
					2	0.004990
	Hook and Jeeve's (HJ)	99.9997	Fig. 8.6	21247	1	0.999990
					2	0.000349
	Steepest descent (SD)	99.9995	Fig. 8.6	870	1	0.999524
					2	0.000286

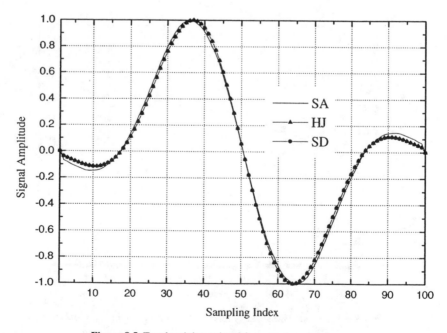

Figure 8.5 Two-baud times signal for a zero-mean case.

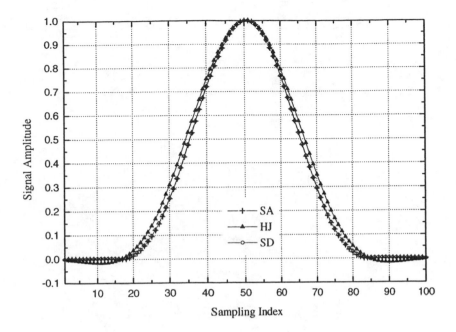

Figure 8.6 Two-baud times signal for a nonzero-mean case.

To further illustrate the application of time limiting and practically band limiting a signal, we consider the electromagnetic scattering from a thin wire of 2m in length and radius 0.01m, as illustrated in Figure 8.7. The wire is illuminated by a Gaussian-shaped pulse, whose derivative is shown in Figure 8.8. The spectrum of this derivative of the pulse is seen in Figure 8.9. The apparent discontinuity at the beginning and at the end of the waveform is a plotting artifact. This pulse will excite the thin wire resulting in an induced current on the structure. The shapes of the transient current at the center of the antenna are plotted in Figure 8.10. If we look at the Fourier transform of the current in Figure 8.11, we observe that the response contains three peaks. Because of the way the structure is excited, we have in the response contributions from the fundamental, third harmonic, and fifth harmonic components of the current. In this case, the resonances are defined by the length of the structure. The response at the third harmonic is 20 dB below that of the fundamental. Suppose our goal is to excite a response at the third harmonic. We would like to have the spectrum-excited response at the third harmonic dominate over the fundamental. To achieve this goal we design a base-band T-pulse as shown in Figure 8.12. We observe that this pulse has its energy concentrated in a very narrow low-frequency band as illustrated in Figure 8.13.

A Dipole Antenna Excited By A Transient Pulse

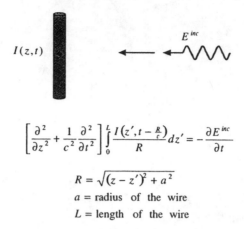

$$\left[\frac{\partial^2}{\partial z^2} + \frac{1}{c^2} \frac{\partial^2}{\partial t^2} \right] \int_0^L \frac{I\left(z', t - \frac{R}{c}\right)}{R} dz' = -\frac{\partial E^{inc}}{\partial t}$$

$$R = \sqrt{(z - z')^2 + a^2}$$

a = radius of the wire

L = length of the wire

Figure 8.7 A wire antenna illuminated by a Gaussian pulse.

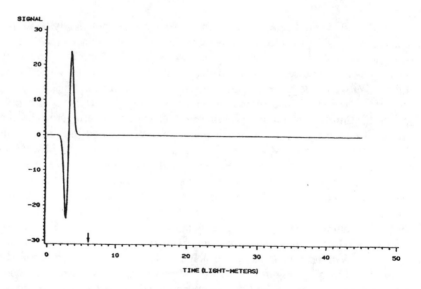

Figure 8.8 Derivative of the incident Gaussian pulse exciting the wire antenna.

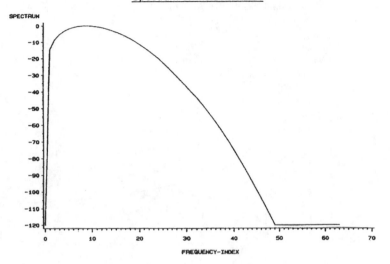

Figure 8.9 Spectrum of the derivative of the incident Gaussian pulse.

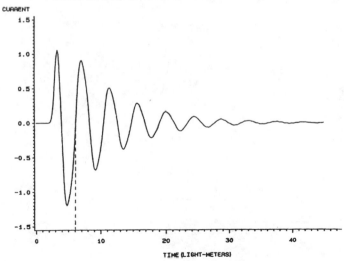

Figure 8.10 Induced current at the center of the wire antenna due to the incident Gaussian pulse.

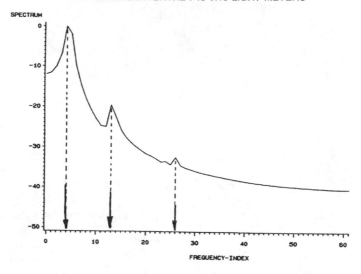

Figure 8.11 Induced current at the center of the wire antenna due to the incident Gaussian pulse.

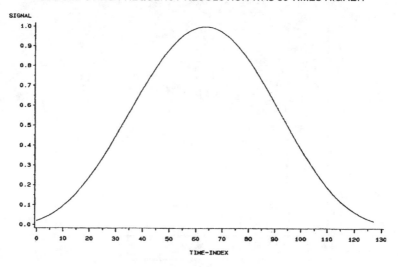

Figure 8.12 Shape of the envelope of the pulse that will excite the wire antenna.

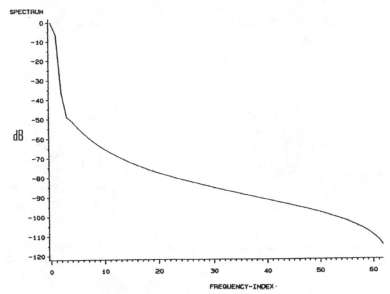

DFT AMPLITUDE SPECTRUM (DB) OF THE BASEBAND SIGNAL
THE FREQUENCY RESOLUTION IS 1/(128T)

Figure 8.13   Spectrum of the envelope of the pulse that will excite the wire antenna.

To transform this energy to the third harmonic frequency of the antenna we modulate this baseband signal with a carrier frequency, which is located near the third harmonic response from the antenna. The modulated carrier is shown in Figure 8.14. Its spectrum is observed in Figure 8.15 and it is seen to be peaked at the third harmonic response. We now excite the wire structure shown earlier in Figure 8.7 by the waveform shown in Figure 8.14. It is seen that the current still oscillates after the incident pulse (of duration 19 light-meters) is gone. The simulated response is calculated using the theory described in [8]. The transient response of the current at the center of the wire due to the modulated T-pulse is shown in Figure 8.16. If we observe the spectrum of this current in Figure 8.17, we notice that the third harmonic response of the structure has been generated without exciting the fundamental. The important point here is that the third harmonic response is approximately 20 dB above the fundamental even though the fundamental response in the broadband excitation as seen in Figure 8.11 is the dominant one. This illustrates that through proper pulse shaping it is possible to induce a response in a narrow band without being camouflaged by the dominant response. For these classes of problems the property of modulation inherent in the Fourier techniques can be advantageous over the principles of dilation and shift.

In short, one can excite some hidden fingerprints of a target, that is, substructure resonances, which may have a lower level of response without inducing the dominant response at the fundamental. Hence, this type of analysis can have significant potential application in target identification.

MODULATED SIGNAL OF DURATION 127T AND CENTER FREQUENCY K=13
THE CORRESPONDING BASEBAND SIGNAL HAS MAXIMUM ENERGY
(99.995%) WITHIN THE BANDWIDTH 2/(128T)
THE COMPUTING FREQUENCY RESOLUTION WAS 30 TIMES HIGHER

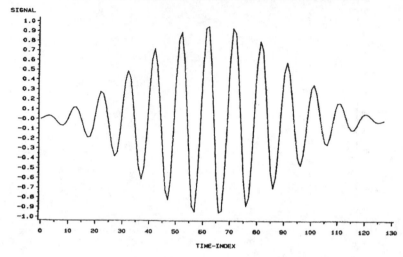

Figure 8.14 A modulated carrier shaped by the T-pulse.

DFT AMPLITUDE SPECTRUM (DB) OF THE MODULATED SIGNAL
THE FREQUENCY RESOLUTION IS 1/(128T)

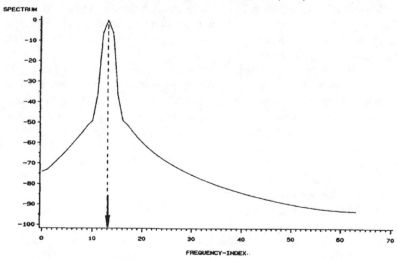

Figure 8.15 A very narrowband spectrum of the zero-mean-modulated T-pulse.

SYNTHESIZED CURRENT RESPONSE AT THE CENTER OF A THIN WIRE
ACTIVATED BY A BANDPASS ELECTRIC FIELD FROM THE BROADSIDE
THE SAMPLING INTERVAL T IS 0.15 LIGHT-METERS

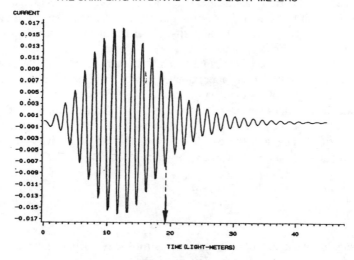

Figure 8.16 Response of the wire due to a modulated pulse.

DFT AMPLITUDE SPECTRUM (DB) OF THE CURRENT RESPONSE (129T THRU 256T)
THE EXCITING PULSE HAS CENTER FREQUENCY 13/(128T)
THE FREQUENCY RESOLUTION IS 1/(128T)
THE SAMPLING INTERVAL T IS 0.15 LIGHT-METERS

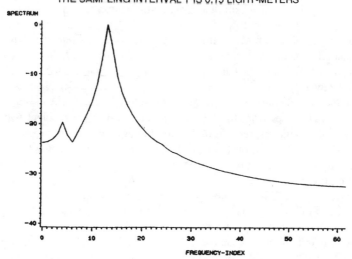

Figure 8.17 Spectrum of the response of the wire irradiated by a zero-mean T-pulse exciting the
second resonance.

The difference between a T-pulse and a wavelet is outlined next. In the T-pulse, a window has been created that is not only limited in time but 99% of its spectrum is focused in a narrow band and in this way, an attempt is made to satisfy the equality in the Heissenberg uncertainty principle (i.e., $\Delta_t \Delta_\omega \to 0.5$). The focused energy within that window can be translated to any frequency of interest by simply translating the spectrum, which is performed by the principle of modulation. Therein lies one of the strengths of the Fourier techniques.

In addition, the expansion in (7.30) can be done in terms of a dilated and shifted version of the T-pulse in an approximate fashion, whereas, for the wavelets, the decomposition by (7.30) is essentially exact and a perfect reconstruction can be done even after the functions have been downsampled [7]. In addition, in (4.30) a desire is to limit the number of nonzero coefficients in (4.30) to improve the efficiency of the decomposition. This is achieved by enforcing the pth derivative and all its lower derivatives are zero at $\omega = 0$, that is,

$$\frac{d^P W}{d\omega^P}(\omega)\Big|_{\omega=0} = 0 \qquad (8.14)$$

Depending on the nature of the requirements for the problem of interest, one can use either of the constraints on the window functions.

Finally we conclude this section by pointing out the differences between the various pulses like the K-pulse and the E-pulse techniques that are widely used in electromagnetic literature. The theory behind a K-pulse is the illustration of the principle that an all-zero function is of finite duration. The existence of a pole in the response makes the response ring for a very long time. So the goal is to excite a particular target by a waveform that contains only zeros and no poles in the transfer function. These zeros correspond to the poles of the target. So since the incident waveform is an all-zero function it is of finite duration, whereas the reflected energy is again an all-zero function because the zeros of the incident field have cancelled the response due to the poles of the target. Therefore, the scattered field is also of finite duration. Hence, if a response from a target due to an incident K-pulse is of finite duration, then the K-pulse is said to be matched to the target. The abbreviation of the K-pulse [9–11] comes from the term *kill pulse* because the incident wave shape is said to have killed the response from the target after some time duration T.

The E-pulse [12, 13], on the other hand, is based on matching the target return with a library of waveforms to see if the target response matches to that of the catalog. Hence, in principle, it is related to the matched filter concept on receive and not on transmit as for the K-pulse concept. Hence, one need not know a priori the characteristics of the target in order to launch the incident pulse on the target. The E-pulse is also related to Prony's method [14].

The principle of the T-pulse is similar to that of a K-pulse in that it is strictly time limited. However, in addition for the T-pulse, most of its energy is simultaneously concentrated in a very narrow band coming close to satisfying Heissenberg's principle of uncertainty. Hence, the goal here is to excite a very narrow spectral band response from the target, which may be its characteristic signature. Also, using the principle of amplitude modulation the energy can be translated to any band of interest. So, with an arbitrary waveform generator followed by an AM modulator with an RF amplifier, it is possible to launch a T-pulse with conventional narrowband antennas.

8.6 CONCLUSION

In this chapter, we have addressed the problem of optimizing the design of optimum discrete finite duration orthogonal Nyquist signals, particularly when the signals last a few baud times. The cost function is formulated in terms of the in-band energy, orthogonality between its shifted versions, and its mean value. We have developed several methods to solve this optimization problem. Both zero-mean and nonzero-mean cases have been discussed. Some of the optimization methods require fewer function evaluations to converge to an optimal solution. Finally, an example has been presented to illustrate the applicability of the T-pulse concept in exciting narrowband substructure resonances for possible target identification.

REFERENCES

[1] D. Slepian and H. Pollak, "Prolate spheroidal wave functions Fourier analysis and uncertainty I," *Bell Syst. Tech. J.*, Vol. 40, Jan. 1961, pp. 43–63.

[2] E. Panayirci and N. Tugbay, "Optimum design of finite duration Nyquist signals," *Signal Processing*, Vol. 7, 1984, pp. 57–64.

[3] P. H. Halpern, "Optimum finite duration Nyquist signals," *IEEE Trans. Communications*, Vol. 27, No. 6, 1979, pp. 884–888.

[4] E. Panayirci, T. Ozugur, and H. Caglar, "Design of optimum Nyquist signals based on generalized sampling theory for data communications," *IEEE Trans. Signal Processing*, Vol. 47, No. 6, 1999, pp. 1753–1759.

[5] Y. Hua and T. K. Sarkar, "Design of optimum discrete finite duration orthogonal Nyquist signals," *IEEE Trans. Acoustics, Speech, and Signal Processing*, Vol. 36, No. 4, 1988, pp. 606–608.

[6] S. Haykin, *Digital Communications*, John Wiley and Sons, New York, 1988.

[7] T. K. Sarkar and C. Su, "A tutorial on wavelets from an electrical engineering perspective, Part 2: The continuous case," *IEEE Antennas Propag. Magazine*, Vol. 40, No. 6, 1998, pp. 36–49.

[8] T. K. Sarkar, W. Lee and S. M. Rao, "Analysis of transient scattering from composite arbitrarily shaped complex structures," *IEEE Trans. Antennas and Propag.*, Vol. 48, No. 10, Oct. 2000, pp. 1625–1634.

[9] E. M. Kennaugh, "The K-pulse concept," *IEEE Trans. Antennas and Propag.*, Vol. 29, March 1981, pp. 327–331.

[10] L. C. Chan, D. L. Moffat, and L. Peters, "A characterization of subsurface radar target," *Proc. IEEE*, Vol. 67, July 1979, pp. 991–1000.

[11] C. W. Chuang and D. L. Moffat, "Natural resonance of radar target via

Prony's method and target discrimination," *IEEE Trans. Aerospace and Electronics,* Vol. AES-12, Sept. 1976, pp. 583–589.

[12] K. M. Chan and D. Westmoreland, "Radar waveform and application for exciting single mode backscatters from a sphere and application to target discrimination," *Radio Science,* Vol. 17, No. 3, May 1982, pp. 574–588.

[13] E. J. Rothwell, K. M. Chen, and D. P. Nyquist, "Extraction of the natural frequencies of a radar target from a measured response using E-pulse techniques," *IEEE Trans. Antennas and Propag.,* Vol. 35, June 1987, pp. 715–720.

[14] Y. Hua and T. K. Sarkar, "A discussion of E-pulse method and Prony's method for radar target resonance retrieval from scattered field," *IEEE Trans. Antennas and Propag.,* Vol. 37, No. 7, July 1989, pp. 944–946.

Chapter 9

OPTIMAL SELECTION OF A SIGNAL-DEPENDENT BASIS AND DENOISING

9.1   INTRODUCTION

In this chapter, we study the utilization of wavelets for characterization of signals, which are represented by waveforms. One of the important issues in the characterization of a waveform is the dimension of the space which that signal spans. In other words, how many independent pieces of information do we need to represent that waveform? The trick here is to either select a basis that would be complete for any type of signals or to develop a signal-independent basis. Of course, in both situations we need the least number of parameters and a computationally efficient way to calculate the various coefficients that weight the various components called the basis functions. The wavelets in this respect provide a viable alternative to the singular value decomposition, as a computationally efficient methodology for developing an optimum basis. However, the methodology of signal characterization based on the singular value decomposition provides more physical insight about the given data.

In addition, all waveforms are somewhat contaminated with noise. The problem of denoising signals is quite an important area in signal processing. Here also, wavelets may provide a computationally efficient way to denoise signals without much computational complexity.

9.2   SELECTION OF AN OPTIMUM BASIS

The goal here is to select the best basis for the representation of a function using minimum computational efforts. Historically, this area has been the domain of the singular value decomposition. However, the computational complexity of a singular value decomposition typically scales as $\theta(N^3)$ where N is the dimensionality of the system and $\theta(\cdot)$ denotes "of the order of." In a singular value decomposition, the system matrix is decomposed into the right- and left-hand singular vectors in addition to the computation of the singular values. These singular vectors are then used to represent the vector of interest. When the system matrix becomes square, then the singular vectors can be related to the eigenvectors as illustrated next.

Consider a D by 1 data vector, which is obtained from a process to be modeled. Assume that it is represented by

$$\{f(n)\} = [f_n, f_{n+1}, f_{n+2}, ..., f_{n+D-1}]^T \tag{9.1}$$

where T denotes the transpose of a matrix. The goal here is to carry out a feature selection methodology, which refers to a process whereby a *data space* is transformed to a *feature space*. The goal of this transformation is such that we would like to represent this vector by a number of effective features and yet retain most of the information. This in principle is achievable through the Karhunen-Loéve expansion [1]. To carry this out numerically, we first expand the data as a linear combination of the eigenvectors $\{v_1\}, \{v_2\}, ..., \{v_M\}$ associated with the eigenvalues $\lambda_1, \lambda_2, ..., \lambda_M$ of the covariance matrix $[R]$ of $\{f(n)\}$. The covariance matrix $[R]$ for the data is defined by

$$[R] = \mathcal{E}[\{f(n)\}\{f(n)\}^H] \tag{9.2}$$

where $\mathcal{E}[\bullet]$ denotes the expected value of a random variable and H denotes the conjugate transpose of a vector. Here, it is assumed that the data sequence $\{f(n)\}$ results from a random realization. Since the covariance matrix can be written in terms of its eigenvalues and eigenvectors, we have

$$[R] = \sum_{i=1}^{D} \lambda_i \, v_i \, v_i^H \tag{9.3}$$

Here D has been assumed to be the dimension of the correlation matrix $[R]$. An exact representation of the finite but random data sequence $\{f(n)\}$ can be made through

$$\{\hat{f}(n)\} = \sum_{i=1}^{D} \alpha_i(n)\{v_i\} \tag{9.4}$$

The goal in this representation which is called the Karhunen-Loéve expansion, is that a vector $\{f(n)\}$ belonging to a wide-sense stationary process of zero-mean and correlation matrix $[R]$ may be expanded in terms of the eigenvectors of $[R]$ which are weighted by the coefficients obtained from the following inner product:

$$\alpha_i(n) = \{v_i\}^H \{f(n)\}, \quad i = 1, 2, ..., D \tag{9.5}$$

having zero mean. Moreover, these coefficients are uncorrelated as

$$\mathcal{E}\,[\alpha_i(n)\,\alpha_j\,(n)] = \mathcal{E}\,[\{\upsilon_i\}^H\,\{f(n)\}\{f(n)\}^H\,\{\upsilon_j\}]$$

$$= \{\upsilon_i\}^H\,[R]\,\{\upsilon_j\} = \lambda_i\,;\quad \text{for } i = j \qquad (9.6)$$

$$= 0\,;\quad \text{otherwise}$$

In a practical situation not all D eigenvalues and eigenvectors of $[R]$ are necessary to represent the function $\{f(n)\}$. Only M dominant eigenvalues may be necessary. Even though such an approximation is quite straightforward, one requires a large computational effort to form the covariance matrix and then to calculate its eigenvalues and eigenvectors, which is a very time-consuming process.

The Weiner filter for example is an optimal estimator generated using a Karhunen-Loéve basis. However, if a signal is from a non-Gaussian process, such as those signals derived from natural images, then a nonlinear estimator will outperform the Weiner filter. This section presents such an estimator, which may be better for non-Gaussian processes.

An alternate methodology, which is less computationally intensive, is to use the wavelet packets where a family of scaling functions and wavelets are constructed by following a binary tree of dilation/translations. Therefore, they are essentially a linear combination/superposition of scaling functions and wavelets satisfying some criteria [2–5]. What that criterion is can be open to discussions. For example, in a conventional wavelet decomposition, usually emphasis is given to the low-frequency components resulting in the wavelet tree of Figure 9.1(a). The significance of each branch of the tree is represented by Figure 9.1(e), where the signal is passed through a low- and a highpass filter. Then the signal is subsampled by a factor of 2, because the effective bandwidth of the filtered signal is half of that of the original signal. This is true for a two-stage decomposition. Hence, in this case the data vector is approximated by a scaling function and wavelets. However, it is also possible to make up the entire tree using an alternate decomposition as shown in Figure 9.1(b). This is identical to the short time Fourier transform. In addition, we could argue that more information is contained in the high-frequency components and follow a tree analogous to a wavelet tree by emphasizing the high frequencies as shown in Figure 9.1(c). These examples illustrate that there may be other ways of approximating the signal as illustrated through Figure 9.1(d), which results in a coarse tree. Wavelet packets make up the entire set of these possible combinations.

As a second example, consider the complete four-level decomposition as shown in Figure 9.2(a). This is analogous to the Fourier transform [2–5]. However, an optimum representation may be a pruned version of the complete tree as shown in Figure 9.2(b). The decision of whether to split or not to split a branch of a tree in a multiresolution decomposition as illustrated through Figure 9.2(a) is determined by the inequalities of the functional M as (defined in the figure) illustrated by Figure 9.3.

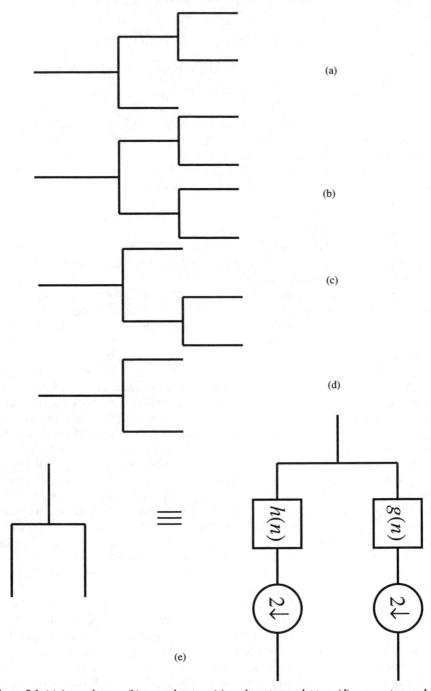

Figure 9.1 (a) A wavelet tree, (b) a complete tree, (c) an alternate wavelet tree, (d) a coarse tree, and (e) the mathematical interpretation for the branches.

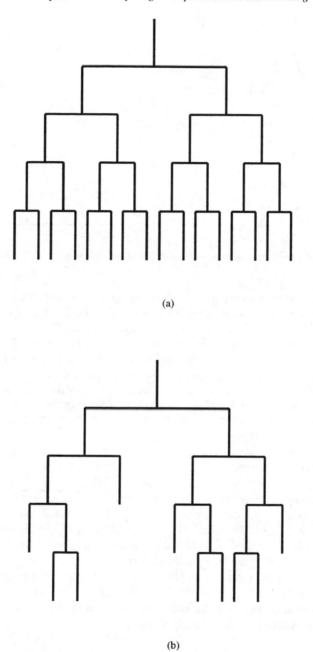

(a)

(b)

Figure 9.2 (a) A complete four-level tree and (b) a pruned four-level tree.

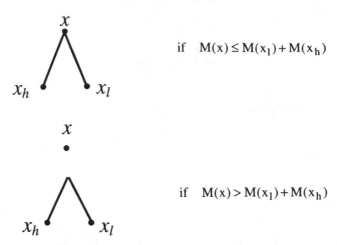

$$\text{if} \quad M(x) \leq M(x_l) + M(x_h)$$

$$\text{if} \quad M(x) > M(x_l) + M(x_h)$$

Figure 9.3 Principles involved in splitting a branch (top) and the decision not to split a branch (bottom).

Here, if the sum of the functionals, which are generated by projecting the function along the different components of the basis at the higher scale (i.e., the sum of the coefficients of the output from the highpass and the lowpass filters), is greater than the original functional at the lower scale (i.e., the sum of the coefficients of the signal), then a decision is made not to further subdivide that portion of the tree into branches, because that would increase the data without any increase in the information content. However, if the magnitude of the functional when the vector is projected to the basis at the higher scale is smaller than the magnitude of the functional at the lower scale, then a decision is made to split that branch (as seen in the lower diagram of Figure 9.3).

Hence, in the conventional Fourier transform where a signal is split into a set of bands with equal bandwidths, the decomposition will provide an entire tree, but the best basis could be a pruned version of that entire tree as shown in Figure 9.2(b). How this procedure is carried out numerically varies for different methods. The key point is that the basis is chosen according to the signal and that the bases need not be orthogonal and may even be linearly dependent. Most importantly, the best basis can be generated in $\theta(D \log D)$ operations. Here we present three algorithms for generating the best possible basis.

The first one is the matching pursuit developed by Mallat and Zhang [2, 6]. The goal here is to iteratively choose the optimal basis. At step k, we define an error in terms of a residual $\Re$ as the difference between the function $\{f(n)\}$ and its approximation by means of the selected bases so that

$$\Re(n) = \{f(n)\} - c_1 \phi_1(n) - c_2 \phi_2(n) - \quad \dots \quad - c_{k-1} \phi_{k-1}(n) \tag{9.7}$$

Out of the remaining bases, we pick an individual basis function $\phi_k(n)$ from the given complete set of bases, so that the largest inner product value for the scalar

$c_k = < \Re; \phi_k >$ is obtained. This is actually a suboptimal selection because we make the current choice for ϕ_k without taking into account how we are going to make the latter choices. We generally terminate the choice after a finite number of steps k (i.e., $k = M$) and then recompute the coefficients c_i to obtain the best fit for the set of basis $\phi_k(n)$; $k = 1, 2, ..., M$. It is said that the error approaches a white noise process as $M \rightarrow \infty$.

The second procedure is called the best basis algorithm developed by Coifman and Wickerhauser [2, 4]. We start the process with a dictionary of bases, which are often orthonormal. For wavelet packets from a binary tree, this method is well adapted. And because of the choice of an orthonormal basis, the computation can be quite fast.

A third method of basis pursuit was developed by Donoho [2, 6–9]. In this procedure an overcomplete dictionary of the bases is selected. Then a nonlinear optimization problem is solved to select the best basis. This method can distinguish between two closely spaced peaks of the function where the previous technique called the matching pursuit has difficulties. This method is also suitable for the situation in which the number of bases is very large along with the dimension of the problem. The selection process for the optimum basis is carried out by minimizing the function

$$\left\| f(n) - \sum_{k=1}^{M} c_k \phi_k(n) \right\|^2 + \lambda M \qquad (9.8)$$

Here the data length is assumed to be N. The number of elements in the dictionary is P, so that $P \gg N$. The dictionary implies that it contains all the possible basis functions that we have at our discretion. It is quite possible that the basis may be linearly dependent or in other words, overcomplete. In other words, we cannot choose a function outside the dictionary. Out of the P basis in the dictionary, we select the best M components. The key in this process is the choice of the parameter λ, which is often selected as $\lambda = \sigma \sqrt{2 \log(P)}$ where it is assumed that the noise contaminating the data is Gaussian with standard deviation σ.

This procedure is also applicable for denoising signals, which we are going to discuss next.

9.3 DENOISING OF SIGNALS THROUGH THE WAVELET TRANSFORM

Typically, in a denoising process the smaller singular values of the system matrix are discarded and the data matrix is reconstructed with the remaining larger singular values to minimize the effects of noise in the data. However, with wavelets a denoising scheme can be implemented with a computational complexity $\theta(N)$, which is much smaller than the $\theta(N^3)$ computations required for the singular value decomposition. In this procedure, the wavelet coefficients are changed, based on the variance of the noise, or the nature of the noise, be it Gaussian or white, and on the number of data samples. By massaging the wavelet coefficients that have been computed for the noisy data, the effect of noise can be minimized. Hence, this method could be complementary to the singular value decomposition in carrying out denoising of a signal in a relatively fast fashion.

Consider a data vector f, which is contaminated by noise. Let us term the noise-contaminated vector as f^{noise} so that

$$f^{\text{noise}}(k) = f(k) + n(k) \quad \text{for} \quad k = 0, 1, \ldots N - 1 \qquad (9.9)$$

where k is the sample value of the variable and n is the noise term. Now if we apply the wavelet transform to (9.9), then we get

$$WT[f^{\text{noise}}(k)] = WT[f(k)] + WT[n(k)] \qquad (9.10)$$

where WT implies the wavelet transform. We consider the noise to be Gaussian in nature. In addition, we assume that σ is the standard deviation of the noise. Furthermore, if we consider $n(k)$ to be an independent identically distributed random variable with a Gaussian probability density function of zero-mean value and a variance of unity, then we obtain

$$\begin{aligned} WT[f^{\text{noise}}(k)] &= WT[f(k)] + WT[\sigma n(k)] \\ &= WT[f(k)] + \sigma\, WT[n(k)] \\ &= WT[f(k)] + \sigma\, w(k) \end{aligned} \qquad (9.11)$$

where $w(k) = WT[n(k)]$. If the wavelet bases are chosen to have a normal distribution, then the wavelet transform of a white Gaussian noise $n(k)$ is a Gaussian white noise process $w(k)$ of the same amplitude [6–8]. Therefore, one can solve for $f(k)$ from

$$f(k) = \{IWT\}\, \{WT[f^{\text{noise}}(k)] - \sigma w(k)\} \qquad (9.12)$$

where $\{IWT\}$ stands for the inverse wavelet transform. In general, we do not know the value of $\sigma w(k)$, so we estimate it by the constant ξ independent of k, yielding the approximate data set $\hat{f}(k)$, which is given by

$$\hat{f}(k) \approx \{IWT\}\, \{WT[f^{\text{noise}}(k)] - \xi\,\} \qquad (9.13)$$

Therefore, the goal is to subtract the noise contribution ξ from each of the noise-contaminated wavelet coefficients, which is defined by $d^{noise}(k) = WT\left[f^{noise}(k)\right]$. A possible implementation of the noise correction can be done by applying the following nonlinear soft thresholding operation to the discrete wavelet coefficients $d^{noise}(k)$. This soft thresholding is carried out by defining the following new modified wavelet coefficients through [6–8]

$$d^{new}(k) = \begin{cases} d(k) - t & \text{if } d(k) > t \\ 0 & \text{if } |d(k)| \leq t \\ d(k) + t & \text{if } d(k) < -t \end{cases} \tag{9.14}$$

Now for a white Gaussian noise, the parameter t is chosen as

$$t = \sigma \sqrt{2 \log(N)} \tag{9.15}$$

where N is the number of data samples. Typically, the noise standard deviation σ is not known, so σ is estimated from

$$\hat{\sigma} = MAV / 0.6745 \tag{9.16}$$

where MAV = mean absolute value of the finest scale wavelet coefficients. However, when noise is white, the threshold t is determined from

$$t = \frac{\sqrt{2 \log(N)}\, \hat{\sigma}}{\sqrt{N}} \tag{9.17}$$

In (9.17) $\hat{\sigma}$ stands for the global noise variance. The concept of denoising then translates into reducing the magnitude of the noisy wavelet coefficients. This wavelet shrinkage methodology therefore acts by setting to zero any coefficients smaller than a "maximum error" bound and retaining the coefficients as opposed to a simple "keep or kill" threshold operation. This guarantees a reconstruction of the data f, which is at least as smooth as f itself (with high probability) and is near optimal in a statistical minimax sense. It is important to note that this choice of the parameters in (9.14) and (9.15) is not unique. There are other ways to select the threshold. An interested reader may find all the details in [3]. Many examples using this technique are also available in [3].

9.4 CONCLUSION

In this chapter, the application of wavelets has been presented to illustrate a computationally efficient way to generate the best/optimal basis in the representation of arbitrary waveforms. In addition, a prescription has also been provided on how to denoise signals by applying a soft threshold to the wavelet coefficients.

REFERENCES

[1] S. Haykin, *Adaptive Filter Theory*, 3rd ed, Prentice Hall, Upper Saddle River, NJ, 1996, p. 175.

[2] G. Strang and T. Nguyen, *Wavelets and Filter Banks*, Wellesley-Cambridge Press, Wellesley, MA, 1996.

[3] A Aldroubi and M. Unser, Eds., *Wavelets in Medicine and Biology*, CRC Press, Boca Raton, FL, 1996.

[4] M. W. Wickerhauser, *Adapted Wavelet Analysis: From Theory to Software*, IEEE Press, New York, 1994.

[5] S. Mallat, *A Wavelet Tour of Signal Processing*, Academic Press, San Diego, CA, 1998.

[6] D. L. Donoho and J. M. Johnstone, "Adapting to unknown smoothness via wavelet shrinkage," *J. Am. Statis. Assoc.*, Vol. 90, 1995, pp. 1200–1224.

[7] D. L. Donoho, "De-noising by soft-thresholding," *IEEE Trans. Information Theory*, Vol. 41, 1995, pp. 613–627.

[8] D. L. Donoho and I. M. Johnstone, "Ideal spatial adaptation via wavelet shrinkage," *Biometrika*, Vol. 81, No. 3, 1994, pp. 425–455.

[9] J. B. Weaver et al., "Filtering noise from images with wavelet transforms," *Magn. Reson. Med.*, Vol. 21, No. 2, 1991, pp. 288–295.

SELECTED BIBLIOGRAPHY

BOOKS

A. Aldroubi and M. Unser, *Wavelets in Medicine and Biology*, CRC Press, Boca Raton, FL, 1996.

M. Ainsworth et al., *Wavelets, Multilevel Methods, and Elliptic PDEs*, Oxford University Press, New York, 1997.

C. K. Chui, *An Introduction to Wavelets*, Academic Press, San Diego, CA, 1992.

C. K. Chui, *Wavelets: A Tutorial in Theory and Applications*, Academic Press, San Diego, CA, 1992.

I. Daubechies, *Ten Lectures on Wavelets*, Capital City Press, Montpelier, VT, 1992.

S. Haykin, *Adaptive Filter Theory,* 3rd ed., Prentice Hall, Upper Saddle River, NJ, 1996.

B. B. Hubard, *The World According to Wavelets: The Story of a Mathematical Technique in the Making*, A K Peters, Ltd., Wellesley, MA, 1996.

A. K. Louis, P. Maab, and A. Reider, *Wavelets: Theory and Applications*, John Wiley & Sons, West Sussex, England, 1997.

Y. Meyer, *Wavelets and Applications*, Masson, Paris, France, 1991.

H. L. Resnikoff and R. O. Wells, *Wavelet Analysis: The Scalable Structure of Information*, Springer-Verlag, New York, 1998.

G. Strang and T. Nguyen, *Wavelets and Filter Banks*, Wellesley-Cambridge Press, Wellesley, MA, 1996.

P. P. Vaidyanathan, *Multirate Systems and Filter Banks*, Prentice Hall, Englewood Cliffs, NJ, 1993.

M. Vetterli and J. Kovačević, *Wavelets and Subband Coding*, Prentice Hall, Upper Saddle River, NJ, 1995.

G. G. Walter, *Wavelets and Other Orthogonal Systems with Applications*, CRC Press, Boca Raton, FL, 1994.

M. V. Wickerhauser, *Adapted Wavelet Analysis from Theory to Software*, A K Peters, Ltd., Wellesley, MA, 1994.

JOURNAL AND CONFERENCE PAPERS

Wavelet Transform

O. M. Bucci et al., "An adaptive wavelet-based approach for non-destructive evaluation applications," *2000 IEEE AP-S Int. Symp. on Antennas & Propagation Digest*, Salt Lake City, UT, July 16–21, 2000, pp. 1756–1759.

L. K. Shi, X. Q. Shen, and W. L. Yan, "A wavelet interpolation Galerkin method for the numerical solution of the boundary problems in 2D static electromagnetic field," *Proc. of the Chinese Society for Electrical Engineering*, No. 9, 2000, pp. 13–16.

L. Tarricone and F. Malucelli, "Efficient linear system solution in moment methods using wavelet expansions," *IEEE Trans. Antennas and Propag.*, Vol. 48, 2000, pp. 1257–1259.

S. Y. Yang et al., "Wavelet-Galerkin method for computations of electromagnetic fields—computation of connection coefficients," *IEEE Trans. Magnetics*, Vol. 36, 2000, pp. 644–648.

Z. Baharav, "Optimal grouping of basis functions," *1999 IEEE AP-S Int. Symp. on Antennas and Propagation Digest*, Orlando, FL, July 11–17, 1999, pp. 340–343.

H. Deng and H. Ling, "Efficient representation of electromagnetic integral equations using pre-defined wavelet packet basis," *1999 IEEE AP-S Int. Symp. Antennas and Propag. Digest*, Orlando, FL, July 11–17, 1999, pp. 336–339.

H. Deng and H. Ling, "Fast solution of electromagnetic integral equations using adaptive wavelet packet transform," *IEEE Trans. Antennas and Propag.*, Vol. 47, 1999, pp. 647–682.

H. Deng and H. Ling, "On a class of predefined wavelet packet bases for efficient representation of electromagnetic integral equations," *IEEE Trans. Antennas and Propag.*, Vol. 47, 1999, pp. 1772–1779.

H. Deng and H. Ling, "Preconditioning of electromagnetic integral equations using pre-defined wavelet packet basis," *Electronics Letters*, Vol. 35, 1999, pp. 1144–1146.

N. Guan, K. Yashiro, and S. Ohkawa, "Wavelet matrix transform approach for the solution of electromagnetic integral equations," *1999 IEEE AP-S Int. Symp. on Antennas and Propagation Digest*, Orlando, FL, July 11–17, 1999, pp. 364–367.

J. M. Huang et al., "Application of mix-phase wavelets to sparsify impedance matrices," *IEICE Trans. on Communications*, Vol. E 82-B, No. 9, 1999, pp. 1688–1693.

G. Kaiser, "Highly focused pulsed-beam wavelets," *Proc. SPIE*, Vol. 3810, 1999, pp. 2–9.

G. W. Pan, M. V. Toupikov, and B. K. Gilbert, "On the use of Coifman intervallic wavelets in the method of moments for fast construction of wavelet sparsified matrices," *IEEE Trans. Antennas and Propag.*, Vol. 47, 1999, pp. 1189–1200.

W. Quan and I. R. Ciric, "A comparative study of wavelet matrix transformations for the solution of integral equations," *1999 IEEE AP-S Int. Symp. on Antennas and Propagation Digest*, Orlando, FL, July 11–17, 1999, pp. 328–331.

W. Quan and I. R. Ciric, "On the semi-orthogonal wavelet matrix transform approach for the solution of integral equations," *1999 IEEE AP-S Int. Symp. on Antennas and Propagation Digest*, Orlando, FL, July 11–17, 1999, pp. 360–363.

Y. L. Sheng and S. Deschenes, "Spatially localized electromagnetic wavelets," *Proc. SPIE*, Vol. 3723 , 1999, pp. 272–276.

L. Tarricone, F. Malucelli, and A. Esposito, "Efficient solution of linear systems in microwave numerical methods," *Applied Computational Electromagnetics Society J.*, Vol. 14, No. 3, 1999, pp. 100–107.

M. Toupikov, G. W. Pan, and B. K. Gilbert, "Weighted wavelet expansion in the method of moments," *IEEE Trans. Magnetics*, Vol. 35, 1999, pp. 1550–1553.

S. Y. Yang and G. Z. Ni, "Wavelet-Galerkin method for the numerical calculation of electromagnetic fields," *Proc. of the Chinese Society for Electrical Engineering*, No. 1, 1999, pp. 56–61.

Q. K. Zhang and N. Ida, "Biorthogonal wavelet analysis on 2D EM scattering from large objects," *IEEE Trans. Magnetics*, Vol. 35, 1999, pp. 1526–1529.

A. J. Blanchard et al., "The use of spline based wavelet filtering to improve classification processing of SAR imagery," *Proc. 1998 IEEE Int. Geoscience and Remote Sensing. Symp.*, 1998, pp. 1757–1759.

H. Deng and H. Ling, "Application of adaptive wavelet packet basis to solving electromagnetic integral equations," *1998 IEEE AP-S Int. Symp. on Antennas and Propagation Digest*, Atlanta, GA, June 22–26, 1998, pp. 1740–1743.

W. L. Golik, "Wavelet packets for fast solution of electromagnetic integral equations," *IEEE Trans. Antennas and Propag.*, Vol. 46, 1998, pp. 618–624.

R. R. Joshi and A. E. Yagle, "Levinson-like and Schur-like fast algorithms for solving block-slanted Toeplitz systems of equations arising in wavelet-based solution of integral equations," *IEEE Trans. Acoustics, Speech, and Signal Processing*, Vol. 46, 1998, pp. 1798–1813.

R. R. Joshi and A. E. Yagle, "Split versions of the Levinson-like and Schur-like fast algorithms for solving block-slanted-Toeplitz systems of equations," *IEEE Trans. Acoustics, Speech, and Signal Processing*, Vol. 46, 1998, pp. 2027–2030.

Y. H. Lee and Y. L. Lu, "Accelerating numerical electromagnetic code computation by using the wavelet transform," *IEEE Trans. Magnetics*, Vol. 34, 1998, pp. 3399–3402.

R. E. Miller and R. D. Nevels, "Investigation of the discrete wavelet transform as a change of basis operation for a moment method solution to electromagnetic integral equations," *1998 IEEE AP-S Int. Symp. on Antennas and Propagation Digest*, Atlanta, GA, June 22–26, 1998, pp. 1258–1261.

G. W. Pan et al., "Use of Coifman intervallic wavelets in 2-D and 3-D scattering problems," *IEE Proc.-Microwaves, Antennas and Propagation*, Vol. 145, No. 6, 1998, pp. 471–480.

Y. Sheng, S. Deschenes, and H. J. Caulfield, "Monochromatic electromagnetic wavelets and the Huygens principle," *Applied Optics*, Vol. 37, No. 5, 1998, pp. 828–833.

L. H. Sibul, L. G. Weiss, and M. J. Roan, "Generalized wavelet transforms and their applications," *Proc. SPIE*, Vol. 3391, 1998, pp. 502–509.

M. S. Wang and A. K. Chan, "Wavelet-packet-based time-frequency distribution and its application to radar imaging," *Int. J. of Numerical Modelling: Electronic Networks, Devices and Fields*, Vol. 11, Jan.–Feb. 1998, pp. 21–40.

M. Werthen and I. Wolff, "A wavelet based time domain moment method for the analysis of three-dimensional electromagnetic fields," *1998 IEEE MTT-S Int. Microwave Symp. Digest*, Baltimore, MD, June 7–12, 1998, pp. 1251–1254.

Z. G. Xiang and Y. L. Lu, "A study of the fast wavelet transform method in computational electromagnetics," *IEEE Trans. Magnetics*, Vol. 34, 1998, pp. 3323–3326.

S. Y. Yang et al., "Wavelet-Galerkin method for computations of electromagnetic fields," *IEEE Trans. Magnetics*, Vol. 34, 1998, pp. 644–648.

Q. K. Zhang and N. Ida, "Application of biorthogonal wavelets on the interval [0,1] to 2D EM scattering," *IEEE Trans. Magnetics*, Vol. 34, 1998, pp. 2728–2731.

A. R. Baghai-Wadji, "An introduction to the fast-MoM in computational electromagnetics," *Electrical Performance of Electronic Packaging*, 1997, p. 231.

R. D. Nevels, J. C. Goswami, and H. Tehrani, "Semi-orthogonal versus orthogonal wavelet basis sets for solving integral equations," *IEEE Trans. Antennas and Propag.*, Vol. 45, 1997, pp. 1332–1339.

L. Rebollo-Neira and J. Fernandez-Rubio, "On wideband deconvolution using wavelet transforms," *IEEE Signal Processing Letters*, Vol. 4, No. 7, 1997, pp. 207–209.

Y. Sheng, S. Deschenes, and H. J. Caulfield, "Monochromatic wavelet and Huygens principle," *Proc. SPIE*, Vol. 3078, 1997, pp. 692–699.

B. Wang, J. C. Moulder, and J. P. Basart, "Wavelets in the solution of the volume integral equation: Application to eddy current modeling," *J. of Applied Physics*, Vol. 81, No. 9, May 1, 1997, pp. 6397–6406.

Z. G. Xiang and Y. L. Lu, "An adaptive wavelet transform for solutions of electromagnetic integral equations," *1997 IEEE AP-S Int. Symp. Antennas and Propag. Digest*, Montreal, Canada, July 14–18, 1997, pp. 1104–1107.

J. C. Yang, K. R. Shao, and J. D. Lavers, "Wavelet-Galerkin method in solving dynamic electromagnetic field problems," *IEEE Trans. Magnetics*, Vol. 33, 1997, pp. 4122–4124.

R. Kastner, G. Nocham, and B. Z. Steinberg, "Multi-region systematic reduced field testing (RFT) for matrix thinning," *1996 IEEE AP-S Int. Symp. on Antennas and Propagation Digest*, Baltimore, MD, July 21–26, 1996, pp. 868–871.

H. D. Kim, H. Ling, and C. K. Lee, "A fast moment method algorithm using spectral domain wavelet concepts," *Radio Science*, Vol. 31, Sept.–Oct. 1996, pp. 1253–1261.

R. D. Nevels and J. C. Goswami, "Current distribution on a scatterer obtained by integral equations with semi-orthogonal and orthogonal wavelet basis sets," *1996 IEEE AP-S Int. Symp. on Antennas and Propagation Digest*, Baltimore, MD, July 21–26, 1996, pp. 340–343.

G. W. Pan, "Orthogonal wavelets with applications in electromagnetics," *IEEE Trans. Magnetics*, Vol. 32, 1996, pp. 975–983.

Y. H. Peng, X. L. Dong, and W. B. Wang, "Applications of wavelet transform in electromagnetic numerical computations," *Acta Electronica Sinica*, Vol. 24, No. 12, 1996, pp. 46–52.

W. Y. Tam, "Weighted wavelet-like basis for electromagnetics," *Proc. 1996 Int. Symp. Antennas and Propag.*, Chiba, Japan, Aug. 1996, Vol.4, pp. 1145–1148.

W. Y. Tam, "Weighted Haar wavelet-like basis for scattering problems," *IEEE Microwave and Guided Wave Letters*, Vol. 6, 1996, pp. 435–437.

G. Wang et al., "Comments on 'On solving first-kind integral equation using wavelets on a bounded interval' [and reply]," *IEEE Trans. Antennas and Propag.*, Vol. 44, 1996, pp. 1306–1307.

M. Werthen and I. Wolff, "A novel wavelet based time domain simulation approach," *IEEE Microwave and Guided Wave Letters*, Vol. 6, 1996, pp. 438–440.

J. C. Goswami, A. K. Chan, and C. K. Chui, "On solving first-kind integral equations using wavelets on a bounded interval," *IEEE Trans. Antennas and Propag.*, Vol. 43, 1995, pp. 614–622.

H. D. Kim, "Application of wavelet transform in electromagnetics," *J. Korean Institute of Telematics and Electronics*, Vol. 32A, No. 9, Sept. 1995, pp. 78–83.

T. K. Sarkar et al., "Design of the mother wavelet: A zero mean pulse of finite support in time-frequency plane," *Proc. European Electromagnetics Int. Symp. on Electromagnetic Environment and Consequences*, 1995, Vol. 2, pp. 1591–1603.

A. Shvartsburg, "Wavelet interactions with dispersive and lossy media," *Proc. SPIE*, Vol. 2242, 1995, pp. 140–146.

L. H. Sibul, "Application of reproducing and invariance properties of wavelet and Fourier-Wigner transforms," *Proc. SPIE*, Vol. 2569, Pt. 1, 1995, pp. 418–428.

L. C. Trintinalia and H. Ling, "Time-frequency representation of wideband radar echo using adaptive normalized Gaussian functions," *1995 IEEE AP-S Int. Symp. on Antennas and Propagation Digest*, Newport Beach, CA, June 18–23, 1995, pp. 324–327.

R. L. Wagner and W. C. Chew, "A study of wavelets for the solution of electromagnetic integral equations," *IEEE Trans. Antennas and Propag.*, Vol. 43, 1995, pp. 802–810.

L. Carin et al., "Dispersive modes in the time domain: analysis and time-frequency representation," *IEEE Microwave and Guided Wave Letters*, Vol. 4, 1994, pp. 23–25.

B. Z. Steinberg and Y. Leviatan, "Periodic wavelet expansions for analysis of scattering from metallic cylinders," *IEEE AP-S Int. Symp. on Antennas and Propagation Digest*, Seattle, WA, June 20–24, 1994, pp. 20–23.

A. J. Blanchard, C. K. Chui, and A. K. Chan, "Spline wavelets applied to SAR information processing," *Proc. 1993 Int. Geoscience and Remote Sensing Symp.*, 1993, pp. 2133–2134.

L. E. Garcia-Castillo, T. K. Sarkar, and M. S. Salazar-Palma, "Wavelets: A promising approach for electromagnetic fields problems," *Proc. of IEEE Electrical Performance of Electronic Packaging*, 1993, pp. 40–42.

Hybrid Methods

T. I. Kosmanis, N. V. Kantartzis, and T. D. Tsiboukis, "A hybrid FDTD-wavelet-Galerkin technique for the numerical analysis of field singularities inside waveguides," *IEEE Trans. Magnetics*, Vol. 36, 2000, pp. 902–906.

D. Sullivan, J. Liu, and M. Kuzyk, "Three-dimensional optical pulse simulation using the FDTD method," *IEEE Trans. Microwave Theory and Techniques*, Vol. 48, 2000, pp. 1127–1133.

M. Walter and I. Wolff, "An algorithm for realizing Yee's FDTD-method in the wavelet domain," *1999 IEEE MTT-S Int. Microwave Symp. Digest*, 1999, Vol. 3, pp. 1301–1304.

Z. Xiang and Y. Lu, "A hybrid FEM/BEM/WTM approach for fast solution of scattering from cylindrical scatterers with arbitrary cross sections," *J. Electromagnetic Waves and Applications*, Vol. 13, 1999, pp. 811–812.

J. C. Yang and K. R. Shao, "Application of interpolation wavelets in difference method for engineering electromagnetic field problems," *Proc. Chinese Society for Electrical Engineering*, Vol. 19, No. 7, 1999, pp. 19–21.

A. Arev, E. Heyman, and B. Z. Steinberg, "A mixed time-frequency-scale analysis of the hybrid wavefront-resonance representation," *Proc. 1998 Int. Symp. on Electromagnetic Theory*, Thessaloniki, Greece, May 25–28, 1998, pp. 608–610.

B. J. Lee, Y. K. Cho, and J. M. Lee, "The application of wavelets to measured equation of invariance," *J. Electrical Engineering and Information Science*, Vol. 3, No. 3, June 1998, pp. 348–354.

C. M. Lin and C. H. Chan, "Monte Carlo simulations for electromagnetic scattering of rough surfaces by a combined wavelet transform and banded-matrix iterative approach/canonical grid method," *Microwave and Optical Technology Letters*, Vol. 19, No. 4, Nov. 1998, pp. 274–279.

C. M. Lin and C. H. Chan, "Combined wavelet transform and banded-matrix iterative approach/canonical grid methods for solving large-scale electromagnetic problems," *1998 IEEE AP-S Int. Symp. on Antennas and Propagation Digest*, Atlanta, GA, June 22–26, 1998, pp. 96–99.

Y. W. Liu, K. K. Mei, and E. K. N. Yung, "Application of discrete periodic wavelets to measured equation of invariance," *IEEE Trans. Antennas and Propag.*, Vol. 46, 1998, pp. 1842–1844.

Z. Xiang, and Y. Lu, "An effective hybrid method for electromagnetic scattering from inhomogeneous objects," *J. Electromagnetic Waves and Applications*, Vol. 12, No. 1, 1998, pp. 91–95.

Y. X. Wang and H. Ling, "Extraction of higher-order modes in open microstrip lines via FDTD and joint time-frequency analysis," *Microwave and Optical Technology Letters*, Vol. 13, No. 6, Dec. 20, 1996, pp. 319–321.

H. D. Kim, H. Ling, and C. K. Lee, "A fast moment method algorithm using spectral domain wavelet concepts," *Radio Science*, Vol. 31, Sept.–Oct. 1996, pp. 1253–1261.

G. Wang, "A hybrid wavelet expansion and boundary element analysis of electromagnetic scattering from conducting objects," *IEEE Trans. Antennas and Propag.*, Vol. 43, 1995, pp. 170–178.

G. Wang and J. C. Hon, "A hybrid wavelet expansion and boundary element method in electromagnetic scattering," *1995 IEEE AP-S Int. Symp. on Antennas and Propagation Digest*, Newport Beach, CA, June 18–23, 1995, pp. 333–336.

G. Wang, G. W. Pan, and B. G. Gilbert, "A hybrid wavelet expansion and boundary element analysis for multiconductor transmission lines in multilayered dielectric media," *IEEE Trans. Microwave Theory and Techniques*, Vol. 43, 1995, pp. 664–675.

L. E. Garce-Castillo, T. K. Sarkar, and M. Salazar-Palma, "An efficient finite element method employing wavelet type basis functions (waveguide analysis)," *Int. J. Computation and Mathematics in Electrical and Electronic Engineering*, Vol. 13, Suppl. A., May 1994, pp. 287–292.

T. K. Sarkar et al., "Utilization of wavelet concepts in finite elements for an efficient solution of Maxwell's equations," *Radio Science*, Vol. 29, 1994, pp. 965–977.

Multiresolution Analysis

G. Carat et al., "An extension of the MRTD technique to the fast computation of the radiation from planar open structures," *2000 IEEE MTT-S Int. Microwave Symp. Digest*, Boston, MA, June 11–16, 2000, pp. 247–250.

K. Y. Choi and W. Y. Tam, "Numerical dispersion of multiresolution time domain method (W-MRTD)." *2000 IEEE AP-S Int. Symp. on Antennas & Propagation Digest*, Salt Lake City, UT, Vol. 1, July 16–21, 2000, pp. 256–259.

K. Kigoshi et al., "Wavelet matrix transform approach for electromagnetic scattering by a dielectric cylinder," *Trans. of the Institute of Electrical Engineers of Japan*, Part A, Vol. 120-A, No. 10, Oct. 2000, pp. 878–884.

G. F. Wang, R. W. Dutton, and J. C. Hou, "A fast wavelet multigrid algorithm for solution of electromagnetic integral equations," *Microwave and Optical Technology Letters*, Vol. 24, No. 2 , Jan. 20, 2000, pp. 86–91.

Z. Chen and J. Zhang, "An efficient eigen-based spatial-MRTD method for computing resonant structures," *IEEE Microwave and Guided Wave Letters*, Vol. 9, 1999, pp. 333–335.

P. Chiappinelli et al., "Multiresolution techniques in microwave tomography and subsurface sensing," *1999 International Geoscience and Remote Sensing Symp.*, Vol. 5, 1999, pp. 2516–2518.

K. Goverdhanam and L. P. B. Katehi, "Applications of Haar wavelet based MRTD scheme in the characterization of 3D microwave circuits," *1999 IEEE MTT-S Int. Microwave Symp. Digest*, Vol. 4 , 1999, pp. 1475–1478.

P. Pirinoli et al., "Multilevel, multiresolution integral equation analysis of printed antennas," *1999 IEEE AP-S Int. Symp. on Antennas and Propagation Digest*, Orlando, FL, July 11–17, 1999, pp. 352–355.

P. Pirinoli, L. Matekovits, and M. S. Garino, "A novel multiresolution approach to the EFIE analysis of printed antennas," *Microwave and Optical Technology Letters*, Vol. 23, No. 1, Oct. 5, 1999, pp. 49–51.

J. L. Sanz et al., "Wavelets applied to cosmic microwave background maps: A multiresolution analysis for denoising," *Monthly Notices of the Royal Astronomical Society*, Vol. 309, No. 3, Nov. 1, 1999, pp. 672–680.

H. Xin and C. H. Liang, "Multiwavelet analysis of 2D EM scattering," *J. Xidian University*, Vol. 26, No. 3, June 1999, pp. 290–292.

M. Fujii and W. J. R. Hoefer, "Multiresolution analysis similar to the FDTD method—derivation and application," *IEEE Trans. Microwave Theory and Tech.*, Vol. 46, 1998, pp. 2463–2475.

B. Z. Steinberg and J. Oz, "A multiresolution approach for the effective properties of complex propagation ducts and effective modes/rays theory," *Proc. 1998 Int. Symp. on Electromagnetic Theory*, Thessaloniki, Greece, May 25–28, 1998, pp. 88–90.

B. Z. Steinberg and J. Oz, "A multiresolution approach for homogenization and effective properties-propagation in complex ducts," *1998 IEEE AP-S Int. Symp. on Antennas and Propagation Digest*, Atlanta, GA, June 22–26, 1998, pp. 871–874.

B. Z. Steinberg and J. J. McCoy, "A multiresolution study of effective properties of complex electromagnetic systems," *IEEE Trans. Antennas and Propag.*, Vol. 46, 1998, pp. 971–981.

E. K. Walton, "Wavelet techniques and other multiresolution techniques for target phenomenology studies," *Advanced Pattern Recognition Techniques Lecture Series*, Sept. 1998, pp. 31–33.

B. Z. Steinberg and J. J. McCoy, "A study of the effective properties of complex scatterers using multiresolution decomposition," *J. Computational Acoustics (Singapore)*, Vol. 5, No. 1, 1997, pp. 1–31.

M. Krumpholz and L. P. B. Katehi, "MRTD: new time-domain schemes based on multiresolution analysis," *IEEE Trans. on Microwave Theory and Techniques*, Vol. 44, 1996, pp. 555–571.

G. Liu et al., "Unsupervised multiresolution segmentation and interpretation of textured SAR image," *Proc. SPIE*, Vol. 2955, 1996, pp. 261–271.

B. Z. Steinberg and Y. Leviatan, "A multiresolution study of two-dimensional scattering by metallic cylinders," *IEEE Trans. Antennas and Propag.*, Vol. 44, 1996, pp. 572–579.

H. D. Kim and H. Ling, "A fast multiresolution moment method algorithm using wavelet concepts," *1995 IEEE AP-S Int. Symp. on Antennas and Propagation Digest*, Newport Beach, CA, June 18–23, 1995, pp. 312–315.

H. D. Kim and H. Ling, "On the efficient representation of the electrodynamic Green's function using multiresolution wavelet concepts," *Microwave and Optical Technology Letters*, Vol. 9, No. 4, July 1995, pp. 183–187.

H. D. Kim and H. Ling, "A fast multiresolution moment-method algorithm using wavelet concepts [EM scattering]," *Microwave and Optical Technology Letters*, Vol. 10, No. 6, Dec. 20, 1995, pp. 317–319.

K. Sabetfakhri and L. P. B. Katehi, "Multiresolution expansions for efficient moment method solution of wave guiding problems," *IEEE AP-S Int. Symp. on Antennas and Propagation Digest*, Seattle, WA, June 20–24, 1994, pp. 24–27.

B. Z. Steinberg, "A multiresolution theory of scattering and diffraction," *IEEE AP-S Int. Symp. on Antennas and Propagation Digest*, Seattle, WA, Vol. 1, June 20–24, 1994, pp. 33–36.

B. Z. Steinberg, "A multiresolution theory of scattering and diffraction," *Wave Motion*, Vol. 19, No. 3, May 1994, pp. 213–232.

Making Dense Matrices Sparse

J. M. Huang et al., "Impedance matrix compression using an effective quadrature filter," *IEE Proc. Microwaves, Antennas and Propagation,*Vol. 147, No. 4, Aug. 2000, pp. 255–260.

A. Laisne et al., "A discrete wavelet transform (DWT)-based compression technique for the computation of Kirchhoff integrals in MR/FDTD," *Microwave and Optical Technology Letters*, Vol. 27, No. 5, Dec. 5, 2000, pp. 312–316.

C. Su and T. K. Sarkar, "Adaptive multiscale moment method (AMMM) for analysis of scattering from perfectly conducting plates," *IEEE Trans. Antennas and Propag.*, Vol. 48, 2000, pp. 932–939.

T. K. Sarkar and K. J. Kim, "Solution of large dense complex matrix equations utilizing wavelet-like transforms," *IEEE Trans. Antennas and Propag.,*Vol. 47, 1999, pp. 1628–1632.

Y. Shifman, Z. Baharav, and Y. Leviatan, "Analysis of truncated periodic array using two-stage wavelet-packet transformations for impedance matrix compression," *IEEE Trans. Antennas and Propag.*, Vol. 47, 1999, pp. 630–636.

C. Su and T. K. Sarkar, "Analysis of scattering from perfectly conducting plates by the use of AMMM," *Progress in Electromagnetics Research*, Vol. 21, 1999, pp. 71–89.

C. Su and T. K. Sarkar, "Adaptive multiscale moment method for solving two-dimensional Fredholm integral equation of the first kind," *Prog. in Electromagnetics Research*, Vol. 21, 1999, pp. 173–201.

Z. Baharav and Y. Leviatan, "Impedance matrix compression (IMC) using iteratively selected wavelet basis," *IEEE Trans. Antennas and Propag.*, Vol. 46, 1998, pp. 226–233.

T. K. Sarkar, C. Su, and M. Salazar Palma, "Solution of large dense complex matrix equations utilizing wavelet-like transforms," *Annales des Telecommunications*, Vol. 53, 1998, pp. 56–67.

Y. Shifman and Y. Leviatan, "Iterative selection of expansion functions from an overcomplete dictionary of wavelet packets for impedance matrix compression," *J. Electromagnetic Waves and Applications*, Vol. 12, No. 11, 1998, pp. 1403–1421.

C. Su and T. K. Sarkar, "Electromagnetic scattering from coated strips utilizing the adaptive multiscale moment method," *Prog. in Electromagnetics Research*, Vol. 18, 1998, pp. 173–208.

C. Su and T. K. Sarkar, "Scattering from perfectly conducting strips by utilizing an adaptive multiscale moment method," *Prog. in Electromagnetics Research*, Vol. 19, 1998, pp. 173–197.

C. Su and T. K. Sarkar, "A multiscale moment method for solving Fredholm integral equation of the first kind," *Prog. in Electromagnetics Research*, Vol. 17, 1997, pp. 237–264.

Z. Baharav and Y. Leviatan, "Impedance matrix compression (IMC) using iteratively selected wavelet basis for MFIE formulations," *Microwave and Optical Technology Letters*, Vol. 12 , No. 3, June 20, 1996, pp. 145–150.

Z. Baharav and Y. Leviatan, "Impedance matrix compression (IMC) using iteratively selected wavelet basis for MFIE formulations," *1996 IEEE AP-S Int. Symp. on Antennas and Propagation Digest*, Baltimore, MD, July 21–26, 1996, pp. 348–351.

Z. Baharav and Y. Leviatan, "Impedance matrix compression using adaptively constructed basis functions," *IEEE Trans. Antennas and Propag.*,Vol. 44, 1996, pp. 1231–1238.

Z. Baharav and Y. Leviatan, "Impedance matrix compression with the use of wavelet expansions [EM scattering]," *Microwave and Optical Technology Letters*, Vol. 12, No. 5, Aug. 5, 1996, pp. 268–272.

S. A. Werness, S. C. Wei, and R. Carpinella, "Experiments with wavelets for compression of SAR data," *IEEE Trans. Geoscience and Remote Sensing*, Vol. 32, 1994, pp. 197–201.

Scattering

F. Berizzi, P. Gamba, and A. Garzelli, "Fractal analysis of ASAR like sea scattering data," *Proc. 2000 IEEE Int. Geoscience and Remote Sensing Symp.*, Vol. 1, 2000, pp. 237–239.

R. S. Chen et al., "Analysis of millimeter wave scattering by an electrically large metallic grating using wavelet-based algebraic multigrid preconditioned CG method," *Int. J. Infrared and Millimeter Waves*, Vol. 21, No. 9, Sept. 2000, pp. 1541–1560.

N. Guan, K. Yashiro, and S. Ohkawa, "Wavelet transform approach on boundary element method for solving electromagnetic scattering problems," *2000 IEEE AP-S Int. Symp. on Antennas & Propagation Digest*, Salt Lake City, UT, July 16–21, 2000, pp. 1834–1837.

L. X. Guo, X. Z. Ke, and Z. S. Wu, "Electromagnetic scattering from a one-dimensional random rough surface using the method of wavelet moment." *Jour. of Xidian University*, Vol. 27, No. 5, 2000, pp. 585–589.

J. L. Leou et al., "Construction of complex-valued wavelets and its applications to scattering problems," *IEICE Trans. on Communications*, Vol. E83-B, No. 6, June 2000, pp. 1298–1307.

J. L. Leou et al., "Minimum and maximum time-localized complex-valued wavelets for scattering problems," *1999 IEEE AP-S Int. Symp. on Antennas and Propagation Digest*, Orlando, FL, July 11–17, 2000, pp. 368–371.

D. Zahn et al., "Numerical simulation of scattering from rough surfaces: A wavelet-based approach," *IEEE Trans. Antennas and Propag.*, Vol. 48, 2000, pp. 246–253.

J. L. Alvarez Perez, S. J. Marshall, and K. Gregson, "Method of moments with wavelet expansions for dielectric surface scattering," *1999 IEEE Int. Geoscience and Remote Sensing Symp.*, Vol. 3, 1999, pp. 1851–1853.

Z. Baharav and Y. Leviatan, "Analysis of scattering by surfaces using a wavelet-transformed triangular-patch model," *Microwave and Optical Technology Letters*, Vol. 21, No. 5, June 5, 1999, pp. 359–365.

L. G. Bruskin et al., "Wavelet analysis of plasma fluctuations in microwave reflectometry," *Rev. of Scientific Instruments*, Vol. 70, No. 1, Pt. 1–2, Jan. 1999, pp. 1052–1055.

Y. W. Cheong et al., "Wavelet-Galerkin scheme of time-dependent inhomogeneous electromagnetic problems," *IEEE Microwave and Guided Wave Letters*, Vol. 9, 1999, pp. 297–299.

N. Guan, K. Yashiro, and S. Ohkawa, "Wavelet matrix transform approach for electromagnetic scattering from an array of metal strips," *IEICE Trans. on Electronics*, Vol. E82-C, No. 7, July 1999, pp. 1273–1279.

H. Ikuno and M. Nishimoto, "Wavelet analysis of scattering responses of electromagnetic waves," *Electronics and Communications in Japan, Part 2 (Electronics)*, Vol. 82, No. 10, Oct. 1999, pp. 35–42.

J. L. Leou et al., "Application of wavelets to scattering problems of inhomogeneous dielectric slabs," *IEICE Trans. on Communications*, Vol. E82-B, No. 9 , Sept. 1999, pp. 1667–1676.

C. M. Lin, C. H. Chan, and L. Tsang, "Monte Carlo simulations of scattering and emission from lossy dielectric random rough surfaces using the wavelet transform method," *IEEE Trans. Geosci. Remote Sens.*, Vol. 37, 1999, pp. 2295–2304.

F. Mattia and T. Le Toan, "Backscattering properties of multi-scale rough surfaces," *J. Electromagnetic Waves and Applications*, Vol. 13, 1999, pp. 493–527.

M. Saillard and G. Soriano, "Rigorous solution of scattering by randomly rough surfaces: Recent contributions to open problems," *Proc. SPIE,* Vol. 3749, 1999, pp. 36–37.

Y. Sheng and S. Dechenes, "Electromagnetic wavelet propagation and diffraction," *Proc. SPIE,* Vol. 3749, 1999, pp. 12–13.

L. T. Thuong, "A comparison of wavelet functions for scattering extraction in centimetre-wave radar," *Proc. Second Australian Workshop on Signal Processing Applications*, Brisbane, Australia, Dec. 1997, pp. 215–218.

Z. Xiang and Y. Lu, "A hybrid FEM/BEM/WTM approach for fast solution of scattering from cylindrical scatterers with arbitrary cross sections," *J. Electromagnetic Waves and Applications*, Vol. 13, 1999, pp. 811–812.

H. Xin and C. H. Liang, "Multiwavelet analysis of 2D EM scattering," *J. Xidian University*, Vol. 26, No. 3, June 1999, pp. 290–292.

Q. K. Zhang and N. Ida, "Biorthogonal wavelet analysis on 2D EM scattering from large objects," *IEEE Trans. Magnetics*, Vol. 35, 1999, pp. 1526–1529.

L. G. Bruskin et al., "Application of wavelet analysis to plasma fluctuation study on GAMMA 10," *Japanese J. Applied Physics, Part 2 (Letters)*, Vol. 37, No. 8A, Aug. 1998, pp. 956–958.

C. A. Guerin and M. Holschneider, "Time-dependent scattering on fractal measures," *Jour. of Mathematical Physics*, Vol. 39, No. 8, Aug. 1998, pp. 4165–4194.

H. Ikuno and M. Nishimoto, "Wavelet analysis of scattering responses of electromagnetic waves," *IEICE Trans. on Electronics*, Vol. J81-C-I, No. 11, Nov. 1998, pp. 609–615.

C. M. Lin and C. H. Chan, "Monte Carlo simulations for electromagnetic scattering of rough surfaces by a combined wavelet transform and banded-matrix iterative approach/canonical grid method," *Microwave and Optical Technology Letters*, Vol. 19, No. 4, Nov. 1998, pp. 274–279.

C. M. Lin, C. H. Chan, and L. Tsang, "Conical diffraction of electromagnetic waves from one-dimensional lossy dielectric rough surfaces by using the wavelet transform," *Proc. 1998 IEEE Int. Geoscience and Remote Sensing. Symp.*, Vol. 1, 1998, pp. 291–293.

H. Ling, "Joint time-frequency analysis of electromagnetic backscattered data," *Proc. SPIE*, Vol. 3391, 1998, pp. 283–294.

G. W. Pan et al., "Use of Coifman intervallic wavelets in 2-D and 3-D scattering problems," *IEE Proc.-Microwaves, Antennas and Propagation*, Vol. 145, No. 6, Dec. 1998, pp. 471–480.

C. Su and T. K. Sarkar, "Electromagnetic scattering from coated strips utilizing the adaptive multiscale moment method," *Prog. in Electromagnetics Research*, Vol. 18, 1998, pp. 173–208.

C. Su and T. K. Sarkar, "Scattering from perfectly conducting strips by utilizing an adaptive multiscale moment method," *Prog. in Electromagnetics Research,* Vol. 19, 1998, pp. 173–197.

K. Takahashi and Y. Miyazaki, "Wavelet analysis of subsurface radar pulse scattering on random surfaces," *Trans. of the Institute of Electrical Engineers of Japan, Part C,* Vol. 118-C, No.1, Jan. 1998, pp. 93–98.

L. T. Thuong and D. T. Nguyen, "Coincident detection approach based on scale-invariance characteristic of scattering locations in wavelet-based range-scale domain," *Electronics Letters,* Vol. 34, No.17, Aug. 20, 1998, pp. 1691–1693.

G. Wang, "Analysis of electromagnetic scattering from conducting bodies of revolution using orthogonal wavelet expansions," *IEEE Trans. Electromagnetic Compatibility,* Vol. 40, 1998, pp. 1–11.

G. W. Wei et al., "An application of distributed approximating functional-wavelets to reactive scattering," *J. Chemical Physics,* Vol. 108, No. 17, May 1998, pp. 7065–7069.

Z. Xiang and Y. Lu, "An effective hybrid method for electromagnetic scattering from inhomogeneous objects," *J. Electromagnetic Waves and Applications,* Vol. 12, No. 1, 1998, pp. 91–95.

Q. K. Zhang and N. Ida, "Application of biorthogonal wavelets on the interval [0,1] to 2D EM scattering," *IEEE Trans. Magnetics,* Vol. 34, 1998, pp. 2728–2731.

L. Carin et al., "Wave-oriented signal processing of dispersive time-domain scattering data," *IEEE Trans. Antennas and Propag.,* Vol. 45, 1997, pp. 592–600.

C. A. Guerin, M. Holschneider, and M. Saillard, "Electromagnetic scattering from multi-scale rough surfaces," *Waves in Random Media,* Vol. 7, No. 3, July 1997, pp. 331–349.

M. Nishimoto and H. Ikuno, "Time-frequency analysis of scattering data using the wavelet transform," *IEICE Trans. on Electronics,* Vol. E80-C, No. 11, Nov. 1997, pp. 1440–1447.

G. Oberschmidt and A. F. Jacob, "Non-uniform wavelets in electromagnetic scattering analysis," *Electronics Letters,* Vol. 33, No. 4, Feb. 13, 1997, pp. 276–277.

B. Z. Steinberg and J. J. McCoy, "A study of the effective properties of complex scatterers using multiresolution decomposition," *J. Computational Acoustics (Singapore),* Vol. 5, No. 1, March 1997, pp. 1–31.

L. T. Thuong and D. T. Nguyen, "Performance of behaviour of scattering mechanisms in time-scale domain against random noise for centimetre wave radar," *Electronics Letters*, Vol. 33 , No. 22, Oct. 23, 1997, pp. 1900–1901.

D. J. Verschuur and A. J. Berkhout, "Estimation of multiple scattering by iterative inversion. II. Practical aspects and examples," *Geophysics*, Vol. 62, No. 5 , Sept.– Oct. 1997, pp. 1596–1611.

G. Wang and B. Z. Wang, "Application of wavelets to the analysis of arbitrary thin-wire loop antennas and scatterers," *Int. J. Numerical Modelling: Electronic Networks, Devices and Fields*, Vol. 10, No. 3, May–June 1997, pp. 193–204.

G. Wang, "Application of wavelets on the interval to numerical analysis of integral equations in electromagnetic scattering problems," *Int. J. Numerical Methods in Engineering*, Vol. 40, No. 1, Jan. 15, 1997, pp. 1–13.

G. Wang, "Application of wavelets on the interval to the analysis of thin-wire antennas and scatterers," *IEEE Trans. Antennas and Propag.*, Vol. 45, 1997, pp. 885–893.

Y. Xu and T. Liu, "The application of intervallic wavelets in electromagnetic scattering," *Chinese J. Electronics*, Vol. 6, No. 4, Oct. 1997, pp. 35–39.

D. Zahn, K. F. Sabet, and K. Sarabandi, "Numerical simulation of scattering from rough surfaces: a wavelet-based approach," *1997 IEEE AP-S Int. Symp. on Antennas and Propagation Digest*, Montreal, Canada, July 14–18, 1997, pp. 1100–1103.

Z. Baharav and Y. Leviatan, "Scattering analysis using fictitious wavelet array sources," *J. Electromagnetic Waves and Applications*, Vol. 10, No. 12, 1996, pp. 1683–1697.

V. Jandhyala, E. Michielssen, and R. Mittra, "Analysis of TM scattering from open two-dimensional perfectly conducting shells using aggregated compactly-supported wavelet bases," *IEE Proc. Microwaves, Antennas and Propagation*, Vol. 143, No. 2 , April 1996, pp. 152–156.

M. Kroon, G. H. Wegdam, and R. Sprik, "Analysis of dynamic light scattering signals with discrete wavelet transformation," *Europhysics Letters*, Vol. 35, No. 8, 1996, pp. 621–626.

Y. Miyazaki and K. Takahashi, "Wavelet analysis of randomly reflected electromagnetic waves of subsurface radar," *Proc. 1996 Int. Symp. on Antennas and Propagation*, Chiba, Japan, Aug. 1996, Vol. 4, pp. 1153–1156.

R. D. Nevels and J. C. Goswami, "Current distribution on a scatterer obtained by integral equations with semi-orthogonal and orthogonal wavelet basis sets," *1996 IEEE AP-S Int. Symp. on Antennas and Propagation Digest*, Baltimore, MD, July 14–18, 1996, pp. 340–343.

W. Y. Tam, "Weighted Haar wavelet-like basis for scattering problems," *IEEE Microwave and Guided Wave Letters*, Vol. 6, 1996, pp. 435-437.

B. Z. Steinberg and Y. Leviatan, "A multiresolution study of two-dimensional scattering by metallic cylinders," *IEEE Trans. Antennas and Propag.*, Vol. 44, 1996, pp. 572–579.

G. Wang, "On the use of orthogonal wavelets on the interval in the moment method [EM scattering]," *Microwave and Optical Technology Letters*, Vol. 11, No. 1, Jan. 1996, pp. 10–13.

G. Wang and J. C. Hou, "Analysis of thin-wire antennas and scatterers using orthogonal intervallic wavelets," *1996 IEEE AP-S Int. Symp. on Antennas and Propagation Digest*, Baltimore, MD, July 14–18, 1996, pp. 344–347.

Z. Baharav and Y. Leviatan, "Resolution enhancement and small perturbation analysis using wavelet transforms in scattering problems," *1995 IEEE AP-S Int. Symp. on Antennas and Propagation Digest,* Newport Beach, CA, June 18–23, 1995, p. 328.

H. D. Kim and H. Ling, "A fast multiresolution moment-method algorithm using wavelet concepts [EM scattering]," *Microwave and Optical Technology Letters*, Vol. 10, No. 6, Dec. 20, 1995, pp. 317–319.

Y. H. Peng et al., "The application of wavelet transform to time-frequency analysis of transient electromagnetic backscatter signals," *Acta Electronica Sinica*, Vol. 23, No. 9, Sept. 1995, pp. 109–111.

Y. C. Tzeng, K. S. Chen, and H. Huang, "Moment method simulation of rough surface scattering based on wavelet representation," *1995 Int. Geoscience and Remote Sensing Symp.*, 1995, pp. 1448–1450.

G. Wang, "A hybrid wavelet expansion and boundary element analysis of electromagnetic scattering from conducting objects," *IEEE Trans. Antennas and Propag.*, Vol. 43, 1995, pp. 170–178.

G. Wang and J. C. Hon, "A hybrid wavelet expansion and boundary element method in electromagnetic scattering," *1995 IEEE AP-S Int. Symp. on Antennas and Propagation Digest*, Newport Beach, CA, June 18–23, 1995, pp. 333–336.

A. Dogariu, J. Uozumi, and T. Asakura, "Wavelet transform analysis of slightly rough surfaces," *Optics Communications*, Vol. 107, No. 1–2, April 1, 1994, pp. 1–5.

O. P. Franza, R. L. Wagner, and W. C. Chew, "Wavelet-like basis functions for solving scattering integral equations," *1994 IEEE AP-S Int. Symp. on Antennas and Propagation Digest*, Seattle, WA, June 20–24, 1994, pp. 3–6.

J. C. Goswami, A. K. Chan, and C. K. Chui, "An analysis of two-dimensional scattering by metallic cylinders using wavelets on a bounded interval," *1994 IEEE AP-S Int. Symp. on Antennas and Propagation Digest*, Seattle, WA, June 20–24, 1994, p. 2.

A. Moghaddar and E. K. Walton, "A data-adaptive time-frequency representation applied to the scattering from a jet aircraft," *1994 IEEE AP-S Int. Symp. on Antennas and Propagation Digest*, Seattle, WA, June 20–24, 1994, pp. 680–683.

L. H. Sibul, L. G. Weiss, and T. L. Dixon, "Characterization of stochastic propagation and scattering via Gabor and wavelet transforms," *J. Computational Acoustics*, Vol. 2, No. 3, Sept. 1994, pp. 345–369.

B. Z. Steinberg and Y. Leviatan, "Periodic wavelet expansions for analysis of scattering from metallic cylinders," *Microwave and Optical Technology Letters*, Vol. 7, No. 6, April 20, 1994, pp. 266–268.

B. Z. Steinberg and Y. Leviatan, "Periodic wavelet expansions for analysis of scattering from metallic cylinders," *1994 IEEE AP-S Int. Symp. Antennas and Propag. Digest*, Seattle, WA, June 20–24, 1994, pp. 20–23.

H. D. Kim and H. Ling, "On the application of fast wavelet transform to the integral-equation solution of electromagnetic scattering problems," *Microwave and Optical Technology Letters*, Vol. 6, No. 3, March 5, 1993, pp. 168–173.

B. Z. Steinberg and Y. Leviatan, "On the use of wavelet expansions in the method of moments (EM scattering)," *IEEE Trans. Antennas and Propag.* , Vol. 41, 1993, pp. 610–619.

M. H. Yaou and W. T. Chang, "Wavelet transform in scattering data interpolation," *Electronics Letters*, Vol. 29, No. 21, Oct. 14, 1993, pp. 1835–1837.

L. B. Wetzel, "A time domain model for sea scatter," *Radio Science*, Vol. 28, No. 2, March–April 1993, pp. 139–150.

H. D. Kim and H. Ling, "Wavelet analysis of electromagnetic backscatter data," *Electronics Letters*, Vol. 28, No. 3, Jan. 30, 1992, pp. 279–281.

H. Ling and H. D. Kim, "Wavelet analysis of backscattering data from an open-ended waveguide cavity," *IEEE Microwave and Guided Wave Letters*, Vol. 2, 1992, pp. 140–142.

A. Tie and L. T. Long, "The character of a scattered wavelet: A spherical obstacle embedded in an elastic medium," *Seismological Research Letters*, Vol. 63, No. 4, Oct.–Dec. 1992, pp. 515–523.

L. F. Bliven and B. Chapron, "Wavelet analysis and radar scattering from water waves," *Naval Research Reviews*, Vol. 41, No. 2, 1989, pp. 11–16.

J. Saniie and N. M. Bilgutay, "Quantitative grain size evaluation using ultrasonic backscattered echoes," *J. Acoustical Society of America*, Vol. 80, No. 6, Dec. 1986, pp. 1816–1824.

J. Saniie and N. M. Bilgutay, "Grain size evaluation through segmentation and digital processing of ultrasonic backscattered echoes," *Proc. IEEE 1984 Ultrasonics Symp.*, Vol. 2, 1984, pp. 847–851.

Inverse Scattering

O. M. Bucci et al., "An adaptive wavelet-based approach for non-destructive evaluation applications," *2000 IEEE AP-S Int. Symp. on Antennas & Propagation Digest*, Salt Lake City, UT, July 16–21, 2000, pp. 1756–1759.

O. M. Bucci et al., "Wavelets in non-linear inverse scattering," *Proc. 2000 IEEE Int. Geoscience and Remote Sensing Symp.*, 2000, pp. 3130–3132.

M. Q. Bao, "Backscattering change detection in SAR images using wavelet techniques," *Proc. 1999 IEEE Int. Geoscience and Remote Sensing Symp.*, 1999, pp. 1561–1563.

A. Dogariu et al., "Fractal roughness retrieval by integrated wavelet transform," *Optical Review*, Vol. 6, No. 4, July–Aug. 1999, pp. 293–301.

M. L. Mittal, V. K. Singh, and R. Krishnan, "Wavelet transform based technique for speckle noise suppression and data compression for SAR images," *Proc. Fifth Int. Symp. on Signal Processing and Its Applications*, 1999, pp. 781–784.

E. Romaneessen et al., "Improved bottom topography in the Elbe estuary using wavelet and active contour methods on SAR images," *Proc. 1999 IEEE Int. Geoscience and Remote Sensing Symp.*, 1999, pp. 1674–1676.

A. J. Van Nevel, B. DeFacio, and S. P. Neal, "Information-theoretic wavelet noise removal for inverse elastic wave scattering theory," *Physical Review E*, Vol. 59, No. 3, Pt. A–B, March 1999, pp. 3682–3693.

L. G. Bruskin et al., "Measurement of plasma density using wavelet analysis of microwave reflectometer signal," *Rev. of Scientific Instruments*, Vol. 69, No. 2, Feb. 1998, pp. 425–430.

F. C. Chen and W. C. Chew, "Development and testing of the time-domain microwave nondestructive evaluation system," *Rev. of Progress in Quantitative Nondestructive Evaluation*, Vol. 1, 1998, pp. 713–718.

Y. Dobashi et al., "A fast volume rendering method for time-varying 3-D scalar field visualization using orthonormal wavelets," *IEEE Trans. Magnetics*, Vol. 34, 1998, pp. 3431–3434.

Z. L. Hu, Y. Z. Chen, and S. Islam, "Multiscaling properties of soil moisture images and decomposition of large- and small-scale features using wavelet transforms," *Int. J. Remote Sensing*, Vol. 19, No. 13, Sept. 10, 1998, pp. 2451–2467.

E. L. Miller, L. Nicolaides, and A. Mandelis, "Nonlinear inverse scattering methods for thermal-wave slice tomography: A wavelet domain approach," *J. Optical Society of America, A*, Vol. 15, No. 6, June 1998, pp. 1545–1556.

M. S. Wang and A. K. Chan, "Wavelet-packet-based time-frequency distribution and its application to radar imaging," *Int. J. Numerical Modelling: Electronic Networks, Devices and Fields*, Vol. 11, No. 1, Jan.–Feb. 1998, pp. 21–40.

L. G. Bruskin et al., "Wavelet application for reflectometry of plasma density profiles," *Japanese J. Applied Physics, Part 2 (Letters)*, Vol. 36, No. 5B, May 15, 1997, pp. 632–633.

C. Castelli and G. Bobillot, "i4D: A new approach to RCS imaging analysis," *19th AMTA Meeting and Symp.*, 1997, pp. 352–357.

T. Doi, S. Hayano, and Y. Saito, "Wavelet solution of the inverse parameter problems," *IEEE Trans. Magnetics*, Vol. 33, No. 2, Pt. 2, 1997, pp. 1962–1965.

A. K. Liu, S. Y. Wu, and W. Y. Tseng, "Wavelet analysis of satellite images for coastal monitoring," *Proc. 1997 IEEE Int. Geoscience and Remote Sensing Symp.*, 1997, pp. 1441–1443.

E. L. Miller and A. S. Willsky, "Multiscale, statistical anomaly detection analysis and algorithms for linearized inverse scattering problems," *Multidimensional Systems and Signal Processing*, Vol. 8, No. 1–2, Jan. 1997, pp. 151–184.

T. L. Olson and L. H. Sibul, "A wavelet detector for model-based imaging," *Proc. SPIE*, Vol. 3068, 1997, pp. 113–123.

G. Saab et al., "Wavelet inverse neutron scattering study of layered metallic NiC-Ti composites," *Rev. of Progress in Quantitative Nondestructive Evaluation*, Vol. 1, 1997, pp. 59–65.

M. X. Wang, J. Li, and L. Ren, "Inversion of inhomogeneous media by wavelets," *J. Southwest Jiaotong University (English Edition)*, Vol. 5, No. 1, May 1997, pp. 85–88.

C. Y. Yu, W. B. Wang, and X. Y. Luo, "On the use of wavelet expansion in reconstruction of conductor profile," *Acta Electronica Sinica*, Vol. 25, No. 3, March 1997, pp. 1–4.

V. C. Chen, "Applications of time-frequency processing to radar imaging," *Proc. SPIE*, Vol. 2762, 1996, pp. 23–31.

R. W. Lindsay, D. B. Percival, and D. A. Rothrock, "The discrete wavelet transform and the scale analysis of the surface properties of sea ice," *IEEE Trans. Geoscience and Remote Sensing*, Vol. 34, 1996, pp. 771–787.

P. Maass and R. Ramlau, "Wavelet-accelerated regularization methods for hyperthermia treatment planning," *Int. J. Imaging Systems and Technology*, Vol. 7, No. 3, Fall 1996, pp. 191–199.

A. Marazzi, P. Gamba, and R. Ranzi, "Rain pattern detection by means of packet wavelets," *1996 Int. Geoscience and Remote Sensing Symp.*, 1996, pp. 266–268.

E. L. Miller and A. S. Willsky, "Wavelet-based methods for the nonlinear inverse scattering problem using the extended Born approximation," *Radio Science*, Vol. 31, No. 1, Jan.–Feb. 1996, pp. 51–65.

W. Qian and L. P. Clarke, "Wavelet-based neural network with fuzzy-logic adaptivity for nuclear image restoration," *Proc. IEEE*, Vol. 84, 1996, pp. 1458–1473.

W. Tobocman, "Application of wavelet analysis to inverse scattering. II," *Inverse Problems*, Vol. 12, No. 4, Aug. 1996, pp. 499–516.

L. H. Sibul, L. G. Weiss, and T. L. Dixon, "New approach to Doppler tomography for microwave imaging," *Int. J. Electronics*, Vol. 78, No. 1, Jan. 1995, pp. 209–218.

G. Z. Chen et al., "Reconstruction of defects from the distribution of current vector potential T using wavelets," *Int. J. Applied Electromagnetics in Materials*, Vol. 5, No. 3, Oct. 1994, pp. 189–199.

E. L. Miller and A. S. Willsky, "A multiscale, decision-theoretic algorithm for anomaly detection in images based upon scattered radiation," *Proc. ICIP-94*, 1994, Vol. 1, pp. 845–849.

D. M. Patterson and B. DeFacio, "Wavelet inversions of elastic wave data for nondestructive evaluation," *Proc. SPIE*, Vol. 2241, 1994, pp. 172–184.

A. E. Yagle, "Reconstruction of layered one-dimensional dielectric media using the wavelet transform," *1994 IEEE AP-S Int. Symp. on Antennas and Propagation Digest*, Seattle, WA, June 20–24, 1994, pp. 12–15.

X. J. Zhu and G. W. Pan, "Theory and application of wavelet based bi-orthonormal decomposition method in the solution of linear inverse problems in electromagnetics," *1994 IEEE AP-S Int. Symp. on Antennas and Propagation Digest*, Seattle, WA, June 20–24, 1994, pp. 8–11.

T. Kikuchi and S. Sato, "Experimental studies on ultrasonic measurements of scattering media by using wavelet transform," *Japanese J. Applied Physics*, Vol. 31, Suppl. 31–1, 1992, pp. 115–117.

J. P. Lefebvre and P. Lasaygues, "Wavelet analysis application to crack detection by ultrasonic echography," *Journal de Physique IV Colloque. (France)*, Vol. 2, No. C1, Pt. 2, April 1992, pp. 637–640.

Target Identification

Y. H. Chen and W. X. Xie, "Detection of radar target in clutter from natural rough surface," *Acta Electronica Sinica*, Vol. 28, No. 7, July 2000, pp. 138–141.

H. Deng and H Ling, "Clutter reduction for synthetic aperture radar images using adaptive wavelet packet transform," *1999 IEEE AP-S Int. Symp. on Antennas and Propagation Digest*, Orlando, FL, July 11–17, 2000, pp. 1780–1783.

C.-P. Lin et al., "Detection of radar targets embedded in sea ice and sea clutter using fractals, wavelets, and neural networks," *IEICE Trans. on Communications*, Vol. E83–B, No. 9, Sept. 2000, pp. 1916–1929.

C. Ohl et al., "Use of minimum phase signals to establish the time reference in transient radar target identification schemes," *IEEE Trans. Antennas and Propag.*, Vol. 48, 2000, pp. 124–125.

D. Carevic, "Clutter reduction and target detection in ground penetrating radar data using wavelets." *Proc. SPIE*, Vol. 3710, Pt. 1–2, 1999, pp. 973–978.

M. McClure, P. Bharadwaj, and L. Carin, "Multiresolution signature-based SAR target detection," *Proc. SPIE*, Vol. 3370, 1998, pp. 318–329.

L. T. Thuong and T. D. Nguyen, "A software package for 2-D scattering feature extraction in wide-band stepped-frequency radar," *Australian J. Intelligent Information Processing Systems*, Vol. 5, Winter 1998, pp. 129–140.

E. K. Walton, "Wavelet techniques and other multiresolution techniques for target phenomenology studies," *Advanced Pattern Recognition Techniques Lecture Series*, Sept. 1998, pp. 31–33.

J. Bertrand and P. Bertrand, "Microwave imaging of time-varying radar targets," *Inverse Problems*, Vol. 13, No. 3, June 1997, pp. 621–645.

H. Luo et al., "Doppler modulated features modeling and radar target recognition," *Proc. SPIE*, Vol. 3069, 1997, pp. 486–493.

T. L. Olson and L. H. Sibul, "A wavelet detector for distributed objects," *Proc. SPIE*, Vol. 3169, 1997, pp. 378–388.

L. T. Thuong, H. Talhami, and D. T. Nguyen, "Target signature extraction based on the continuous wavelet transform in ultra-wideband radar," *Electronics Letters*, Vol. 33, No. 1, Jan. 2, 1997, pp. 89–91.

L. T. Thuong, "Neural network applied to narrow-band noise target signatures for automatic target recognition," *Proc. Second Australian Workshop on Signal Processing Applications*, Brisbane, Australia, Dec. 1997, pp. 219–222.

G. Thomas, B. C. Flores, and A. Martinez, "ISAR imaging of moving targets via the Gabor wavelet transform," *Proc. SPIE*, Vol. 3161, 1997, pp. 90–101.

N. C. Yen and L. R. Dragonette, "Wave packet decomposition for acoustic target recognition," *Proc. SPIE*, Vol. 3069, 1997, pp. 436–445.

T. L. Dixon and L. H. Sibul, "A parameterized Hough transform approach for estimating the support of the wideband spreading function of a distributed object," *Multidimensional Systems and Signal Processing*, Vol. 7, No. 1, Jan. 1996, pp. 75–86.

N. Ehara, I. Sasase, and S. Mori, "Moving target detection by quadrature mirror filter," *Electronics and Communications in Japan, Part 1 (Communications)*, Vol. 79, No. 4, April 1996, pp. 55–62.

C. P. Lin, M. Sano, and M. Sekine, "Detection of targets embedded in sea ice clutter by means of MMW radar based on fractal dimensions, wavelets, and neural classifiers," *IEICE Trans. on Communications*, Vol. E79–B, No. 12, Dec. 1996, pp. 1818–1826.

M. Nishimoto and H. Ikuno, "Time-frequency analysis of radar target echo by using wavelet transform," *Proc. 1996 Int. Symp. on Antennas and Propagation*, Chiba, Japan, Sept. 24–27, 1996, Vol. 4, pp. 1149–1152.

Y. Su et al., "Ultra-wideband radar signal detection, estimation and experiment," *Proc. SPIE*, Vol. 2845, 1996, pp. 343–349.

L. Carin, L. B. Felsen, and C. Tran, "Model-based object recognition by wave-oriented data processing," *Proc. SPIE*, Vol. 2485, 1995, pp. 62–73.

T. L. Dixon, "Wavelet processing applied to the estimation of continuously distributed objects," *Proc. SPIE*, Vol. 569, Pt. 1, 1995, pp. 164–174.

D. M. Drumheller et al., "Identification and synthesis of acoustic scattering components via the wavelet transform," *J. Acoustical Society of America*, Vol. 97, No. 6, June 1995, pp. 3649–3656.

N. Ehara, I. Sasase, and S. Mori, "Moving target detection by quadrature mirror filter," *Trans. of IEICE (Japan) B-II*, Vol. J78B-II, No. 5, May 1995, pp. 401–407.

E. J. Rothwell et al., "A radar target discrimination scheme using the discrete wavelet transform for reduced data storage," *IEEE Trans. Antennas and Propag.*, Vol. 42, 1994, pp. 1033–1037.

C. Chambers et al., "Wavelet processing of ultra wideband radar signals," *IEE Colloquium on Antenna and Propagation Problems of Ultrawideband Radar Digest*, 1993, No. 004.

I. Jouny, "Description and classification of ultra-wideband radar targets using wavelets," *Digital Signal Processing*, Vol. 3, No. 2, April 1993, pp. 78–88.

H. D. Kim and H. Ling, "Wavelet analysis of radar echo from finite-size targets," *IEEE Trans. Antennas and Propag.*, Vol. 41, 1993, pp. 200–207.

P. Moulin, "A wavelet regularization method for diffuse radar-target imaging and speckle-noise reduction," *J. Mathematical Imaging and Vision*, Vol. 3, No. 1, March 1993, pp. 123–134.

I. Jouny, "Wavelet decomposition of UWB radar signals," *1992 IEEE AP-S Int. Symp. on Antennas and Propagation Digest*, Chicago, IL, July 20–25, 1992, pp. 1132–1135.

H. D. Kim and H. Ling, "Analysis of electromagnetic backscattering data using wavelets," *1992 IEEE AP-S Int. Symp. on Antennas and Propagation Digest*, Chicago, IL, July 20–25, 1992, pp. 1877–1880.

Electromagnetic Compatibility

I. Jouny and M. G. Amin, "Mitigation of broadband coherent interference using subbanding," *1996 IEEE AP-S Int. Symp. on Antennas and Propagation Digest*, Baltimore, MD, July 21–26, 1996, pp. 1200–1203.

S. Maslakovic et al., "Excising radio frequency interference using the discrete wavelet transform," *Proc. IEEE-SP Int. Symp. on Time-Frequency and Time-Scale Analysis*, 1996, pp. 349–352.

Wireless Communication

X. M. Jiang and M. X. Li, "Application of wavelet transform and adaptive noise cancelling in ultrasonic closed-crack detection," *Acta Acustica*, Vol. 25, No. 2, March 2000, pp. 97–102.

M. Martone, "Wavelet-based separating kernels for array processing of cellular DS/CDMA signals in fast fading," *IEEE Trans. Communications*, Vol. 48, 2000, pp. 979–995.

M. Martone, "Wavelet-based separating kernels for sequence estimation with unknown rapidly time-varying channels," *IEEE Workshop on Signal Processing Advances in Wireless Communications*, 1999, pp. 255–258.

H. Ujiie, K. Maruo, and T. Shimura, "Improvement of holdover characteristics of GPS timing and frequency reference for CDMA system using wavelet transform," *Trans. of Institute of Electrical Engineers of Japan, Part C*, Vol. 119–C, No. 7, July 1999, pp. 802–809.

ABOUT THE AUTHORS

TAPAN KUMAR SARKAR received a B. Tech. from the Indian Institute of Technology, Kharagpur, India, in 1969, an M.Sc.E. from the University of New Brunswick, Fredericton, Canada, in 1971, and an M.S. and a Ph.D. from Syracuse University in Syracuse, New York, in 1975.

Dr. Sarkar has worked for the General Instruments Corporation and the Rochester Institute of Technology in New York. He was also a research fellow at Harvard University. He is currently a professor in the Department of Electrical and Computer Engineering at Syracuse University. His current research interests deal with numerical solutions of operator equations arising in electromagnetics and signal processing with application to system design. He received one of the "best solution" awards in 1977 at the Rome Air Development Center (RADC) Spectral Estimation Workshop. He has authored or coauthored more than 210 journal articles and numerous conference papers, has written chapters in 28 books, and has authored 10 books, including *Iterative and Self-Adaptive Finite-Elements in Electromagnetic Modeling* (Artech House, 1998).

Dr. Sarkar is a registered professional engineer in the State of New York. He received the Best Paper Award of the *IEEE Transactions on Electromagnetic Compatibility* in 1979 and the National Radar Conference in 1997. He received the College of Engineering Research Award in 1996 and the chancellor's citation for excellence in research at Syracuse University in 1998. He was an associate editor for feature articles in the *IEEE Antennas and Propagation Society Newsletter*, and he was the technical program chairman for the 1988 IEEE Antennas and Propagation Society International Symposium and URSI Radio Science Meeting. Dr. Sarkar is a distinguished lecturer for the Antennas and Propagation Society. He is on the editorial boards of the *Journal of Electromagnetic Waves and Applications* and *Microwave and Optical Technology Letters*. He has also been appointed as the U.S. Research Council Representative to many URSI general assemblies. He was the chairman of the Intercommission Working Group of International URSI on Time Domain Metrology from 1990 to 1996. He is a fellow of the IEEE and a member of Sigma Xi and the International Union of Radio Science Commissions A and B. He received the title Docteur Honoris Causa from Université Blaise Pascal, Clermont Ferrand, France, in 1998, and *the friend of the city (Clermont Ferrand)* in 2000.

MAGDALENA SALAZAR-PALMA received an *Ingeniero de Telecomunicación* and a Ph.D. from the Universidad Politécnica de Madrid in Madrid, Spain, where she is a *Profesor Titular* in the Signals, Systems, and Radiocommunications Department at the Escuela Técnica Superior de Ingenieros de Telecomunicación. She has taught courses on electromagnetic field theory, microwave and antenna theory, circuit networks and filter theory, analog and digital communication systems theory, numerical methods for electromagnetic field problems, and related laboratories.

Dr. Salazar-Palma has developed her research within the Microwave and Radar Group in the areas of electromagnetic field theory, computational and numerical methods for microwave structures, passive components, and antenna analysis; design, simulation, optimization, implementation, and measurements of hybrid and monolithic microwave integrated circuits; and network and filter theory and design. She has frequently been a visiting professor in the Electrical Engineering and Computer Science Department at Syracuse University.

In addition, Dr. Salazar-Palma has authored one book and has contributed several chapters and articles to books and papers in international journals, conferences, symposiums, and workshops, as well as a number of national publications and reports. She has delivered a number of invited presentations, lectures, and seminars and has lectured in several short courses, some of them in the frame of European Community Programs. Dr. Salazar-Palma has also participated in many projects and contracts financed by international and national institutions and companies. She has been a member of the technical program committee of several international symposiums and has acted as reviewer for several international scientific journals, symposiums, and publishers. Dr. Salazar-Palma has assisted the National Board of Research in project evaluation and has also served on several evaluation panels of the Commission of the European Communities. She has acted as the topical editor for compilation of references of the triennial *Review of Radio Science*. Dr. Salazar-Palma is a member of the editorial boards of three scientific journals, a registered engineer in Spain, and a senior member of the IEEE. She has served as vice chairman and chairman of IEEE Microwave Theory and Techniques Society/Antennas and Propagation Society (MTT-S/AP-S) Spain joint chapter and chairman of IEEE Spain section. She is a member of IEEE Region 8 Nominations and Appointments Committee, IEEE Ethics and Member Conduct Committee, and IEEE Women in Engineering Committee (WIEC). She acts as a liaison between the IEEE Regional Activities Board and the IEEE WIEC. Dr. Salazar-Palma has also received two individual research awards from national institutions.

MICHAEL C. WICKS received a BSEE from Rensselaer Polytechnic Institute in 1981, and a Ph.D. from Syracuse University in 1995, both in electrical engineering. His interests include adaptive radar signal processing, wideband radar, ground penetrating radar, and knowledge-base applications to sensor signal processing. He is a principal engineer in the Air Force Research Laboratory Radar Signal Processing Branch. He is a member of the Association of Old Crows and a fellow of the IEEE.

RAVIRAJ ADVE received his B. Tech. in electrical engineering from the Indian Institute of Technology, Bombay, in 1990, and his Ph.D. from Syracuse University in 1996. His dissertation received the Syracuse University Doctoral Prize for Outstanding Dissertation Research. He has worked for Research Associates for Defense Conversion (RADC), Inc., on contract with the Air Force Research Laboratory in Rome, New York. He joined the faculty at the University of Toronto in 2000.

Dr. Adve's research interests include practical adaptive signal processing algorithms for smart antennas with applications in wireless communications and radar. His research areas include nonstatistical, direct data domain adaptive processing and the applications of numerical electromagnetics to adaptive signal processing. He has also investigated the applications of signal processing techniques to numerical and experimental electromagnetics.

ROBERT J. BONNEAU obtained a B.S.E.E in 1988 and an M.S.E.E. in 1989 from Cornell University, and an M.S. in 1996 and a Ph.D. in 1997 from Columbia University, all in electrical engineering. He was a postdoctoral research associate at Columbia University in 1988 and was a research electrical engineer at the Sensors Directorate of Air Force Research Laboratory in Rome, New York. Currently, he is the program manager at the Advanced Technology Office of DARPA. He has made substantial contributions improving the technology of image compression. His work has provided a solution to the communications problem of transferring large quantities of image data across limited bandwidth connections by prioritizing data according to Markov models of objects of interest in scenes. Thus, background information is transferred at low resolution, while foreground information is transferred at high resolution, thereby increasing the amount of compression achievable by a factor of 2 to 4 over the best conventional methods. In wireless applications Dr. Bonneau's work with multiresolution Markov models has enabled efficient design of new pseudo-random noise sequences for transferring more information with less interference across a highly partitioned spectrum. Additionally, this work has led to improved spectrum allocation as well as the spatial reutilization of existing spectrum. He has published many journal and conference papers and holds three patents.

RUSSELL D. BROWN received a B.S.E.E. in 1972 and an M.S.E.E. in 1973 from the University of Maryland and a Ph.D. in 1995 from Syracuse University, all in electrical engineering. From 1973 to 1978, he served as a communications electronics officer for the U.S. Air Force. Since 1978, he has performed research as a principal engineer in the Radar Signal Processing Branch of the Air Force Research Laboratory. He is a fellow of the IEEE and a member of the Association of Old Crows.

LUIS-EMILIO GARCÍA-CASTILLO received an *Ingeniero de Telecomunicación* and a Ph.D. from the Universidad Politécnica de Madrid in 1992 and 1998, respectively. His Ph.D. thesis received two prizes from the Colegio Oficial de Ingenieros de Telecomunicación in Spain and the Universidad Politécnica de Madrid. Dr. García-Castillo has frequently been a visiting scholar in the Electrical Engineering and Computer Science Department at Syracuse University and a *Profesor Titular de Escuela Universitaria* at the Universidad Politécnica de Madrid. Dr. García-Castillo is currently a *Profesor Titular de Universidad* at the Universidad de Alcalá in Madrid, Spain. His research activities and interests are focused in the application of numerical methods and signal processing techniques to electromagnetic problems. Other research areas are network theory and filter design. He has coauthored the book *Iterative and Self-*

Adaptive Finite-Elements in Electromagnetic Modeling (Artech House, 1998), contributed chapters and articles in books, and written articles in international journals and papers in international conferences, symposiums, and workshops as well as a number of national publications and reports. He currently leads a National Plan for Research of Spain project. Dr. García-Castillo has also participated in several projects and contracts, financed by international and national institutions and companies.

YINGBO HUA received a B.E. from the Nanjing Institute of Technology (currently Southeast University), Nanjing, China, in 1982, and an M.S. and a Ph.D. from Syracuse University, in 1983 and 1988, respectively. He won the Chinese Government Scholarship for Overseas Study and the Syracuse University Graduate Fellowship. Dr. Hua has also worked as a teaching assistant, research assistant, summer graduate course lecturer, and research fellow at Syracuse University.

Dr. Hua has worked as a lecturer, senior lecturer, associate professor and reader, and both undergraduate and graduate study coordinator at the University of Melbourne in Australia. In addition, he was a project leader for statistical signal processing at the Australian Cooperative Research Center for Sensor Signal and Information Processing. Dr. Hua has been a visiting faculty member with the Hong Kong University of Science and Technology. He is currently a professor of electrical engineering at the University of California in Riverside, California.

Dr. Hua is an author and coauthor of many journal articles, several book chapters, and 130 conference papers in the fundamental areas of estimation, detection, system identification, and fast algorithms, with applications in communications, remote sensing, and medical data analysis. He is the coeditor of *Signal Processing Advances in Wireless and Mobile Communications* (Prentice Hall, 2001). Dr. Hua has served as an associate editor of *IEEE Transactions on Signal Processing* and *IEEE Signal Processing Letters*. He is an invited reviewer for more than 16 international journals and is an invited speaker, session chair, and organizer of many international conferences. In the IEEE Signal Processing Society's Technical Committees, Dr. Hua was an elected member of the Underwater Acoustic Signal Processing Committee, and is currently an elected member of the Sensor Array and Multichannel Signal Processing and the Signal Processing for Communications Committees. He is a fellow of the IEEE.

ZHONG JI received both a B.S. and an M.S. in electronic engineering from Shandong University in China, in 1988 and 1991, respectively. He received a Ph.D. from Shanghai Jiao Tong University in China in 2000. He was a teacher at Shandong University from 1991 to 1997. He is currently a postdoctoral fellow at Syracuse University. His research interests are in the areas of propagation models for wireless communications and smart antenna.

KYUNGJUNG KIM received a B.S. from Inha University in Korea and an M.S. from Syracuse University. He is currently working toward a Ph.D. in the

Department of Electrical Engineering at Syracuse University. He was a research assistant at Syracuse University from 1998 to 2001 and has been a graduate fellow since 2001. His current research interests include digital signal processing related to adaptive antenna problems and wavelet transform.

JINHWAN KOH is a professor in the Department of Electronics and Electrical Engineering at Kyungpook National University in Taegu, Korea. He received a B.S. in electronics from Inha University in Korea, in 1991, and an M.S. and a Ph.D. in electrical engineering from Syracuse University in 1999. Dr. Koh has worked for Hyundai Electronics, where he developed many test tools for TTL and ECL. He has also worked as a research assistant in the Computational Electromagnetics Laboratory at Syracuse University. He received the All University Doctoral Prize from Syracuse University in 2000. His current research interests include wireless communications, digital signal processing related to adaptive antenna problems, and data restoration.

SERGIO LLORENTE-ROMANO received an *Ingeniero de Telecomunicación* from the Universidad Politécnica de Madrid in 2001. He has worked in the Microwave and Radar Group at the Departamento de Señales, Sistemas y Radiocomunicaciones (SSR), at the Universidad Politécnica de Madrid, where he carried out his master's thesis on techniques for the simulation and design of filters and diplexers in waveguide technology and where he is currently working toward a Ph.D. He was a visiting scholar in the Department of Electrical Engineering and Computer Science at Syracuse University, where he worked on the use of wavelets basis functions for matrix compression applications. His research activities and interests are in the area of numerical methods applied to electromagnetic problems related to the design of microwave devices.

RAFAEL RODRIGUEZ-BOIX received a Licenciado and a Ph.D. in physics from the University of Seville in Spain, in 1985 and 1990, respectively. Dr. Boix is an associate professor in the Electronics and Electromagnetics Department at the University of Seville. He has been a visiting scholar at the Electrical Engineering Department of the University of California at Los Angeles and at the Electrical and Computer Engineering Department at Syracuse University. His current research interest is focused on the fast numerical analysis of planar passive microwave circuits, planar periodic transmission lines, frequency selective surfaces, and printed circuit antennas. Dr. Boix is also engaged in the study of the effects of complex substrates (e.g., anisotropic dielectrics, magnetized ferrites, and chiral materials) on the performance of planar microwave circuits and printed antennas. Dr. Boix has authored and coauthored more than 30 papers in international refereed journals. Dr. Boix is on the editorial board of *IEEE Transactions on Microwave Theory and Techniques*.

CHAOWEI SU received a B.S in 1981 and an M.Sc. in 1986 in applied mathematics from Northwestern Polytechnical University (NPU). In 1982, he joined the Department of Applied Math at NPU. He was appointed as an associated professor in 1993 and as a full professor at NPU in 1996. He has been a visiting scholar at SUNY at Stony Brook, New York, supported by the Grumman Fellowship from December 1990 to June 1992. From 1996 to 2000, he has been a visiting professor at Syracuse University. He was a senior research fellow at the Wireless Communication Center at the City University of Hong Kong in 2001. Mr. Su is currently working for Integrated Engineering Software. He is the author of *Numerical Methods in Inverse Problems of Partial Differential Equations and Their Applications* (NPU Press, 1995). His main research interests are in the areas of numerical methods of electromagnetic scattering and inverse scattering and in inverse problems of partial differential equations.

Index

Recent Titles in the Artech House Electromagnetic Analysis Series

Tapan K. Sarkar, Series Editor

Wavelet Applications in Engineering Electromagnetics,
Tapan K. Sarkar, Magdalena Salazar-Palma, and
Michael C. Wicks

For further information on these and other Artech House titles,
including previously considered out-of-print books now available through our
In-Print-Forever® (IPF®) program, contact:

Artech House
685 Canton Street
Norwood, MA 02062
Phone: 781-769-9750
Fax: 781-769-6334
e-mail: artech@artechhouse.com

Artech House
46 Gillingham Street
London SW1V 1AH UK
Phone: +44 (0)20 7596-8750
Fax: +44 (0)20 7630 0166
e-mail: artech-uk@artechhouse.com

Find us on the World Wide Web at:
www.artechhouse.com